SAFETY, HEALTH, AND ENVIRONMENTAL PROTECTION

McGraw-Hill Series in Water Resources and Environmental Engineering

CONSULTING EDITOR

George Tchobanoglous, *University of California—Davis*

Bailey and Ollis: *Biochemical Engineering Fundamentals*
Bouwer: *Groundwater Hydrology*
Canter: *Environmental Impact Assessment*
Chanlett: *Environmental Protection*
Chow, Maidment, and Mays: *Applied Hydrology*
Crites and Tchobanoglous: *Decentralized Wastewater Management Systems*
Davis and Cornwell: *Introduction to Environmental Engineering*
de Nevers: *Air Pollution Control Engineering*
Eckenfelder: *Industrial Water Pollution Control*
La Grega, Buckingham, and Evans: *Hazardous Waste Management*
Linsley, Franzini, Freyburg, and Tchobanoglous: *Water Resources and Engineering*
McGhee: *Water Supply and Sewerage*
Mays and Tung: *Hydrosystems Engineering and Management*
Metcalf & Eddy, Inc.: *Wastewater Engineering: Collection and Pumping of Wastewater*
Metcalf & Eddy, Inc.: *Wastewater Engineering: Treatment, Disposal, Reuse*
Peavy, Rowe, and Tchobanoglous: *Environmental Engineering*
Sawyer and McCarty: *Chemistry for Environmental Engineering*
Schroeder, Eweis, Ergas, Chang: *Bioremediation Principles*
Tchobanoglous, Theisen, and Vigil: *Integrated Solid Waste Management: Engineering Principles and Management Issues*
Wentz: *Hazardous Waste Management*
Wentz: *Safety, Health, and Environmental Protection*

SAFETY, HEALTH, AND ENVIRONMENTAL PROTECTION

Charles A. Wentz

Boston, Massachusetts   Burr Ridge, Illinois   Dubuque, Iowa
Madison, Wisconsin   New York, New York   San Francisco, California   St. Louis, Missouri

WCB/McGraw-Hill

A Division of The **McGraw·Hill** *Companies*

SAFETY, HEALTH, AND ENVIRONMENTAL PROTECTION

Copyright © 1998 by The McGraw-Hill Companies, Inc. Printed in the United States of America. Except as permitted under the United States Copyright Act of 1976, no part of this publication may be reproduced or distributed in any form or by any means, or stored in a data base or retrieval system, without the prior written permission of the publisher.

This book is printed on acid-free paper.

1 2 3 4 5 6 7 8 9 0 DOC/DOC 9 0 9 8 7

ISBN 0-07-069310-2

Executive editor: *Eric Munson*
Project manager: *Carrie Sestak*
Production supervisor: *Heather D. Burbridge*
Designer: *Larry J. Cope*
Senior photo research coordinator: *Keri Johnson*
Photo researcher: *Debra Hershkowitz*
Compositor: *Ruttle, Shaw & Wetherill, Inc.*
Art studio: *Fine Line Illustrations*
Typeface: *10/12 Times Roman*
Printer: *R. R. Donnelley & Sons Company*

Library of Congress Cataloging-in-Publication Data
Wentz, Charles A.
    Safety, health, and environmental protection / Charles A. Wentz.
      p.   cm.
    ISBN 0-07-069310-2
    Includes bibliographical references and index.
    1. Industrial safety. 2. Industrial hygiene. I. Title.
T55.W47 1998
363.11dc—21   97-16349

http://www.mhhe.com

CONTENTS

PREFACE

The importance of the safety, health, and environmental fields has risen dramatically in recent years. Today, industry, government, the public, and academia take a much more serious approach to this subject, than a few decades ago. The concern to protect people, property, and the environment has identified this area as needing greater emphasis in academic programs at colleges and universities.

The purpose of *Safety, Health, and Environmental Protection* is to bring together all aspects of this subject into a single, comprehensive textbook. This book has three major components: hazard identification and characterization; safety, health, and environmental management; and the protection of safety, health, and the environment. These components have been integrated to bring together workplace and environmental hazards to facilitate their management and control.

This textbook has been primarily written for senior undergraduate students who are knowledgeable about the fundamentals of science, mathematics, and engineering. An undergraduate course based on this textbook should satisfy the safety component requirements of the Accreditation Board for Engineering and Technology (ABET). The book is also suitable for a beginning graduate level course in engineering and science curriculums. Additionally, managers, engineers, and scientists will find it a useful reference in dealing with the complexities of safety, health, and environmental programs.

The format of the book allows the course material to be presented in a single semester. The numerous problems and questions at the end of each chapter reinforce the material in a logical, systematic fashion. Every chapter should be covered to achieve the primary goal of the book, which is to satisfy the ABET requirements.

The author would like to thank Gloria Pederson, who was particularly helpful in the manuscript preparation of this textbook.

Charles A. Wentz

ABOUT THE AUTHOR

Charles A. Wentz is the president of International Scientific Management, Inc., and president of Wentz Health Care, Inc. Dr. Wentz was the Associate Dean of the Petroleum and Petrochemical College at Chulalongkorn University in Bangkok, Thailand. He was formerly a manager at Argonne National Laboratory in the fields of hazardous waste management and safety engineering. Dr. Wentz has been manager of waste management at the University of North Dakota Energy Research Center and Professor of Engineering Management where he was responsible for environmental research projects. He has taught safety, health, and environmental management to both undergraduate and graduate students, as well as in continuing education programs for professional societies. His prior industrial experience included the presidency of ENSCO in Arkansas, the Presidency of Newpark Waste Treatment in New Orleans, and various executive positions with Phillips Petroleum in oil shale, chemicals, polymers, federal government relations, and investor relations. He received his B.S. and M.S. degrees in chemical engineering from the University of Missouri-Rolla, his Ph.D. in chemical engineering from Northwestern University, and an M.B.A. from Southern Illinois University. He has been chairman of numerous technical sessions at national AlChE meetings. He has organized and presented conferences and workshops dealing with safety, health, and the environment. He is a CSP and a registered professional engineer. He is a member of a number of business, honorary, and technical organizations, as well as an author of technical papers and books on environmental and safety matters.

INTRODUCTION TO REGULATIONS AND STANDARDS

Safety must be paramount in industrial processes and plant facilities that involve hazardous chemicals. [19] Better engineering and safer design and operation might have prevented the disasters that occurred at Bhopal and Chernobyl, in which thousands of people were killed and thousands more were injured. [14] While both these accidents happened outside the United States, there have been numerous serious incidents at U.S. plants.

The worst industrial accident in the history of the United States occurred in 1947, when a French fertilizer ship carrying ammonium nitrate exploded in Texas City, killing 576 people and injuring another 5000. [10] In 1981 a Dow Chemical polyethylene plant in Freeport, Texas, exploded, killing six people. In 1985 a Warren petroleum plant at Mont Belvieu, Texas, exploded, killing two people. In 1989, at a Phillips Petroleum polyethylene plant near Houston, a valve in the process line that contained volatile hydrocarbons, was accidentally opened to the atmosphere; 23 people were killed in the resulting explosion. [15,16,17]

In 1990, at ARCO Chemical's complex in Channelview, Texas, a 900,000-gallon waste storage tank blew up, killing 17 people. Workers had been installing a new compressor and were unaware that a gauge measuring the amount of oxygen in the tank had malfunctioned. Volatile hydrocarbons in the wastewater ignited when the compressor was started up. [1]

In 1991 an explosion at an Albright & Wilson America's phosphorus chemicals plant in Charleston, South Carolina, killed six people. [21] At the time of the explosion, workers were mixing chemicals in the reactor in the special-products unit to manufacture a flame retardant. One of these chemicals was moisture-sensitive and flammable. The plant had been shut down the previous week and was just coming back on stream; this may have contributed to the explosion.

In 1992 a liquified petroleum gas pipeline exploded in Texas, killing one person and igniting a blaze that burned out of control for several hours. Investigators believe that leaking liquified petroleum gas had collected in a low-lying ravine and may have been ignited by sparks from a vehicle or the pilot light on a gas stove at a nearby home.

Because of disasters such as these, engineers, scientists, and managers need to be well-informed about potential hazards in process technology, operation, and design. Technology and production methods have improved tremendously in order to keep pace with a growing market demand for high-quality and improved products for the public. Hazard identification, safety design, and operating procedures must also keep pace with these improvements. Safety must be emphasized throughout the development, construction, start-up, and operation of manufacturing facilities that involve hazardous chemicals and processes. [12,20] Even if a company has built and successfully operated numerous identical manufacturing facilities in other locations, there are no grounds for complacency. A good safety record will be of little consolation to the relatives and friends of someone who has been killed in an industrial accident.

HAZARD IDENTIFICATION AND CHARACTERIZATION

There are many hazards in process plants. The chemicals used at the facility may be hazardous because they are flammable, explosive, reactive, radioactive, or toxic. [11] Usually these chemical hazards are in the form of mixtures, making it more difficult to identify the severity of the potential hazard. [15] Mechanical and electrical hazards as well as industrial noise further complicate plant process operation. All of these hazards, no matter how complex, must be adequately identified through proper characterization of the hazards in order to arrive at a satisfactory solution to safety problems.

CONSEQUENCE OF HAZARDS

Every day in the United States many thousands of accidents are caused by some failure of people, equipment, chemicals, or the surrounding environment. These unplanned and unforeseen events result in loss of life, injury, and property damage. Even apparently minor personal injuries can later result in fatalities or permanent disabilities. Many accidents are avoidable and can often be traced to poor management policies, procedures, and decisions.

Hazard analysis should be performed for each task at an operating facility in order to assess the risks involved. Job safety analysis is often used to study the basic steps of the job, identify the hazards, and prescribe appropriate safety controls. Both hazard analysis and job safety analysis can provide personnel with information that is useful in anticipating and reducing the risks in process and plant operation.

Accidents have direct, indirect, and root causes. A *direct* cause is one that is attributable to equipment failure or unsafe operating conditions. For example, faulty ventilating equipment may allow volatile gases to build up, resulting in an explosion. *Indirect* causes are usually not as readily apparent and can generally be tied to some human shortcoming or failure. Finally, the basic, or *root,* cause is usually the result of poor management safety policies, procedures, or decisions. Good accident investigations should not be witch-hunts conducted merely to fix blame; their real objective should be future accident prevention.

There are many direct causes of accidents (see Table 1-1). Physical and energy sources are directly related to process and plant operations. Hazardous material must be stored and used at an operating facility so as to reduce the risk of accident. [3]

Table 1-2 on page 4 lists indirect causes of many industrial accidents. Inattentive or improperly trained personnel may perform the job in an unsafe manner. Unsafe operating conditions are usually the result of inadequate planning, faulty process design, poor plant layout, or unprofessional operation of the process and plant facility. These same conclusions also apply to laboratory safety. [18]

Direct and indirect causes of accidents can usually be traced back to root causes: poor management practices and inappropriate or inadequate safety policies and procedures (see Table 1-3 on page 4). An industrial safety program must have the complete support and commitment of upper management to succeed. All of these basic causes of accidents are the result of management's failure to adequately plan, organize, implement, motivate, and control with the proper emphasis on safety throughout the entire organization.

Future accidents can be reduced by eliminating one or more potential causes. Accident investigations should be conducted whenever there is a lost-time injury or a fatality. The direct cause of the accident determines *what* happened; indirect causes explain *how* it happened. The root cause of the accident points to *why* it happened. We all learn from experience. A proper investigation can prevent the recurrence of accidents. An accident investigator should study each event and the sequence of events that led to the accident. Often the accident investigation will identify potentially unsafe conditions not only at the site of the accident but elsewhere at the operating facility.

TABLE 1-1
DIRECT CAUSES OF ACCIDENTS

Physical and energy sources	Hazardous materials
Mechanical: equipment, tools, moving parts	Compressed or liquified gas
Electrical: poor insulation, high voltage	Corrosive, flammable, radioactive, or explosive materials
Thermal: conduction, convection, radiation heat transfer	Oxidizers Poisons
Noise: intensity, frequency	Dust
Radiation: ionizing, nonionizing	

TABLE 1-2
INDIRECT CAUSES OF ACCIDENTS

Unsafe acts	Unsafe operating conditions
Failure to use proper personal protective equipment	Workplace congestion
Improper equipment or chemical placement	Poor housekeeping
Failure to secure equipment	Fire and explosion hazards
Making safety devices inoperable	Hazardous air contaminants
Operating equipment without authority	Poor ventilation
Operating equipment improperly	Excessive radiation
Operating defective equipment	Excessive noise
Servicing equipment as it is operating	Inadequate warning systems
Failure to warn personnel	Poor illumination
Horseplay	Defective equipment or tools
Drugs or alcohol	Inadequate guards and supports
Smoking in a prohibited area	Off-specification supplies and chemicals
Using an unsafe position	Electrical power failure
	Instrumentation failure
	Control valve failure
	Equipment failure

The recommended procedures for investigating an accident are shown in Table 1-4. It is important that management be supportive of any accident investigation. Management must also be supportive of the recommendations of the investigation team in order to reduce or eliminate a recurrence.

Safety has to be an integral part of any process plant. Safe facilities that meet performance and economic requirements are in the best interest of all concerned parties. Unless personnel know and understand the principles behind process and

TABLE 1-3
ROOT CAUSES OF ACCIDENTS

Safety policy not defined and communicated
Responsibility, authority, and accountability not assigned
Emphasis on production, rather than safety
Lack of direct communication with management
Unsafe design and selection of equipment, chemicals, process, and facilities
Inadequate safety inspection procedures
Insufficient procedures and safety training for normal and emergency situations
Inadequate employee selection, supervision, and rewards

TABLE 1-4
STEPS INVOLVED IN AN ACCIDENT INVESTIGATION

1 Define the scope and management support of the accident investigation.
2 Select the investigation team.
3 Present the preliminary situation to the team.
4 Visit and inspect the accident site.
5 Interview victims, witnesses, and on-the-scene personnel.
6 Determine if anything different or unusual occurred prior to the accident.
7 Trace the most probable sequence of events, and determine the direct, indirect, and root causes of the accident.
8 Develop any reasonable alternatives for the accident's occurrence.
9 Review the initial findings of the investigative team with all the participants in the investigation.
10 Prepare a written summary and include recommendations for preventing a recurrence.

plant safety, there is an increased risk of accidents that could result in injuries, illness, fatalities, and loss of property.

CONTROL OF HAZARDS

Engineering or management controls can eliminate many hazards from the workplace. For example, innovative process design can reduce or eliminate hazardous air contaminants. Or less hazardous materials can be substituted. Finally, a well-designed ventilation system may be the best solution to the problem.

Exhaust ventilation may remove harmful contaminants from the air. The best ventilation systems have exhaust fans that pull the contaminated air out rather than forcing it through ductwork. The contaminant may be removed and the air recirculated, or new air could be added at the point where the contamination is taking place.

Cyclones, electrostatic precipitators, and bag filters are useful engineering controls in removing solid particulate contaminants from air. Wet scrubbers are often used to wash the air of particulate and gaseous contaminants.

Changing work practices and rotating job assignments are examples of management controls to keep employee exposures below the permissible exposure limit (PEL). Industrial noise can cause chronic hearing problems. Exposure may be controlled through equipment design or by locating employees in the facility so that the PEL of 90 decibels, or the action level of 85 decibels, on the A scale for an 8-hour time-weighted average is not reached. The absolute sound intensity varies inversely with the square of the distance from the sound source. As a rule of thumb, doubling the distance from the noise source results in a fourfold reduction in sound intensity.

If the hazard cannot be eliminated from the work environment, personal protective equipment is mandatory. Hard hats, safety shoes, and safety glasses are common in process industries. The degree and type of protection needed requires an understanding of the hazard and appropriate testing of the equipment to ensure that its comfort, appearance, and protection meet the needs of the employees. [50] All protective equipment should meet the performance specification standards of organizations such as the American National Standards Institute (ANSI) and the National Institute for Occupational Safety and Health (NIOSH). Protective equipment is a major business expense, but so are lost-time accidents.

In any operation, the best approach to controlling hazards is to bring together personnel, materials, and equipment in a well-balanced safety system. Such a system will result in cost-effective and efficient production and performance. Safe operations need to take into account both present and potential hazards and hazards that may act independently or synergistically with other hazards.

Uncontrolled hazards may result in tragic accidents, usually at significant cost to the operation or business. The future operation of the facility itself may be in jeopardy. Damage to equipment, process materials and products, and property can be substantial; there will usually be hidden costs and losses in operation efficiencies. Government regulatory agencies may scrutinize the operation, levy fines, and restrict or even close the facilities. Both public and employee relations will suffer as a result. For all of these reasons, hazards must be reduced.

SAFETY AND HEALTH LEGISLATION AND REGULATIONS

The Occupational Safety and Health Act was created on December 29, 1970. The purpose of the act is to reduce work-related injuries, illnesses, and death and, incidentally, to cut resulting costs (lost wages and productivity, medical expenses, disability compensation). Occupational health and safety standards are developed jointly by the Department of Labor Occupational Safety and Health Administration (OSHA) and the National Institute for Occupational Safety and Health (NIOSH), which is part of the Department of Health and Human Services.

The Department of Labor is responsible for collecting, compiling, and analyzing statistics on all disabling or serious work-related injuries and illnesses, whether or not they involve loss of time from work. No statistics are kept on minor injuries that do not involve medical treatment, loss of consciousness, restriction of work or motion, or transfer to another job. NIOSH also carries out research and develops criteria for health and safety standards.

The Department of Labor sets OSHA standards and regulations based on its own research and the recommendations of NIOSH. The individual states are responsible for administering and enforcing OSHA regulations at the local level.

When OSHA was formed, there were already established federal standards for occupational safety and health, as well as safety and health standards that had been adopted and promulgated by nationally recognized standards-producing organizations such as the American National Standards Institute (ANSI), the National Fire Protection Association (NFPA), and the American Society for Testing and Materials (ASTM). Rather than reinventing the wheel, for two years OSHA was allowed by rule to promulgate any of these established federal standards or national consensus standards as an OSHA standard. Most of the OSHA standards of the early 1970s were promulgated by this mechanism.

OSHA headquarters are located at 200 Constitution Avenue NW, Washington, DC 20210; telephone: 202-523-8151. There are 10 OSHA administrative regions (see Fig. 1-1). The addresses and telephone numbers of OSHA regional offices are listed in Table 1-5 on page 8.

ENFORCEMENT OF OSHA STANDARDS

OSHA can conduct workplace inspections, issue citations, and propose penalties for alleged violations. If the employer contests the violation, it is adjudicated by the Occupational Safety and Health Review Commission (Review Commission). Employers may also appeal adjudications through the courts.

FIGURE 1-1
OSHA REGIONS IN THE UNITED STATES.

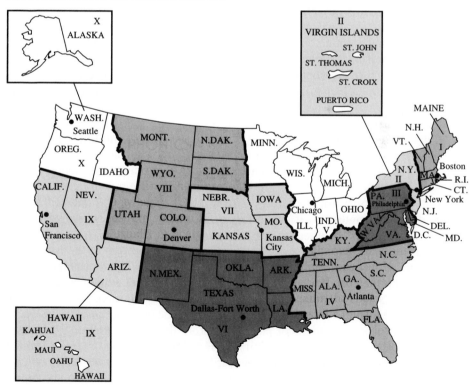

TABLE 1-5
OSHA REGIONAL OFFICES

Region	Address	Telephone
1	133 Portland St., Boston, MA 02114	617-565-7159
2	201 Varick St., New York, NY 10014	212-337-2378
3	3535 Market St., Philadelphia, PA 19104	215-596-1201
4	1375 Peachtree St. NE, Atlanta, GA 30367	404-347-3573
5	230 S. Dearborn St., Chicago, IL 60604	312-353-2057
6	555 Griffin Sq. Bldg., Dallas, TX 75202	214-767-4731
7	911 Walnut St., Kansas City, MO 64106	426-867-5861
8	1961 Stout St., Denver, CO 80294	303-844-3061
9	71 Stevenson St., San Francisco, CA 94105	415-744-6670
10	1111 Third Ave., Seattle, WA 98101	206-553-5930

OSHA compliance safety and health officers are authorized to enter any workplace at reasonable times and within reasonable limits to make inspections and question personnel. If the officer is refused entry, OSHA legal personnel can institute a compulsory inspection. Generally, OSHA officers are not required to give employers advance notice of inspections. Both the employer and the employees may have a representative present during the inspection. Most OSHA inspections are triggered by written employee complaints; the employee is allowed to remain anonymous. If the OSHA officer concludes that workplace conditions could cause death or serious physical harm, the employer and employees are informed of the imminent danger. [23]

OSHA citations for alleged violations or notice of de minimis violations, which have no direct or immediate relationship to safety or health, are issued to the employer after the inspection. The citation must describe the alleged violation and fix a reasonable time for the abatement of the alleged violation. The citation does not constitute a finding that the violation occurred unless the employer does not contest or, if contested, unless the citation is affirmed by the Review Commission.

The employer must post the citation in a prominent location in the workplace until the violation has been abated or for 3 working days, whichever is later. If the citation is contested, the employer may also post such a notice in the same prominent location. Notices of de minimis violations need not be posted. [24]

Employers have 15 working days from receipt of the citation to notify OSHA of their intent to contest citations or penalties before the Review Commission. Employees have 15 working days to notify OSHA that the period of time fixed for abatement in the citation is unreasonable. [25]

Employers must keep records of occupational injuries and illnesses and report them to OSHA on OSHA Form 200 within 6 working days of their occurrence. The employer must post an annual summary on OSHA Form 102 of recordable occupational injuries and illnesses. Recordable occupational injuries and illnesses are defined by OSHA as any of the following [26]:

1 Fatalities
2 Lost workday cases
3 Nonfatal cases without lost workdays, resulting in job transfer or termination, physician-directed medical treatment, but not first aid, loss of consciousness, restriction of work, or restriction of motion.

In addition to Form 200, the employer must complete OSHA Form 101, which is a supplementary record, within the same 6 working days. Records of all of these forms must be retained by the employer for at least 5 years following the end of the 5 years to which they relate.

Workplace fatalities or multiple hospitalization accidents are treated as serious, special-case situations, requiring more stringent OSHA reporting by the employer. Within 48 hours after a workplace fatality or after the hospitalization of five or more employees, the employer must notify OSHA orally or in writing.

COMPLIANCE AND ETHICAL BEHAVIOR

Both employers and employees must comply with the Occupational Safety and Health Act and with all OSHA standards and regulations. Most of the burden of compliance falls on employers because they must maintain a safe and healthy work environment.

Employers may request a variance from an OSHA standard by stipulating:

1 The reasons why the employer is unable to comply with the standard
2 The steps taken to protect employees against the hazard
3 An estimated future compliance date
4 That employees have been notified of the situation

Employers may also request a variance from an OSHA standard by stipulating that:

1 The employer is going to use a manner, technique, or method that differs from the requirements of the standard
2 The different manner, technique, or method would be at least as safe and healthful as required by the standard
3 The employees have been notified of the situation

Exemptions or variances may also be obtained from OSHA to avoid serious impairment of the national defense. Modifications of OSHA rules or orders may be granted for hardship based upon the grounds for relief on a case-by-case basis.

OSHA STANDARDS FOR INDUSTRY

OSHA standards and regulations are found in 29 Code of Federal Regulations (29 CFR 1900–1999). Industrial occupational safety and health standards are found in 29 CFR 1910. [22] The major categories are shown in Table 1-6.

Walking-Working Surfaces

Under 29 CFR 1910, pt. D, all places of employment must be kept clean, orderly, and sanitary. Aisles, loading docks, and passageways should be unobstructed and have safe clearances. Guardrails and covers should be provided wherever appropriate to protect personnel from open tanks, pits, vats, and ditches. Floor and wall openings should be guarded with railings and covers as appropriate.

Industrial stairs must have a minimum width of 22 inches, carry a moving concentrated load of at least 1000 pounds, and be 30° to 50° to the horizontal. Fixed ladders should be designed at a minimum for a single concentrated live load of 200 pounds; the clear length of rungs should be at least 16 inches, with no more than 12 inches between rungs.

TABLE 1-6
INDUSTRIAL OCCUPATIONAL SAFETY AND HEALTH STANDARDS

Section	Safety subject
D	Walking-working surfaces
E	Means of egress
G	Occupational health and environmental control
H	Hazardous materials
I	Personal protective equipment (PPE)
J	General environmental controls
K	Medical and first aid
L	Fire protection
N	Materials handling and storage
O	Machinery and machine guarding
P	Hand and portable powered tools and other handheld equipment
Q	Welding, cutting, and brazing
R	Special industries
S	Electrical
T	Commercial diving operations
Z	Toxic and hazardous substances

Scaffolds, which are used whenever work cannot be done safely from the ground or solid construction, must be on a firm foundation. Scaffolds should be of sufficient size and strength to safely support the intended use. Most of the OSHA walking-working surface standards have been developed from ANSI standards.

Means of Egress

In 29 CFR 1910, pt. E, means of egress is defined as a continuous and unob-structed way of exit travel from any point in a building or workplace to a public way. Means of egress includes the way of exit access, the exit itself, and the way of exit discharge.

The building must allow prompt escape of occupants in case of fire or other emergencies. Multiple safeguards must be provided for exits to have free egress whenever a building is occupied. Exits or routes to exits must be clearly visible to occupants.

In buildings of one to three stories, exit doors must have a 1-hour fire resistance rating at least. In buildings of more than three stories, a 2-hour fire resistance rat-ing is required. Exits must be at least 30 to 44 inches wide, depending on the num-ber of occupants. The way of exit access must be at least 28 inches wide. The way of exit discharge should be directly to the street or to open space with safe access to a public way. Exits and access to exits must be marked by a readily visible sign, illuminated by at least 5 foot-candles. An employee emergency escape plan and fire prevention plan must be in place. This entire OSHA standard has been promul-gated from the NFPA 101-1970, life safety code. The NFPA is located at 470 At-lantic Avenue, Boston, MA 02210.

Occupational Health and Environmental Control

OSHA standards for ventilation, noise exposure, and radiation in the workplace are listed in 29 CFR 1910, pt. G.

Ventilation Blast-cleaning enclosures must be ventilated with a continuous inflow of air during the blasting operation to promptly clear the dust-laden air from the enclosure. Abrasive-blasting respirators must be worn by all of the operators working inside blast-cleaning rooms.

Grinding, polishing, or buffing requires hoods connected to exhaust systems. Dust or dirt particles from these operations must flow into the hood and not into the breathing zone of the operator. The required air exhaust for these operations is a minimum of 220 ft^3/min and depends on the size of equipment.

Spray-paint booths must be constructed of noncombustible materials and venti-lated with 50 to 200 ft^3/min airflow velocities, depending upon the operating con-ditions. The total air volume exhausted through a spray booth must dilute vapors to

≤25 percent of their lower explosive limit. Table 1-7 is helpful in determining the dilution air volume required in spray-finishing operations for some common solvents. [27]

Example 1-1

Determine the dilution volume of air in cubic feet required to dilute 0.8 gallons of toluene solvent per minute to 25 percent of the lower explosive limit (LEL).

Solution

$$Dilution\ volume\ = \left(\frac{100 - LEL}{0.25\ LEL} \right) \left(\frac{volume\ vapor \times}{gal/min} \right)$$

where LEL = 1.4 percent at 70°F
vapor volume = 30.4 ft³/gal at 70°F

$$Dilution\ volume = 4\left(\frac{100 - 1.4}{1.4} \right)(30.4 \times 0.8)$$
$$= 6851$$

The required air ventilation is 6851 ft³/min.

Uncontaminated air must be supplied to a spray booth. During emergencies employees must wear respirators if the oxygen content of the workplace air is less than 19.5 volume percent.

Noise Exposure   Occupational noise exposure is regulated by OSHA under 29 CFR 1910.95. In the workplace employees must be provided with protection when sound levels measured on the A scale of a standard sound level meter exceed OSHA permissible noise limits (see Table 1-8). [28]

When employees are subjected to sound levels that are ≥85 dBA on an 8-hour time-weighted-average (TWA) basis, the employer must administer a hearing conservation program that includes monitoring, notification, testing, protection, training, and record keeping for all employees in danger of exposure. Monitoring includes developing a sampling strategy to identify employees to be included in the hearing conservation program. Each employee being monitored must be notified of the results. Employees may observe the monitoring by the employer. The employer must establish and maintain an audiometric testing program that is performed by a qualified person at no cost to the employees. Testing includes a baseline audiogram within 6 months of the employee's first exposure and annual audiograms thereafter. The annual audiogram must be evaluated for a standard threshold shift of 10 dB or more at 2000, 3000, or 4000 hertz (Hz) in either ear. Hearing protectors must be provided by the employer as part of the hearing conservation program. The em-

TABLE 1-7
VAPOR VOLUMES AND LOWER EXPLOSIVE LIMITS FOR SELECTED SOLVENTS

Solvent	Vapor volume of liquid at 70°F (ft³/gal)	Lower explosive limit by volume of air at 70°F (%)
Acetone	44.0	2.6
Amyl acetate (iso)	21.6	1.0*
Amyl alcohol (n)	29.6	1.2
Amyl alcohol (iso)	29.6	1.2
Benzene	36.8	1.4*
Butyl acetate (n)	24.8	1.7
Butyl alcohol (n)	35.2	1.4
Butyl cellosolve	24.8	1.1
Cellosolve	33.6	1.8
Cellosolve acetate	23.2	1.7
Cyclohexanone	31.2	1.1*
1, 1-Dichloroethylene	42.4	5.9
1, 2-Dichloroethylene	42.4	9.7
Ethyl acetate	32.8	2.5
Ethyl alcohol	55.2	4.3
Ethyl lactate	28.0	1.5*
Methyl acetate	40.0	3.1
Methyl alcohol	80.8	7.3
Methyl cellosolve	40.8	2.5
Methyl ethyl ketone	36.0	1.8
Naphtha (VM&P) (76° Naphtha)	22.4	0.9
Naphtha (100° Flash) Safety Solvent—Stoddard Solvent	23.2	1.0
Propyl acetate (n)	27.2	2.8
Propyl acetate (iso)	28.0	1.1
Propyl alcohol (n)	44.8	2.1
Propyl alcohol (iso)	44.0	2.0
Toluene	30.4	1.4
Turpentine	20.8	0.8
Xylene (o)	26.4	1.0

*At 212°F.

TABLE 1-8
OSHA PERMISSIBLE NOISE EXPOSURE LIMITS.

Daily Exposure Duration (h)	Maximum Sound Level (dBA)
8	90
6	92
4	95
3	97
2	100
1.5	102
1	105
0.5	110
≤0.25	115

ployer must institute a training program in the use and care of hearing protectors for all employees who are exposed to an 8-h TWA ≥85 dBA. The hearing protectors must attenuate 8-h TWA to at least 90 dBA and for employees with a standard threshold shift down to at least 85 dBA. The audiometric test records must be retained by the employer for the duration of the affected employee's employment.

Radiation   Ionizing radiation includes alpha, beta, gamma, and x-rays, but not sound or radio waves, or visible, infrared, or ultraviolet light. It is regulated by OSHA under 29 CFR 1910.96 (see Table 1-9).

Employers must monitor the workplace to assure compliance with Table 1-9. Personnel monitoring equipment like dosimeters or badges must be supplied to each potentially affected employee. Each radiation area must be conspicuously posted with signage bearing the radiation symbol shown in Figure 1-2. This radiation symbol is magenta or purple on a yellow background. The caution signage must also contain the appropriate warning description. [29]

Nonionizing radiation in the radio-frequency region is regulated by OSHA under 29 CFR 1910.97. The OSHA radiation protection guidelines for radio frequency or electromagnetic radiation, averaged over 0.1 hour, are shown in Table 1-10.

TABLE 1-9
OSHA EXPOSURE LIMITS TO IONIZING RADIATION

Human exposure	Maximum limit (rems/quarter)
Whole body, head and trunk, active blood-forming organs, eye lens, gonads	1.25
Skin of whole body	7.5
Hands and forearms, feet and ankles	18.75

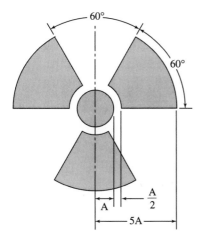

FIGURE 1-2
OSHA IONIZING RADIATION HAZARD WARNING
SYMBOL.

The guidelines apply to both continuous or intermittent radiation and to either whole-body or partial-body radiation exposure. The warning symbol for nonionizing radiation is a red isosceles triangle above an inverted black isosceles triangle with aluminum letters and border (see Fig. 1-3 on page 16). [30]

Hazardous Materials

OSHA standards for hazardous materials are in 29 CFR 1910, pt. H. Compressed gases are regulated by OSHA in accordance with standards of the Compressed Gas Association. [31] Cylinders and other compressed-gas containers must be inspected to verify they are in safe condition and must have pressure relief devices. The compressed gases under this standard include acetylene, hydrogen, oxygen, and nitrous oxide. [32]

The storage, piping, valves, fittings, loading, dispensing, and processing of flammable and combustible liquids is covered by 29 CFR 1910.106. Explosives and blasting agents are under 29 CFR 1910.109. The storage and handling of liquified petroleum gases, usually propane or normal butane, are regulated with 29 CFR

TABLE 1-10
OSHA NONIONIZING RADIATION PROTECTION GUIDELINES

Type of exposure	Guidelines for 0.1 h
Power density	$\leq 10 \dfrac{mW}{cm^2}$
Energy density	$\leq 1 \dfrac{mW \cdot h}{cm^2}$

FIGURE 1-3
OSHA RADIO-FREQUENCY RADIATION HAZARD WARNING
SYMBOL.

1910.110. The storage and handling of anhydrous ammonia is under 29 CFR
1910.111.

Process safety management of hazardous chemicals, under 29 CFR 1910.119,
seeks to prevent or minimize the consequences of catastrophic releases of chemicals into the workplace and surrounding areas. These chemical releases may result
in toxic, fire, or explosion hazards.

Hazardous waste treatment, disposal, and site remediation are covered under 29
CFR 1910.120. These OSHA standards are also used in emergency response situations with hazardous wastes.

Personal Protective Equipment

OSHA standards for personal protective equipment (PPE) are in 29 CFR 1910, pt.
I. These standards cover protective equipment for eyes, face, head and extremities,
protective clothing, respirators, and shields. All PPE must be maintained in a sanitary and reliable condition. Generally, PPE is provided by the employer for use by
the employees. Whenever PPE is provided by employees for themselves, the employer is still responsible for assuring its adequacy, maintenance, and sanitation.

The OSHA standards for eye and face, respiratory, head, foot, and electrical device protection are in 29 CFR 1910.132–137.

General Environmental Controls

OSHA standards for general environmental controls for permanent places of employment are in 29 CFR 1910, pt. J. Potable water, toilet and washing facilities, change rooms, and clothes-drying facilities must be provided. Food handling and the consumption of food and beverages on the premises must be done under sanitary conditions in accordance with sound hygienic principles. [33]

Red and yellow colors in the workplace should be used to mark physical hazards. Red should be the basic identification color for fire protection equipment, safety cans for flammable liquids, danger signs, and emergency stop devices for machinery. Red lights should be provided at barricades. Yellow should be used as the basic color to designate caution. [34]

Danger signs should be red, black, and white. Caution signs should be yellow with black letters. Safety instruction signs should be white with black letters or green with white letters. [35]

Machines and equipment must have lockout and tagout devices so that they don't start up unexpectedly while being maintained or repaired. [36] A *lockout device* is a key or combination lock that holds an energy-isolating device in a safe position and prevents the energization of a machine or equipment until it is removed.

A *tagout device* is a prominent warning device or sign (e.g., Do Not Start or Do Not Operate) that can be securely fastened to an energy-isolating device to indicate that the equipment being controlled should not be operated until the tagout device is removed.

Lockout and tagout devices must be durable and substantial to withstand adverse weather and operating conditions, as well as to prevent removal without the use of excessive force.

Fire Protection

OSHA standards for fire protection are listed in 29 CFR 1910, pt. L. These cover private fire departments, fire suppression equipment, and fire detection and alarm systems. These standards are based primarily on NFPA and ANSI.

OSHA has designated the following fire classifications:

- Class A fires involve ordinary combustible materials like paper, wood, cloth, and some rubber and plastic materials.
- Class B fires involve flammable or combustible liquids, flammable gases, greases and similar materials, and some rubber and plastic materials.
- Class C fires involve energized electrical equipment where safety to the employee requires the use of electrically nonconductive extinguishing media.
- Class D fires involve combustible metals like magnesium, titanium, zirconium, sodium, lithium, and potassium.

The employer is responsible for the organization and management of the fire department or fire brigade. Employees who are expected to do interior structural fire fighting must be physically capable of performing these duties. The employer must provide quality training and education to all members of the fire brigade. Protective clothing must be supplied by the employer. [37]

The employer must provide portable fire extinguishers that are mounted, located, and identified so that they are readily accessible to employees. Carbon tetrachloride or chlorobromomethane extinguishing agents are not allowed because of toxicity and environmental concerns. The maximum employee travel distance to portable fire extinguishers for Class A fires is 75 feet (22.9 m); for Class B fires, 50 feet (15.2 m); for Class C and Class D fires, 75 feet (22.9 m). The employer must inspect equipment monthly, check for maintenance annually, and test in accordance with the type of extinguisher. [38]

Automatic fire sprinkler systems must be properly maintained; this includes an annual main-drain flow test and an opening of the test valve every other year to ensure the sprinkler system works properly. The minimum vertical clearance between sprinklers and material below them must be at least 18 inches (45.7 cm). [39]

The employer must install and maintain an automatic fire detection system. The system should actuate related fire extinguishment systems in time to control or extinguish a fire within 30 seconds. [40]

The employee fire alarm system must provide a warning for the necessary emergency action and sufficient reaction time for the safe escape of employees from the affected work area. The employee alarm must be distinctive and recognizable as a signal to evacuate the work area or to perform the actions of the emergency action plan. [41]

Materials Handling and Storage

OSHA standards for materials handling and storage are listed in 29 CFR 1910, pt. N. Safe clearances for mechanical handling equipment must be provided for aisles, loading docks, turns, and passage. Bags, containers, and bundles must be stacked so that they are stable and secure against sliding or collapse. All storage areas must have good housekeeping practices to minimize hazards from tripping, fire, explosion, or pests. The employer must provide appropriate clearance limit signs. [42]

Machinery and Machine Guarding

OSHA standards for machine guarding are listed in 29 CFR 1910, pt. O. Machine operators and other personnel in the machine area must be protected from hazards created by the point of operation, pinch point, rotating parts, flying chips, and sparks. Machine guards should be affixed to the machine, or secured elsewhere if attachment to the machine is not possible. The guard cannot become an accident hazard in itself.

The point of operation, where the work is actually being performed on the material being processed, must have a guard to prevent any part of the body of the operator from being in the danger zone during the operating cycle. Special hand tools for placing and removing material may be provided for supplemental protection in addition to the guards. The types of machines that often require point-of-operation guarding include power presses and saws, milling machines, jointers, forming rolls and calendars, cutters, and shears. [43]

Handheld Equipment

OSHA standards for handheld equipment are listed in 29 CFR 1910, pt. P. The employer is responsible for the safe condition of the tools and equipment used by employees. Compressed air may only be used for cleaning purposes when the pressure is less than 30 parts per square inch (psi) and with chip guarding and personal protective equipment.

Portable powered tools must have guarding, as described in 29 CFR 1910.243. This applies to circular saws, belt sanding machines, pneumatic powered tools, grinders, fasteners, power lawnmowers, and jacks.

Welding, Cutting, and Brazing

OSHA standards for welding, cutting, and brazing materials and operations are listed in 29 CFR 1910, pt. Q. The OSHA standards are based on NFPA Standard 51B, Standard for Fire Prevention in Use of Cutting and Welding Processes.

The basic precautions for fire prevention in welding or cutting work include (1) isolation or guarding of all movable fire hazards in the vicinity of the work to be performed, (2) providing fire extinguishing equipment for instant use if necessary, and (3) inspection and authorization prior to proceeding with the work. Protective equipment such as goggles, helmets, and hand shields must be provided by the employer. The lenses in the goggles must have the proper shade number based on the type of welding operation. [44]

Hazardous materials may be used in welding fluxes, filler metals, and coatings. These hazardous materials may produce fumes and gases from welding and brazing that are hazardous to health. Caution and warning labels should be used in welding and brazing whenever appropriate. [44] Ventilation must be provided at least at the rate of 2000 ft^3/min (57 m^3/min), except where local exhaust hoods and booths or airline respirators are provided. Where hoods are provided, the airflow rate in the hood must be at least 100 linear feet per minute in the zone of welding. The hazardous materials used may include toxic metals, anions, and gases.

Special precautions are required for oxygen-fuel gas welding and cutting. [23] Oxygen cylinders must not be stored near flammable materials like acetylene, oil, and grease. They should be stored outside at least 20 feet from flammable materials or separated by a 5-foot barrier with a fire resistance rating of at least 0.5 hours. Fuel gas cylinders should be stored, shipped, and used with the valve end up.

Arc welding machines must be able to carry their rated load with rated temperature rises where the cooling air does not exceed 40°C and the altitude does not exceed

3300 feet (1006 m). Hazardous gases, dust, and light rays can be produced by arc welding. The maximum alternating current (AC) requirement for manual arc welding is 80 volts; for automatic arc welding machines, 100 volts. For direct current (DC) the maximum voltage is 100 volts for either manual or automatic machines. [45]

Electrical

OSHA electrical system standards are listed in 29 CFR 1910, pt. S. Numerous ANSI and NFPA standards were used to develop the OSHA standards. All electrical systems that use electric power and light for employees in the workplace are covered by these standards.

Electrical equipment must be free from recognized hazards related to installation and intended use, insulation, durability, mechanical strength, and heat and arcing effects. [46]

Employees must be protected from electrical shock or other injuries when work is being performed near energized electrical equipment. While any employee is exposed to de-energized electrical equipment, the circuits energizing the parts must be locked out or tagged out, preferably both. [47] Insulated personal protective equipment that is safe and reliable must be provided by the employer. Head, eye, and body PPE should be nonconductive or properly insulated. Tools or handling equipment that may come in contact with energized conductors or circuits should be insulated or nonconductive. Safety signs, tags, and barricades should be used when appropriate to warn employees about electrical hazards. [48]

Toxic and Hazardous Substances

OSHA standards for workplace hazards for toxic substances and communication to employees about these hazards are listed in 29 CFR 1910, pt. Z. These standards cover many toxic air contaminants, asbestos, vinyl chloride, arsenic, lead, benzene, acrylonitrile, ethylene oxide, and formaldehyde. [49] The 8-hour TWA, STEL, and ceiling values are tabulated in 29 CFR 1910.1000.

The hazards of all appropriate produced or imported chemicals must be transmitted to employers and employees. Comprehensive hazard communication programs as required in 29 CFR 1910.1200 must be followed to comply with this standard. Included in this hazard communication program are material safety data sheets, container labeling, and employee training.

Chemical laboratories are a special situation for occupational exposure to hazardous chemicals. The OSHA standards for such workplaces are found in 29 CFR 1910.1450. These OSHA standards interact greatly with 29 CFR 1910.1200 for communication and training of laboratory employees. A chemical hygiene plan is included in the laboratory safety standard. Under 29 CFR 1910.1450 the employer must provide all employees who work with hazardous chemicals with the opportunity to have medical examinations and consultation appropriate for the chemical exposure in the laboratory.

ENVIRONMENTAL LEGISLATION AND REGULATIONS

Since 1970 numerous federal laws and regulations have been enacted in response to public concern about protecting the environment. [52] The following laws (and their subsequent amendments) are generally considered to be the main driving forces in the environmental field in the United States. [51]

- National Environmental Policy Act of 1970 (NEPA)
- Clean Air Act of 1970 (CAA)
- Federal Water Pollution Control Act of 1972 (FWPCA)
- Safe Drinking Water Act of 1974 (SDWA)
- Toxic Substances Control Act of 1976 (TSCA)
- Resource Conservation and Recovery Act of 1976 (RCRA)
- Comprehensive Environmental Response, Compensation, and Liability Act of 1980 (CERCLA)

The National Environmental Policy Act requires an environmental impact statement (EIS) for every project that may affect the environment and needs a federal permit.

The Clean Air Act provided guidelines and standards to control air pollution. The reduction of undesirable air emissions is its major emphasis. The Federal Water Pollution Control Act, also known as the Clean Water Act, was enacted to control pollution of lakes, rivers, and streams caused by improper discharge of hazardous waste. Both of these laws have resulted in many long-term research and commercial programs designed to advance federal environmental objectives.

The Safe Drinking Water Act was enacted to ensure public drinking water supplies will be maintained at a high level of quality.

The Toxic Substances Control Act provided a framework to regulate the introduction of chemicals into the marketplace. The premanufacturing and premarketing tests and notification required by TSCA initiated many new research and commercial fields, particularly with respect to human health and environmental effects. In addition, TSCA amendments have led to environmental regulations for polychlorinated biphenyls (PCBs) and asbestos.

The Resource Conservation and Recovery Act was necessary to control the disposal of hazardous and nonhazardous waste and to reduce its generation. RCRA has become the primary focus for regulating hazardous waste generation from industry in the United States. The Pollution Prevention Act of 1990 added further emphasis to this area of environmental protection.

The Comprehensive Response, Compensation, and Liability Act was enacted to clean up high-priority hazardous waste sites generated by past disposal practices. CERCLA legislation led to the Superfund program, which provides funding for cleanup of hazardous wastes sites that endanger human health and the environment.

The regulation of municipal solid waste disposal is generally under RCRA. If a municipal solid waste landfill endangers human health and the environment, then it may come under CERCLA regulation.

Environmental legislation is regulated by the U.S. Environmental Protection Agency (EPA), which is headquartered at 401 M Street SW, Washington, DC 20460; telephone: 202-260-2090. There are 10 EPA regions in the United States (see Fig. 1-4). The headquarters location for each of the EPA regions is listed in Table 1-11.

All of the EPA regulations are found in 40 Code of Federal Regulations (40 CFR), which is updated annually by the Office of the Federal Register. All EPA proposed and final regulations that have general applicability and legal effect appear in the Federal Register, which is published on each federal workday.

Air Quality

The U.S. Environmental Protection Agency under the Clean Air Act of 1970 has identified eight major air pollutants that are potentially hazardous to public health and

FIGURE 1-4
U.S. ENVIRONMENTAL PROTECTION AGENCY, REGIONS.

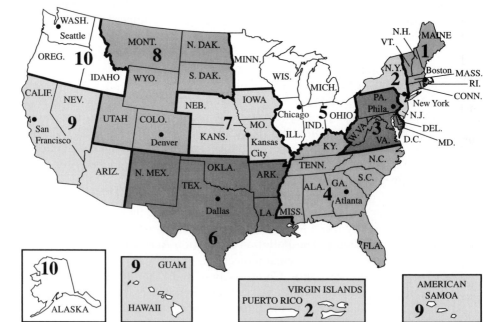

TABLE 1-11
EPA REGIONAL HEADQUARTER LOCATIONS

Region	Address
1	John F. Kennedy Federal Bldg., Boston, MA 02203
2	26 Federal Plaza, New York, NY 10278
3	841 Chestnut St., Philadelphia, PA 19107
4	345 Courtland St. NE., Atlanta, GA 30365
5	230 S. Dearborn St., Chicago, IL 60604
6	1445 Ross Ave., Dallas, TX 75202
7	726 Minnesota Ave., Kansas City, KS 66101
8	999 18th St., Denver, CO 80202
9	215 Fremont St., San Francisco, CA 94105
10	1200 6th Ave., Seattle, WA 98101

welfare (see Table 1-12 on page 24). The Clean Air Act empowered the EPA to establish national ambient air quality standards (NAAQS) (see Table 1-13 on page 25).

In order to control air quality in the United States, EPA imposed limits for the prevention of significant deterioration (PSD) in those areas of the country that were already cleaner than required by the NAAQS. Under PSD there are three classes of air quality areas:

- Class I: pristine areas that are subject to highly restrictive controls
- Class II: areas of moderate industrial growth
- Class III: areas of major industry activity.

For each of these classifications the EPA set numerical limits to indicate the maximum permissible increment of air quality degradation in that area from new stationary facilities. The EPA specified that each major new industrial facility must install the best available control technology (BACT) to limit air emissions. All major stationary sources and major modifications located in attainment areas are subject to PSD regulations.

The areas in the United States that failed to attain compliance with NAAQS have been designated nonattainment areas. Any new industrial facility may be constructed in a nonattainment area only if that facility has met certain stringent air quality control conditions. If new industrial sources are locating in a nonattainment area, reductions in air emissions from other sources in the region have to more than offset these air pollution discharges.

In order to allow more flexibility and creativity in reduction of air quality pollutants, EPA has instituted a "bubble policy." A company may develop the most efficient, cost-effective way to control a plant's air emissions so long as the sum of the total air pollution from the plant does not exceed the sum of the current limits on the individual point sources at the plant. This EPA bubble policy also covers more

TABLE 1-12
EIGHT MAJOR AIR POLLUTANTS

Pollutant	Health and environmental concern	Environmental sources
Particulates	Respiratory and visual irritant	Dust, combustion, and minerals processes
Sulfur dioxide	Respiratory irritant; vegetation damage	Combustion and minerals processes
Carbon monoxide	Cardiovascular, nervous, and pulmonary systems	Automobiles, combustion, minerals processes, natural sources
Nitrogen dioxide	Respiratory illness and lung damage	High temperature combustion and natural sources
Ozone	Respiratory irritant; vegetation damage	Atmospheric reactions
Lead	Retardation and brain damage	Combustion, minerals processes, and natural sources
Hydrocarbons	Respiratory and visual irritant	Automobiles, combustion, and natural sources
Photochemical oxidants	Respiratory and visual irritant; vegetation damage	Atmospheric reaction

than one plant within the same area, thereby allowing the company flexibility to make air pollution control choices within the total bubble.

EPA also has an offset policy. A new facility or a modification of the existing facility can meet the air quality regulations in a nonattainment area if another air pollution source in the area institutes additional air quality control actions, thereby offsetting increased emissions from the new facility.

In 1990, Clean Air Act Amendments (CAAA) identified 189 hazardous air pollutants (HAP); these pollutants also come under stringent EPA air quality regulations (see Table 1-14 on pages 26–27). HAP emissions are severely restricted.

Any facility that emits ≥ 10 tons per year of a single HAP, or ≥ 25 tons per year of any combination of HAPs, is defined as a major source. Each major source is required to meet maximum achievable control technology (MACT) requirements. For new sources, MACT has been defined as the best available control system. For existing sources, MACT has been defined as at least as good as the average of the best 12 percent of the sources in the category. [2]

Example 1-2

A continuous round-the-clock industrial operation emits 1500 grams of methyl ethyl ketone (MEK) per hour into the atmosphere. Is this facility a major source of hazardous air pollutants under the CAAA of 1990?

TABLE 1-13
NATIONAL AMBIENT AIR QUALITY STANDARDS

Pollutant	Primary (health-related)		Secondary (welfare-related)	
	Averaging time	Concentration	Averaging time	Concentration
Particulates	Annual geometric mean 24-h	50 µg/m³	Same as primary	Same as primary
Sulfur dioxide	Annual arithmetic mean 24-h	80 µg/m³ (0.03 ppm) 365 µg/m³ (0.14 pm)	3-h	1300 µg/m³ (0.50 ppm)
Carbon monoxide	8-h 1-h	10 mg/m³ (9 ppm) 40 mg/m³ (35 ppm)	Same as primary Same as primary	Same as primary Same as primary
Nitrogen dioxide	Annual arithmetic mean	100 µg/m³ (0.053 ppm)	Same as primary	Same as primary
Ozone	1-h	235 µg/m³ (0.12 ppm)	Same as primary	Same as primary
Lead	Maximum quarterly average	1.5 µg/m³	Same as primary	Same as primary

Source: U.S. Environmental Protection Agency

TABLE 1-14
HAZARDOUS AIR POLLUTANTS IDENTIFIED IN THE CLEAN AIR ACT AMENDMENTS
OF 1990

Acetaldehyde	2,4-D, salts and esters	Ethylene thiourea
Acetamide	1,1 dichloro-2, 2-bis (*p*	Ethylidene dichloride (1,1-
Acetonitrile	chlorophenyl) ethylene	dichloroethane)
Acetophenone	Diazomethane	Formaldehyde
2-Acetylaminofluorene	Dibenzofurnas	Heptachlor
Acrolein	1,2-Dibromo-3-	Hexachlorobenzene
Acrylamide	chloropropane	Hexachlorobutadiene
Acrylic acid	Dibutylphthalate	Hexachlorocyclopentadiene
Acrylonitrile	1,4-Dichlorobenzene (p)	Hexachloroethane
Allyl chloride	3,3-Dichlorobenzidene	Hexamethylene-1,6-
4-Aminobiphenyl	Dichloroethyl ether (bis [2-	diisocyanate
Aniline	chloroethyl] ether)	Hexamethylphosphora-
o-Anisidine	1,3-Dichloropropene	mide
Asbestos	Dichlorvos	Hexane
Benzene (including	Diethanolamine	Hydrazine
benzene from gasoline)	N,N-Diethylaniline (N,N-	Hydrochloric acid
Benzidine	dimethylaniline)	Hydrogen fluoride
Benzotrichloride	Diethyl sulfate	(hydrofluoric acid)
Benzyl chloride	3,3-Dimethoxybenzidine	Hydrogen sulfide
Biphenyl	Dimethyl aminoazobenzene	Hydroquinone
Bis (2-ethylhexyl) phthalate	3,3-Dimethyl benzidine	Isophorone
(DEHP)	Dimethyl carbamoyl	Lindane (all isomers)
Bis (chloromethyl) ether	chloride	Maleic anhydride
Bromoform	Dimethyl formamide	Methanol
1,3-Butadiene	1,1-Dimethyl hydrazine	Methoxychlor
Calcium cyanamide	Dimethyl phthalate	Methyl bromide
Caprolactam	Dimethyl sulfate	(bromomethane)
Captan	4,6-Dinitro-*o*-cresol, and	Methyl chloride
Carbaryl	salts	(chloromethane)
Carbon disulfide	1,4-Dinitrophenol	Methyl chloroform (1,1,1,-
Carbon tetrachloride	2,4-Dinitrotoluene	trichloroethane)
Carbonyl sulfide	1,4-Dioxane (1,4-	Methyl ethyl ketone (2-
Catechol	diethyleneoxide)	butanone)
Chloramben	1,2-Diphenylhydrazine	Methyl hydrazine
Chlordane	Epichlorohydrin (2-chloro-	Methyl iodide
Chlorine	2, 3-epoxypropane)	(iodomethane)
Chloroacetic acid	1,2-Eposybutane	Methyl isobutyl ketone
2-Chloroacetophenone	Ethyl acrylate	(hexone)
Chlorobenzene	Ethyl benzene	Methyl isocyanate
Chlorobenzilate	Ethyl carbamate (ure-	Methyl methacrylate
Chloroform	thane)	Methyl tert butyl ether
Chloromethyl methyl	Ethyl chloride	4,4-Methylene bis (2-
ether	(chloroethane)	chloroaniline)
Chloroprene	Ethylene dibromide	Methylene diphenyl
Cresols/Cresylic acid	(dibromoethane)	diisocyanate (MDI)
(isomers and mixture)	Ethylene dichloride (1,2-	4,4'-Methylenedianiline
0-Cresol	dichloroethane)	Naphthalene
m-Cresol	Ethylene glycol	Nitrobenzene
p-Cresol	Ethylene imine (aziridine)	4-Nitrobiphenyl
Cumene	Ethylene oxide	4-Nitrophenol

TABLE 1-14
HAZARDOUS AIR POLLUTANTS IDENTIFIED IN THE CLEAN AIR ACT AMENDMENTS
OF 1990 (*CONTINUED*)

2-Nitropropane	Styrene	Xylenes (isomers and
N-Nitroso-N-methylurea	Styrene oxide	mixture)
N-Nitrosodimethylamine	2,3,7,8-Tetrachlorodibenzo-	o-Xylenes
N-Nitrosomorpholine	p-dioxin	m-Xylenes
Parathion	1,1,2,2-Tetrachloroethane	p-Xylenes
Pentachloronitrobenzene	Tetrachloroethylene	Antimony compounds
(quintobenzene)	(perchloroethylene)	Arsenic compounds
Pentachlorophenol	Titanium tetrachloride	(inorganic including
Phenol	Toluene	arsine)
p-Phenylenediamine	2,4-Toluene diamine	Beryllium compounds
Phosgene	2,4-Toluene diisocyanate	Cadmium compounds
Phosphine	o-Toluidine	Chromium compounds
Phosphorus	Toxaphene (chlorinated	Cobalt compounds
Phthalic anhydride	camphene)	Coke oven emissions
Polychlorinated biphenyls	1,2,4-Trichlorobenzene	Cyanide compounds
(arochlors)	1,1,2-Trichloroethane	Glycol ethers
1,3-Propane sulfone	Trichloroethylene	Lead compounds
beta-Propiolactone	2,4,5-Trichlorophenol	Manganese compounds
Propionaldehyde	2,4,6-Trichlorophenol	Mercury compounds
Propoxur (baygon)	Triethylamine	Fine mineral fibers
Propylene dichloride (1,2-	Trifluralin	Nickel compounds
dichloropropane)	2,2,4-Trimethylpentane	Polycyclic organic matter
Propylene oxide	Vinyl acetate	Radionuclides (including
1,2-Propylenimine (2-	Vinyl bromide	radon)
methyl aziridine)	Vinyl chloride	Selenium compounds
Quinoline	Vinyllidene chloride (1,1-	
Quinone	dichloroethylene)	

Solution

MEK is listed in Table 1-14, which means the regulatory criteria is $\geq$ 10 tons MEK per year.

$$\text{Emissions} = \text{g/h} \times 24 \text{ h} \times 365 \text{ days}$$
$$= 13{,}140{,}000 \text{ g/year}$$
$$= 14.5 \text{ tons/year (1 ton} = 2000 \text{ lb, or } 908{,}000 \text{ g)}$$

This is a major source of MEK, which is a hazardous air pollutant.

The CAAA of 1990 also added a program to further control sulfur dioxide (SO_2) emissions by 10 million tons per year. Utilities must reduce SO_2 emissions to 2.5 pounds per million Btu coal, with a further reduction to 1.2 pounds per million Btu coal by the year 2000. The utilities may adopt their own strategy to achieve this reduction, such as switching to low-sulfur coal, coal washing, or flue gas desulfurization.

Water Quality

The Clean Water Act of 1972 is the primary federal legislation affecting wastewater treatment. It established a national pollutant discharge elimination system (NPDES) to regulate the discharge of pollutants into the nation's waters through a permit that must be renewed every 5 years. The NPDES permit stipulates regulatory limits for pollutants in the effluent from a facility and requirements for monitoring and recording. The restrictions from the NPDES permit are generally based on national effluent guidelines but may be modified on a local case-by-case basis.

The Clean Water Act and its subsequent amendments control acceptable levels of toxic chemicals in water through standards based on the 129 priority pollutants shown in Table 1-15. These priority pollutants should be particularly targeted in the design of wastewater treatment plants.

Industrial and municipal wastewater treatment plants have their own individual NPDES permit. Municipal plants must practice at least secondary sewage treatment prior to discharge into a receiving river or stream. This secondary treatment requirement restricts the effluent to 30 mg/L suspended solids and 30 mg/L biochemical oxygen demand (BOD) and requires chlorination to destroy pathogens. [13]

Additional, more restrictive standards are placed on certain receiving streams based on the usage category, for example, receiving bodies of water used for potable water or swimming. Rivers used only for navigation may not require such elaborate treatment prior to discharge. A treatment plant that discharges into a quality-limited stream or river may be forced to produce an effluent that is superior to the industry standard. Quality-limited receiving bodies may periodically fail to meet required standards for designated parameters, such as temperature, dissolved oxygen, BOD, or suspended solids.

As municipal sewage treatment plants generate high-quality effluent, they may be permitted to recycle into the local water supply for reuse, provided proper safeguards are in place to protect public health. Often sewage plant effluent is used to irrigate farms, parks, and roadways. This type of reuse captures the nutrient value of effluent and lessens the impact on the receiving body of water.

The characterization of wastewater is the most important first step in developing both regulatory compliance and cost-effective treatment facilities. Domestic sewage and industrial waste are point sources of waste discharge that are treated and conveyed to a single discharge point and into the receiving water. These point sources are contrasted with nonpoint sources, which are urban and agricultural runoff and involve multiple discharge points.

The goal of the Clean Water Act has been to create fishable and swimmable water in the United States. Biodegradable organic substances compete for the dissolved oxygen in water because they can be oxidized in the receiving water; hence the need to remove these organics prior to discharge. Chemical oxygen demand (COD), total organic carbon (TOC), and biochemical oxygen demand (BOD) are the primary analytical tests that are used to determine organic compounds in wastewaters and their potential effect on receiving waters.

The organic substances in wastewater are often removed by biodegradation, which requires sufficient nutrients and trace metals and salts. Nitrogen and phos-

TABLE 1-15
PRIORITY POLLUTANTS FOR WATER QUALITY MANAGEMENT

Priority pollutant	Type of chemical substance
Acenaphthene	Aromatic
Acenaphthylene	Aromatic
Acrolein	Organic
Acrylonitrile	Organic
Aldrin	Pesticide
Anthracene	Aromatic
Antimony	Metal
Arsenic	Metal
Asbestos	Mineral
Beryllium	Metal
Benzene	Aromatic
Benzidine	Substitute aromatic
Benzo [a] anthracene	Aromatic
3,4-Benzofluoranthane	Aromatic
Benzo [k] fluoranthane	Aromatic
Benzo [ghi] perylene	Aromatic
Benzo [e] pyrene	Aromatic
e-BHC-α	Pesticide
b-BHC-β	Pesticide
r-BHC (lindane)-δ	Pesticide
g-BHC-δ	Pesticide
Bis (2-chloroethoxy) methane	Chlorinated ether
Bis (2-chloromethyl) ether	Chlorinated ether
Bis (chloromethyl) ether	Chlorinated ether
Bis (2-chloroisopropyl) ether	Chlorinated ether
Bis (2-ethylhexyl) phthalate	Phthalate ester
Bromoform	Chlorinated alkane
4-Bromophenyl phenyl ether	Chlorinated ether
Butyl benzyl phthalate	Phthalate ester
Cadmium	Metal
Carbon tetrachloride	Chlorinated alkane
Chlordane	Pesticide
Chlorobenzene	Chlorinated alkane
Chlorodibromonethane	Chlorinated alkane
Chloroethane	Chlorinated ether
2-Chloroethyl vinyl ether	Chlorinated ether
Chloroform	Chlorinated alkane
2-Chlorophenol	Phenol
4-Chlorophenyl phenyl ether	Chlorinated esther
2-Chlorophythalene	Chlorinated aromatic
Chromium	Metal
Chrysene	Aromatic
Copper	Metal
Cyanide	Miscellaneous
4,4-DDD	Pesticide
4,4-DDE	Pesticide
4,4-DDT	Pesticide
Dibenzo [a,h] anthracene	Chlorinated aromatic
1,3-Dichlorobenzene	Chlorinated aromatic

(continued)

TABLE 1-15
PRIORITY POLLUTANTS FOR WATER QUALITY MANAGEMENT *(CONTINUED)*

Priority pollutant	Type of chemical substance
1,4-Dichlorobenzene	Chlorinated aromatic
3,3-Dichlorobenzidene	Substituted aromatic
Dichlorbromothane	Chlorinated alkane
Dichlorodifluoromethane	Chlorinated alkane
1,1-Dichloroethane	Chlorinated alkane
1,2-Dichloroethylene	Chlorinated alkane
1,1-Dichloroethylene	Chlorinated alkane
2,4-Dichlorophenol	Phenol
1,3-Dichloropropane	Chlorinated alkane
1,2-Dichloropropylene	Chlorinated alkane
Dieldrin	Pesticide
Diethyl phthalate	Phthalate ester
2,4-Dimethyl phenol	Phenol
Dimethyl phthalate	Phthalate ester
Di-*n*-Butyl phthalate	Phthalate ester
4,6-Dinitro-*o*-cresol	Phenol
2,4-Dinitrophenol	Phenol
2,4-Dinitrotoluene	Substituted aromatic
2,6-Dinitrotoluene	Substituted aromatic
Di-*N*-octyl phthalate	Phthalate ester
1,21-Diphenyl hydrazine	Substituted aromatic
A-Endosulfan-α	Pesticide
B-Endosulfan-ß	Pesticide
Endosulfan sulfate	Pesticide
Endrin	Pesticide
Endrin aldehyde	Pesticide
Ethylbenzene	Aromatic
Fluoranthene	Aromatic
Fluorene	Aromatic
Haphthalene	Aromatic
Heptachlor	Pesticide
Heptachlor epoxide	Pesticide
Hexachlorobenzene	Chlorinated aromatic
Hexachlorobutadiene	Chlorinated alkane
Hexachloroethane	Chlorinated alkane
Indeno [1,2,3-c,d] pyrene	Aromatic
Isophorone	Organic
Lead	Metal
Mercury	Metal
Methyl bromide	Chlorinated alkane
Methyl chloride	Chlorinated alkane
Methylene chloride	Chlorinated alkane
Nickel	Metal
Ntirobenzene	Substituted aromatic
2-Nitrophenol	Phenol
4-Nitrophenol	Phenol
n-Nitrosodimethylamine	Organic
n-Nitrosodi-*N*-propylamine	Organic
n-Nitrosodiphenylamine	Organic

TABLE 1-15
PRIORITY POLLUTANTS FOR WATER QUALITY MANAGEMENT *(CONTINUED)*

Priority pollutant	Type of chemical substance
Para-chlor-meta-cresol	Phenol
PCB-1016	Chlorinated biphenyl
PCB-1221	Chlorinated biphenyl
PCB-1232	Chlorinated biphenyl
PCB-1248	Chlorinated biphenyl
PCB-1254	Chlorinated biphenyl
PCB-1260	Chlorinated biphenyl
Pentachlorophenol	Phenol
Phenanthane	Aromatic
Phenol	Phenol
Pyrene	Aromatic
Selenium	Metal
Silver	Metal
2,3,7,8-Tetrachlorodibenzo-*p*-dioxin	Chlorinated organic
1,1,2,2,-Tetrachloroethane	Chlorinated alkane
Tetrachloroethylene	Chlorinated alkane
Thallium	Metal
Toluene	Aromatic
1,2-*trans*-Dichloroethylene	Chlorinated alkane
1,2,4-Trichlorobenzene	Chlorinated aromatic
1,1-Trichloroethane	Chlorinated alkane
1,1,2-Trichloroethane	Chlorinated alkane
Trichloroethylene	Chlorinated alkane
Trichlorofluoromethane	Chlorinated alkane
2,4,6-Trichlorophenol	Phenol
Vinyl chloride	Chlorinated phenol

phorus are the two primary nutrients that must be available for effective biodegradation processes to succeed. Iron, potassium, calcium, magnesium, cobalt, molybdenum, chlorides, and sulfur are also desirable so long as they are present in trace quantities in the wastewater stream. Excessive amounts of either nutrients or trace elements in wastewater must normally be controlled because they will adversely affect the receiving stream of the discharge effluent from the waste treatment facility.

Microorganisms must be present in order for biological treatment processes to be used on the wastewater stream. If the microorganisms contain high levels of pathogens, the water effluent from the treatment facility may be unfit for swimming and unsafe for fishing. Some marine life, such as shellfish, may concentrate the pathogens in their bodies, thereby increasing their toxicity, particularly if they are eaten raw.

The effluent from wastewater treatment facilities should have pH values between 6 and 8 to minimize the impact on the receiving stream and the environment. Low pH values create corrosive conditions in wastewater effluents, while high pH values will reduce the solubility of many metal ions, causing them to precipitate. Both of these phenomena are undesirable in protecting the environment.

Most industrial process waters and wastewater streams are warmer than the water of the receiving body. Effluents from wastewater treatment facilities should be maintained below 37°C if the wastewaters are to be biologically treated. Above this temperature the microorganism population can be severely depleted, thereby reducing the effectiveness of the treatment process. Increases in water effluent temperatures can have a significant effect on the marine life in the vicinity of the discharge. In some species of marine life such as shellfish and certain game fish, warmer effluent water can enhance their growth. Other game fish may be adversely affected by the discharge of warm-water effluents. It is desirable to maintain effluent discharge temperatures from wastewater treatment plants between 10°C and 30°C to optimize and possibly enhance protection of the environment.

Oxygen is critical for aquatic life in receiving waters. Since the EPA has set minimum dissolved oxygen levels in waters at 5 mg/L maximum dissolved oxygen levels are desirable in effluents from wastewater treatment facilities that are being discharged. The solubility of oxygen in water is approximately 9 mg/L at 20°C and 1 atmosphere (atm) of pressure.

All of these wastewater-related parameters may enter into consideration in the NPDES permit for the discharge of treatment plant effluents at industrial facilities.

Solid Waste

Solid waste generally consists of municipal trash, refuse, construction debris, and the like which has been discarded from residences and businesses. [4] Solid waste must be disposed of in an environmentally acceptable manner to reduce potential problems from fire and disease and for appearance's sake. The management of solid waste is regulated under the Solid Waste Disposal Act of 1965, the Resources Recovery Act of 1970, and the Resource Conservation and Recovery Act of 1976. Conservation of natural resources through the management, reuse, and recovery of solid waste is a highly desirable goal that has been used effectively in some areas, for example, the recycling of aluminum cans. Landfill space for disposal of solid waste is becoming limited, as is the capacity of the municipal trash incinerators, which greatly reduce the volume of solid waste.

Hazardous Waste

The Resource Conservation and Recovery Act of 1976 established a comprehensive federal regulatory program to manage hazardous waste. This program is based on a cradle-to-grave concept whereby generators identify their wastes as hazardous and transporters deliver the hazardous waste to the facility that the generator selects for treatment, storage, and disposal (TSD). Generators, transporters, and TSD facilities all must obtain EPA identification numbers; this allows the EPA to track and regulate the cradle-to-grave framework of hazardous waste management. [51]

Wastes may be characterized as hazardous because they possess certain characteristics, because the EPA lists them as being hazardous, or because generators declare them to be hazardous. [5]

There are four characteristic tests that may be used to define a waste as hazardous: ignitability, corrosivity, reactivity, and toxicity. The EPA selected these four characteristics because they had a basis in protecting human health and the environment, were reasonably simple and inexpensive for the generator to perform, were based on standard testing protocols, and did not involve biomedical testing, which was too time consuming and subjective.

The ignitability characteristic for liquids is based on a flash point below 60°C, and for solids causing ignition at standard conditions. These are D001 hazardous wastes.

The corrosivity characteristic for aqueous wastes is based on pH ≤ 2 or ≥ 12.5, or corrosion of steel at a rate ≥ 0.25 inches per year. These are D002 hazardous wastes.

The reactivity characteristic designates wastes that are unstable, react violently with water or air, form explosive mixtures with water, generate certain deadly gases with water, or are capable of detonation. These are D003 hazardous wastes.

The toxicity characteristic measures the ability for the waste to generate toxic leachate in a landfill that could potentially harm groundwater for drinking purposes. The toxicity characteristic leaching procedure (TCLP) has replaced the extraction procedure (EP) for this test protocol. The extract regulatory limits for hazardous waste designation under this protocol are shown in Table 1-16 on page 34.

The ignitability characteristic defines wastes as hazardous if they could cause a fire during the transport, storage, or disposal of the waste. The corrosivity characteristic is used to define hazardous wastes that can react dangerously with other wastes, corrode steel containers easily, or cause toxic contaminants to migrate from certain wastes. The reactivity characteristic was chosen because unstable waste can pose an explosive problem or release hazardous constituents at any stage in the cradle-to-grave waste management system. The toxicity characteristic for hazardous waste was designated to identify wastes that were likely to leach hazardous concentrations of particular hazardous constituents into the groundwater as a result of improper management in landfills and other disposal facilities. The toxicity characteristic leaching procedure is used in the toxicity characteristic test method to simulate the leaching actions that occur in landfill. If the extract concentrations of the hazardous constituents shown in Table 1-16 exceed the maximum allowable concentrations listed, the waste is deemed to be hazardous by the toxicity characteristic.

Example 1-3

A liquid hydrocarbon waste has a 65°C flash point. Is this a hazardous waste?

Solution

This waste is not hazardous by the ignitability characteristic because the flash point is greater than 50°C. The waste should be checked for the other characteristics and compared to the hazardous waste lists before declaring it to be nonhazardous.

The key regulatory document that is used to track the cradle-to-grave disposal of hazardous waste is the EPA hazardous waste manifest (Fig. 1-5 on page 35). The uniform information required on the manifest by the federal EPA must be completed by the generator, transporter, and TSD facility for the cradle-to-grave regula-

TABLE 1-16
HAZARDOUS WASTE TOXICITY CHARACTERISTIC CONSTITUENTS

EPA Hazardous Waste Number	Contaminant	Regulatory level note (mg/L)
D004	Arsenic	5.0
D005	Barium	100.0
D018	Benzene	0.5
D006	Cadmium	1.0
D019	Carbon tetrachloride	0.5
D020	Chlordane	0.03
D021	Chlorobenzene	100.0
D022	Chloroform	6.0
D007	Chromium	5.0
D023	*o*-Cresol	200.0
D024	*m*-Cresol	200.0
D025	*p*-Cresol	200.0
D026	Cresol	200.0
D016	2,4-D	10.0
D027	1,4-Dichlorobenzene	7.5
D028	1,2-Dichloroethane	0.5
D029	1,1-Dichloroethylene	0.7
D030	2,4-Dinitrotoluene	0.13
D031	Heptachlor (and its epoxide)	0.008
D032	Hexachlorobenzene	0.13
D033	Hexachlorobutadiene	0.5
D034	Hexachloroethane	3.0
D008	Lead	5.0
D003	Lindane	0.4
D009	Mercury	0.2
D014	Methoxychlor	10.0
D035	Methyl ethyl ketone	200.0
D036	Nitrobenzene	2.0
D037	Pentachlorophenol	100.0
D038	Pyridine	5.0
D010	Selenium	1.0
D011	Silver	5.0
D039	Tetrachloroethylene	0.7
D015	Toxaphene	0.5
D040	Trichloroethylene	0.5
D041	2,4,5-Trichlorophenol	400.0
D042	2,4,6-Trichlorophenol	2.0
D017	2,4,5-TP (Silvex)	1.0
D043	Vinyl chloride	0.2

UNIFORM HAZARDOUS WASTE MANIFEST	1. Generator's US EPA ID No.	Manifest Document No.	2. Page 1 of	Information in the shaded areas is not requested by Federal law.	
3. Generator's Name and Mailing Address				A. State Manifest Document Number	
				B. State Generator's ID	
4. Generator's Phone ( )					
5. Transporter 1 Company Name	6.	US EPA ID Number		C. State Transporter's ID	
				D. Transporter's Phone	
7. Transporter 2 Company Name	8.	US EPA ID Number		E. State Transporter's ID	
				F. Transporter's Phone	
9. Designated Facility Name and Site Address	10.	US EPA ID Number		G. State Facility's ID	
				H. Facility's Phone	

11. US DOT Description (Including Proper Shipping Name, Hazard Class, and ID Number)	12. Containers		13. Total Quantity	14. Unit Wt/Vol	I. Waste No.
	No.	Type			
G a.					
E b.					
N c.					
d.					

J. Additional Descriptions for Materials Listed Below	K. Handling Codes for Wastes Listed Above

15. Special Handling Instructions and Additional Information

16. Generator's Certification: I hereby declare that the contents of this consignment are fully and accurately described above by proper shipping name and are classified, packed, marked, and labeled, and are in all respects in proper condition for transport by highway according to applicable international and national government regulations.

Unless I am a small quantity generator who has been exempted by statute or regulation from the duty to make a waste minimization certification under Section 300(b) of RCRA, I also certify that I have a program in place to reduce the volume and toxicity of waste generated to the degree I have determined to be economically practicable and I have selected the method of treatment, storage, or disposal currently available to me which minimizes the present and future threat to human health and the environment.

Printed/Typed Name	Signature	Month	Day	Year

17. Transporter 1 Acknowledgement of Receipt of Materials

Printed/Typed Name	Signature	Month	Day	Year

18. Transporter 2 Acknowledgement of Receipt of Materials

Printed/Typed Name	Signature	Month	Day	Year

19. Discrepancy Indication Space

20. Facility Owner or Operator: Certification of receipt of hazardous materials covered by this manifest except as noted in item 19.

Printed/Typed Name	Signature	Month	Day	Year

EPA Form 8700-22 (Rev. 4-85) Previous edition is obsolete.

FIGURE 1-5
EPA HAZARDOUS WASTE MANIFEST.

tory framework of managing hazardous waste. Individual states have the right to require additional information over and above the uniform information requirements. The waste minimization certification in the hazardous waste manifest that the generator must sign requires that industrial operations that generate hazardous waste be designed to reduce the volume and toxicity of the waste to the maximum extent that is economically practicable. This requirement further stipulates that the generator has selected the EPA treatment storage and disposal method that minimizes the threat to human health and the environment in the disposal of the waste.

The use of the RCRA characteristic test in identifying hazardous waste has been supplemented by publication of official lists of designated hazardous wastes, as follows:

1 *F-listed wastes.* These are nonspecific source wastes commonly produced by manufacturing and industrial processes. Examples of F-listed wastes include spent halogenated solvents used in degreasing and wastewater treatment sludge from electroplating processes. [6]

2 *K-listed wastes.* These are wastes from specifically identified industries such as wood preserving, petroleum refining, and organic chemical manufacturing. K wastes typically include sludges, still bottoms, wastewaters, spent catalysts, and production process residues. [7]

3 *P- and U-listed wastes.* These consist of specific commercial chemical products or manufactured chemical intermediates such as chloroform, creosote, sulfuric acid, hydrochloric acid, and certain pesticides. [8]

Any waste mixture that contains a listed hazardous waste should be considered a hazardous waste and managed accordingly. This requirement applies regardless of what percentage of the waste mixture is composed of the listed hazardous waste. This controversial mixture rule has been tested from time to time in legal proceedings and is still under legal scrutiny.

Pollution Prevention

In the Hazardous and Solid Waste Amendments (HSWA) of 1984, Congress stated the objective of the national waste policy: "to promote the protection of health and the environment and to conserve valuable material and energy resources." The Congress also declared: "wherever feasible the generation of hazardous waste is to be reduced or eliminated as expeditiously as possible." Pollution prevention has encouraged process substitutions, materials recovery, recycling, reuse, and treatment. [9]

With this legislation the U.S. government has made pollution prevention and waste reduction the ideal environmental management objective. While all concerned parties have universally embraced this concept, in actuality it has not been vigorously implemented yet by many organizations.

Industry has generally not taken advantage of many opportunities to reduce waste. Reducing waste involves more than buying a black box, reading the directions, and plugging it in. Even a simple step toward waste reduction can seem difficult to a company with few technical resources. While government and the public want to *reduce* waste, there are even greater pressures to meet existing government-imposed requirements to *control* waste. The attention industry must give to hazardous waste treatment limits the resources that can be devoted to waste reduction. Government actions often send an ambiguous message to waste generators. Often, environmental protection emphasizes control and cleanup of hazardous substances after they have been generated. Such hazardous industrial wastes are often destroyed by subsequent pollution control methods, but they may also be put back into the land, water, or air to eventually disperse and migrate. [51]

In practice, waste reduction is frequently subordinated to pollution control, even though reducing waste can be the most effective way to prevent environmental risk. The domination of pollution control over waste reduction has occurred over a long period of time, and it will not be reversed overnight.

The HSWA amendments require all generators of hazardous waste to certify on the hazardous waste manifest (Fig. 1-5) that they have a program in place to reduce the volume and toxicity of their waste to the maximum degree that the generator has determined to be economically practical. This environmental legislation should be foremost in the minds of people engaged in the design of plant processes and related manufacturing facilities. In the conversion of raw materials to finished products, there are numerous opportunities to eliminate or reduce the generation of wastes. The historic disposal of these wastes has resulted in massive quantities of pollutants in the form of wastewater, air emissions, and solid waste that are recycled into the environment.

BIBLIOGRAPHY

1 Ainsworth, S., "Arco Agrees to Pay Record Fine Following Blast," *C&E News,* January 14, 1991.
2 Buonicore, A. J., and W. T. Davis, *Air Pollution Engineering Manual,* Van Nostrand Reinhold, New York, 1992.
3 Fawcett, H. H., *Hazardous and Toxic Materials,* John Wiley, New York, 1984.
4 40 CFR 261.2, July 1, 1994.
5 40 CFR 261.20–261.24, July 1, 1994.
6 40 CFR 263.31, July 1, 1994.
7 40 CFR 261.32, July 1, 1994.
8 40 CFR 261.33, July 1, 1994.
9 U.S. Congress, *Hazardous and Solid Waste Amendments of 1984,* 99th Cong., 1984.
10 Hoversten, P., "Big Blasts in Texas," *USA Today,* April 8, 1992.
11 Kavianian, H. R., and C. A. Wentz, *Occupational and Environmental Safety Engineering and Management,* Van Nostrand Reinhold, New York, 1990.
12 Kolaczkowski, S. T., and B. D. Crittenden, *Management of Hazardous and Toxic Wastes in the Process Industries,* Elsevier Applied Science, London, 1987.
13 Kremmer, F. N., *The Nalco Water Handbook,* 2d ed., McGraw-Hill, New York, 1988.
14 Long, J. R., "Standard for Material Safety Data Sheets in the Offing," *Chemical and Engineering News,* May 18, 1992.
15 *Oil and Gas Journal,* pp. 46–47, April 30, 1990.
16 *Oil and Gas Journal,* p. 36, May 28, 1990.
17 *Oil and Gas Journal,* pp. 28–32, July 16, 1990.
18 OSHA, "Occupational Exposures to Hazardous Chemicals in Laboratories," *Federal Register,* 55, no. 21 (30 January 1990): 3300–3335.
19 Seton, W., et al., "Hazmat Storage," *Chemical Engineering,* pp. 118–119, June 1992.
20 Slote, L., *Handbook of Occupational Safety and Health,* John Wiley, New York, 1987.
21 Thayer, A., "Charleston Explosion," *C&E News,* June 24, 1991.
22 29 CFR 1910, July 1, 1994.
23 29 CFR 1903, July 1, 1994.
24 29 CFR 1903.16, July 1, 1994.
25 29 CFR 1903.17, July 1, 1994.
26 29 CFR 1904, July 1, 1994.

27 29 CFR 1910.94, July 1, 1994.
28 29 CFR 1910.95, July 1, 1994.
29 29 CFR 1910.96, July 1, 1994.
30 29 CFR 1910.97, July 1, 1994.
31 29 CFR 1910.101, July 1, 1994.
32 29 CFR 1910.102–1910.105, July 1, 1994.
33 29 CFR 1910.141, July 1, 1994.
34 29 CFR 1910.144, July 1, 1994.
35 29 CFR 1910.145, July 1, 1994.
36 29 CFR 1910.147, July 1, 1994.
37 29 CFR 1910.156, July 1, 1994.
38 29 CFR 1910.157, July 1, 1994.
39 29 CFR 1910.159, July 1, 1994.
40 29 CFR 1910.164, July 1, 1994.
41 29 CFR 1910.165, July 1, 1994.
42 29 CFR 1910.176, July 1, 1994.
43 29 CFR 1910.212, July 1, 1994.
44 29 CFR 1919.252, July 1, 1994.
45 29 CFR 1910.254, July 1, 1994.
46 29 CFR 1910.303, July 1, 1994.
47 29 CFR 1910.333, July 1, 1994.
48 29 CFR 1910.335, July 1, 1994.
49 29 CFR 1910.1000–1910. 1048, July 1, 1994.
50 U.S. Department of Labor, Mine Safety, and Health Administration, *Safety Manual No. 10,* Washington, D.C., 1987.
51 Wentz, C.A., *Hazardous Waste Management,* 2d ed. McGraw-Hill, New York, 1995.
52 Whitehead, L., "Planning Considerations for Industrial Plants Emphasizing Occupational and Environmental Health and Safety Issues," *Applied Industrial Hygiene,* 2(2): 1987.

PROBLEMS

1 ARCO Chemical Company agreed to a major revamping of its workplace safety programs at its facilities nationwide and to pay a record $3.4 million fine for alleged workplace violations at its Channelview, Texas, chemicals complex because of an explosion that killed 17 people and charred an area the size of a city block. At the time of the fine it was the largest ever paid to OSHA. ARCO plans to spend an additional $36 million to add safety measures and rebuild the facility. ARCO has stated that although it doesn't agree with all of the OSHA citations, it believes "its best interests are better served by focussing on improving workplace safety, rather than by contesting or litigating the differences it may have with OSHA." Do you believe that ARCO did the right thing by not contesting the OSHA fine?

2 When the polyethylene plant of Phillips Petroleum Company exploded and killed 23 people, OSHA fined Phillips $5,666,200. Phillips contested all the citations issued by OSHA and also disputed the legality of OSHA's method of determining the number of violations. Phillips also stated that this action in no way contradicts its concern for employee and community safety. Do you believe that Phillips did the right thing in contesting the OSHA fine and violations while it proceeded with the rebuilding of the polyethylene plant?

3 Why are leaks and fugitive emissions major safety and occupational health concerns when handling toxic chemicals at a plant facility?

4 A storage warehouse containing flammable liquids burned to the ground, destroying several nearby structures. Three lives were lost, 10 people were injured and there was about $10 million in property damage. List the possible direct and indirect causes of this accident.

5 You have just accepted the job of managing a process facility that has a poor safety record including numerous lost-time accidents and many OSHA citations. Top management of the organization has given you complete freedom to turn this facility around. What steps would you take immediately to eliminate the unsafe conditions and unsafe acts within the operating facility?

6 How many OSHA regional headquarters are there and where are they located?

7 How many days does an employer have to post an OSHA citation for an alleged violation in the workplace?

8 An employer receives an OSHA citation for an alleged violation on Monday morning, June 3. By what date must the employer notify OSHA of the intent to contest the citation before the Review Commission?

9 A tragic accident occurs in the workplace, resulting in one fatality, five employees being off the job for a week, and three employees on restricted duty. How many occupational injuries and illnesses should be recorded by the employer on OSHA Form 200?

10 How many cubic feet (ft^3) of dilution air are required to dilute 0.4 gallons of methyl ethyl ketone solvent per minute to 25 percent of the lower explosive limit?

11 A plant operates 24 hours per day, five days per week. The operation emits 4 lb/h of ethylene glycol into the atmosphere. Is this plant a major source of HAP?

12 A round-the-clock operation emits 2 lb/h butadiene, 1.7 lb/h hexane, and 2.2 lb/h acrylonitrile. Is this plant a major source of HAP?

13 You are the manager of a large coal-fired utility plant in central Illinois. The plant uses high sulfur bituminous coal from a nearby Illinois coal mine. You made a major capital expenditure in 1988 in flue gas desulfurization equipment, but your SO_2 emissions are still greater than 2.5 pounds per million Btu coal. An economic evaluation of additional scrubbing equipment, to achieve the 1.2 pounds per million Btu coal by the year 2000, is financially unfeasible. Switching to an out-of-state low-sulfur coal source would achieve the regulatory standard and close the local coal mine. There are 450 employees at the local mine in a rural area of Illinois. What would you do?

14 The flash point of a liquid waste is 45°C. Is this waste hazardous?

15 The pH of an aqueous waste is 13. Is this waste hazardous?

16 A drum of hazardous waste explodes while being transported from the generator to the disposal facility. Characterize the hazard designation for the waste.

17 A waste contains lead and cadmium. The TCLP extract from this waste measured 3 mg/L lead and 1.5 mg/L cadmium. Is this waste hazardous?

18 A storage tank at an industrial facility contains 4 million gallons of wastewater to be processed through the on-site wastewater treatment plant prior to discharge of the treated effluent. You inadvertently add 1 gallon of a listed hazardous waste to the contents of the storage tank. The tank is then mixed thoroughly. Can you legally process the entire storage tank contents through the wastewater treatment plant?

2

RISK PERCEPTION, ASSESSMENT, AND MANAGEMENT

There is some risk in every decision or action. [32] This risk is present in all industrial, government, and public situations. It is also in our own lifestyles and personal decisions and actions. In order to appraise risk and safety, quantitative methods are preferred because they are more disciplined and objective than purely subjective conclusions. Even quantitative methods often involve some degree of subjectivity that introduces uncertainty.

RISK PERCEPTION

The following example illustrates the changes in public perception that have occurred over the past several decades.

- *1960:* A logger sells a truckload of lumber for $100. His cost of production is four-fifths of this price. What is his profit?
- *1970 (new math):* A logger exchanges a set L of lumber for a set M of money. The cardinality of set M is 100, and each element is worth $1. Make 100 dots representing the elements of the set M. The set C of the costs of production contains 20 fewer points than set M. Represent the set C as a subset of M, and answer the following question: What is the cardinality of the set P for profits?
- *1980:* A logger sells a truckload of wood for $100. His cost of production is $80, and his profit is $20. Your assignment: Underline the number 20.
- *1990 (outcome-based education):* By cutting down beautiful forest trees, a logger makes $20. What do you think of this way of making a living? (Topic for class participation: How did the forest birds and squirrels feel?)

The perception of risk depends a great deal on our personal situation. We take numerous risks daily with little, if any, concern. Yet we become highly concerned about other less serious risks because of our personal perception of an activity, chemical substance, or process operation.

News media announcements about cancer-causing chemicals, catastrophic accidents, crime, drugs, regional conflicts, and weather-related tragedies make the world seem to be a hazardous place. Everyone would like to live in a risk-free environment, but is this really an attainable goal?

In the United States the average life expectancy during the past century has risen from about 50 years to 70 years of age. Improvements in our standard of living and our working conditions have been mainly responsible for this change. The sum of all our risks now must be less than the risks were about 100 years ago. Some of the risks eliminated were quite large by the standards of today. As we become more knowledgeable about the remaining hazards, we are more conscious of a number of smaller risks that may have been present in our lives for extended periods of time.

Our perception of both the type and magnitude of risks has a great deal to do with our attitude about these risks. Our personal risks begin each day when we get out of bed in the morning. As electric lights and appliances are turned on, do we think about the 500 people who are electrocuted each year? As we shower to remove harmful contaminates, what about the many chemicals in the soap and how they affect our skin? Are the bleaching agents for our white clothes completely safe? As we walk down the stairs, do we recall that 12,400 people die each year from falls, 6200 of them in their homes? [1] Does the caffeine in coffee or tea adversely affect our nervous system or cause cancer? Is it better to risk heart disease by eating sugar or to risk cancer by eating saccharin? Should we eat meat and risk cancer or be a vegetarian? Should we eliminate peanut butter and peanuts, which contain aflatoxin, a potent carcinogen? [15]

Most choices about commuting to work are heavily influenced by the travel distance and the viable alternatives: by car, walking, train, bicycle, or bus. There are health, convenience, economic, accident, and societal issues in travel, all of which involve risks. We take our choice for travel for granted because we are accustomed to the risks, which have become part of a daily routine. Traveling into major metropolitan areas brings exposure to air pollution. So does exposure at home or work to tobacco smoke, which causes about 40 percent of all cancers and kills about 15 percent of all Americans.

Drinking a glass of tap water has risks. Has the added chlorine killed all of the cholera and typhus microbes in the water supply? Has the chlorine produced chlorinated compounds, which are carcinogens?

Homes and offices are often constructed of brick and concrete blocks that contain radioactive minerals and the derived radon gas, a carcinogen. A chest x-ray for tuberculosis or cancer detection introduces a cancer risk.

Airplane travel is commonly used for long trips without concern for cosmic radiation hazards from outer space. Wood-burning fireplaces cause air pollution that could be avoided, but what about the risks to coal miners who produce coal for coal-fired electric utilities or the risk of explosion in natural gas transmission lines. The risk of nuclear power plants has been widely reported by the news media.

The public perception of risk is often vastly different from the risk perception of industry and statisticians. While the public is frequently exposed to cases of accidental death, there is more public concern about a risky chemical hazard presence. For example, what would be the local public reaction if Union Carbide wanted to construct a pesticide plant similar to the one in Bhopal, India? [25]

In 1992 there were 83,000 accidental deaths in the United States (see Table 2-1). This was the lowest number since 1922, when the population was less than half the 1992 population level. The 40,300 motor vehicle deaths in 1992 was the lowest level since 1961. [1] As Table 2-1 shows, accidental deaths and disabling injuries are much more common at public locations and at home than in the workplace. If we are to continue to reduce accidental deaths in the future, considerable attention must be focused on the public and on home exposure to safety, health, and environmental hazards.

The leading cause of accidental death among people below age 80 is from motor vehicle accidents (see Fig. 2-1). The accidental death rate for falls increases dramatically with age for senior citizens. Accidental death rates usually peak in the late teens, mainly from motor vehicle deaths.

There are many other significant causes of accidental death: poisoning, drowning, fires, suffocation, and firearm accidents (see Table 2-2 on page 44).

Most deaths in the United States are the result of heart disease, cancer, and stroke (see Fig. 2-2 on page 45). Deaths from these causes become particularly ev-

TABLE 2-1
ACCIDENTAL DEATHS IN THE UNITED STATES IN 1992

Classification	Deaths	Death rate per 100,000 people	Disabling injuries
Public			
Motor vehicle	37,100		2,000,000
Other	17,900		5,800,000
Subtotal	55,000	21.6	7,800,000
Home			
Motor vehicle	200		<10,000
Other	19,300		6,000,000
Subtotal	19,500	7.6	6,000,000
Work			
Motor vehicle	3,000		200,000
Other	5,500		3,100,000
Subtotal	8,500	3.3	3,300,000
All Classes			
Motor vehicle	40,300		2,200,000
Other	42,700		14,900,000
Total	83,000	32.5	17,100,000

Source: National Safety Council.

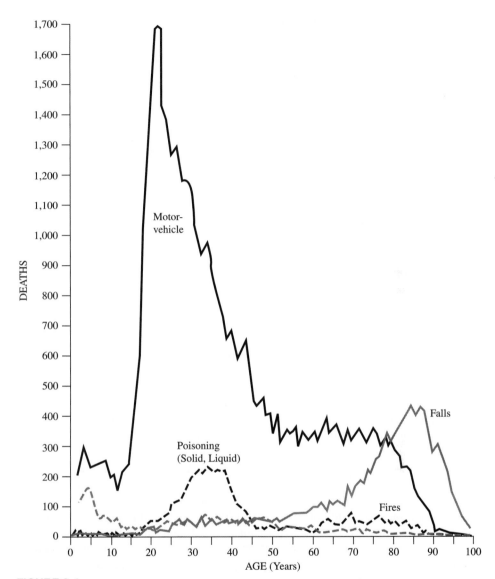

FIGURE 2-1
CAUSES OF ACCIDENTAL DEATH IN 1989 BY AGE. (*Source:* Used by permission of the
National Safety Council, Itasca, Illinois.)

TABLE 2-2
CAUSES OF ACCIDENTAL DEATHS IN THE UNITED STATES IN 1992

Type of accident	Deaths	Death rate per 100,000 people
Motor vehicle	40,300	15.8
Falls	12,400	4.9
Poisoning by solids and liquids	5,200	2.0
Drowning	4,300	1.7
Fires and burns	4,000	1.6
Suffocation by ingestion	2,700	1.1
Firearms	1,400	0.5
Poisoning by gases and vapors	700	0.3
Other	12,000	4.6
Total	83,000	32.5

Source: National Safety Council.

ident after age 40 and continue to escalate as we grow older. Former Surgeon General C. Everett Koop in 1988 declared tobacco products containing nicotine to be addictive. [12]

There are many other common causes of death in the United States (see Table 2-3 on page 46). Most of these health-related causes result from risks in our lifestyles, homes, or occupational environments. The identification of the precise risk or risks for these causes is a complex problem because of the severity and length of exposure to a wide variety of risks during a lifetime. As medical science becomes more knowledgeable about health-related risks, life expectancies will increase even more. For example, tobacco use has been linked to many causes of death. The risk for the cause of AIDS deaths will remain the subject for considerable research for the foreseeable future. In 1993, AIDS became the leading cause of death for men ages 25 to 44 years, greatly surpassing accidents, which are in second place.

Too much fat and oil in our diet is hazardous to our health. The more common types of fatty acids are saturated, polyunsaturated, and monounsaturated. All of these contain about 9 calories per gram, but blood cholesterol level can be significantly affected by the type of fatty acid. Polyunsaturated fatty acids reduce blood cholesterol levels, while saturated fatty acids tend to increase it. Monounsaturated fatty acids tend to reduce low-density lipoprotein (LDL), or bad cholesterol. Trans-fatty acids tend to raise LDL levels. [13]

The incidence rate is commonly used to measure and compare industrial occupational injuries and illnesses.

$$\text{Incidence rate} = \frac{(\text{total injuries and illnesses} \times 200,000) \text{ or } (\text{total lost workdays} \times 200,000)}{\text{total hours worked by all employees during period}}$$

The 200,000 constant is based on 100 full-time equivalent workers working 40 hours a week, 50 weeks a year. The 1992 incidence rates for selected industry sec-

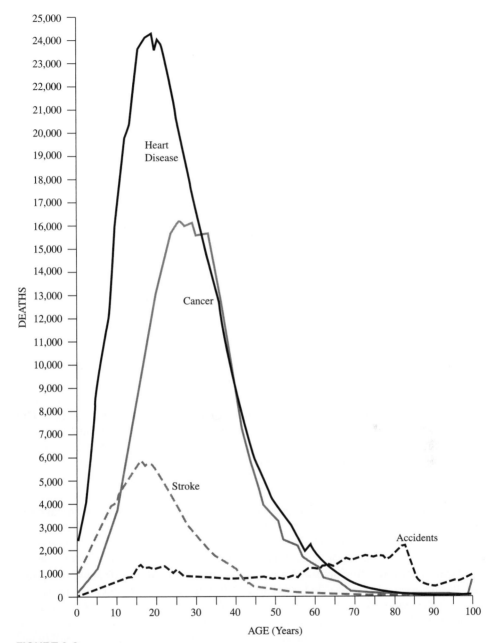

FIGURE 2-2
LEADING CAUSES OF DEATH IN 1989 BY AGE. (*Source:* Used by permission of the
National Safety Council, Itasca, Illinois.)

TABLE 2-3
THE MAJOR CAUSES OF DEATH IN THE UNITED STATES IN 1990

Cause of death	Deaths	Death rate per 100,000 people
Heart disease	720,058	289.5
Cancer	505,322	203.2
Stroke	144,088	57.9
Accidents	91,983	37.0
Chronic obstructive pulmonary disease	86,679	34.9
Pneumonia	79,513	32.0
Diabetes	47,664	19.2
Suicide	30,906	12.4
Cirrhosis of the liver	25,815	10.4
AIDS	25,188	10.1
Homicide	24,932	10.0
Nephritis, nephrosis	20,764	8.3
Septicemia	19,169	7.7
Atherosclerosis	18,047	7.3
Other	308,335	123.9
Total	2,148,463	863.8

Source: National Safety Council.

tors is shown in Figure 2-3. Based upon these statistics the trucking, transit, railroad, primary nonferrous metals, motor vehicle, ship-building, and government sectors have the highest incidence rates. The lowest incidence rates are found in the textile, chemical, aircraft, coal-mining and petroleum sectors of industry. The normal public perception is probably much different than these data indicate.

The more common industrial trauma disorders are sprains, strains, bruises, lacerations, and fractures to the back, shoulder, fingers, wrist, and knee. Overexertion by lifting, falls, and bodily contact with objects are the main causal events; accidents normally involve containers, floors, materials, machinery, tools, or vehicles. [44]

Incidence rates based on the type of industry are shown in Table 2-4 on page 48. Cumulative trauma disorders (CTDs) in industry have increased dramatically in recent years. CTDs develop gradually over time from repeated stresses to a body part. Industries with the highest CTD risk are those with repetitive movements, awkward positions, or vibration for prolonged periods of time. Manufacturing has by far the greatest risk of CTD. Agriculture and manufacturing have the most risk of skin disorders. Manufacturing also has fairly high-risk levels for respiratory problems. The mining industry leads in risk of lung disease from dust exposure.

We cannot legislate or ban all known risks out of existence. The commonly accepted policy for identifying risk is to quantitatively measure the risk and then assign an acceptable limit. Most government and industrial authorities assess an increased risk of death of one in a million as acceptable. If the risk of dying in a given year is increased, then the risk of dying from another cause is decreased in later years. Increased risk shortens the life expectancy; for example, life shortened

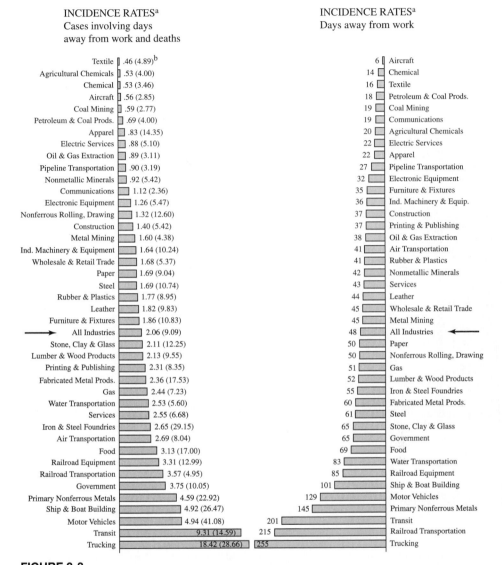

FIGURE 2-3
SELECTED INDUSTRIAL OCCUPATIONAL INJURY AND ILLNESS INCIDENCE RATES
FOR 1992. [a]Incidence rates per 100 full-time employees, using 200,000 employee hours as
the equivalent. [b]Rates in parentheses are Total Recordable Cases. (*Source:* Used by permission of the National Safety Council, Itasca, Illinois.)

TABLE 2-4
OCCUPATIONAL ILLNESS AND INCIDENCE RATE BY TYPE OF INDUSTRY IN 1991

	Agriculture	Mining	Construction	Manufacturing	Transportation	Trade	Finance	Services	Total
Occupational illness number (thousands)									
Trauma disorder	0.8	0.6	1.2	185.5	5.5	12.2	5.1	12.7	223.6
Skin disorder	3.6	0.2	2.2	31.8	2.8	4.4	0.6	12.6	58.2
Respiratory, toxic agents	0.3	0.0	0.8	8.8	1.1	1.6	0.7	4.8	18.3
Physical agent disorder	0.2	0.1	1.0	12.0	1.1	1.1	0.1	2.5	18.2
Poisoning	0.4	0.1	0.4	3.3	0.3	0.4	0.1	1.7	6.7
Lung disease, dust	0.1	0.5	0.2	1.0	0.2	0.2	0.0	0.3	2.5
Other	0.7	0.2	1.1	12.8	2.4	5.3	1.7	16.6	40.8
Total	6.1	1.7	7.0	255.2	13.4	25.2	8.3	51.3	368.3
Occupational illness rate (per 10,000 full-time workers)									
Trauma disorder	7.9	8.2	3.1	104.6	10.3	6.2	8.6	6.1	29.7
Skin disorder	34.5	2.9	5.6	17.9	5.3	2.3	1.0	6.0	7.7
Respiratory, toxic agents	2.7	0.6	2.1	5.0	2.1	0.8	1.2	2.3	2.4
Physical agent disorder	1.9	1.3	2.6	6.8	2.1	0.6	0.2	1.2	2.4
Poisoning	4.2	1.4	0.9	1.9	0.6	0.2	0.2	0.8	0.9
Lung disease, dust	0.5	7.3	0.4	0.5	0.3	0.1	0.0	0.1	0.3
Other	7.0	2.6	2.8	7.2	4.5	2.7	2.8	7.9	5.4
Total	58.9	24.4	17.5	144.0	25.3	12.9	14.1	24.5	49.0

Source: National Safety Council.

TABLE 2-5
ACTIVITIES THAT INCREASE THE RISK OF DEATH BY 1 PER MILLION

Activity	Cause of death
Smoking 1.4 cigarettes	Cancer, heart disease
Drinking 0.5 liters wine	Cirrhosis of liver
One hour in coal mine	Black lung disease
Traveling 6000 miles in jet	Cancer, cosmic radiation
Two months of second-hand smoke	Cancer, heart disease
Eating 40 tbsp peanut butter	Cancer, aflatoxin
Drinking 1.25 cases saccharin diet soda	Cancer, saccharin
Five years living near nuclear power plant	Cancer, radiation
Eating 100 charcoal-broiled steaks	Cancer, benzopyrene
Living 2 months in stone or brick home	Cancer, radon
Having one chest x-ray	Cancer, radiation
Drinking chlorinated water 1 year	Cancer, chlorine

Source: R. Wilson, *Technology Review,* February 1979.

30 years by an accident or 15 years by cancer. A risk of 1×10^{-6} shortens life expectancy by 30×10^{-6} years (16 minutes) for an accident and 15×10^{-6} years (8 minutes) for cancer. There are many familiar activities that increase the risk of death by 1×10^{-6} (see Table 2-5). [15] Most, if not all of these activities, are perceived by the public to be either risk-free or to possess an inconsequential risk at most.

These risks are not uniformly distributed throughout the population. There is no way of identifying exactly which individuals will die of cancer, be killed in a car wreck or skiing accident, or die from an industrial work situation. [10] In order to quantify these various risks, all involved participants are lumped together and are assumed to be equally at risk.

During the past several decades public concern has shifted from visible and common safety, health, and environmental concerns—such as auto emissions, smoking, raw sewage, and property damage—to the effects of very low concentrations of toxic pollutants.

People may underestimate or overestimate the magnitude and acceptability of the risk based upon their perception, as shown in Table 2-6 on page 50. [36] If the public or a special-interest group perceives the risk to be unacceptable, then the risk will be unacceptable. These perceptions apply to industry, government, and the public. Employees usually identify with a standard operating procedure based on how they were taught to do it or the way they have observed coworkers doing it. [5]

The news media and environmental groups have played major roles in making people fearful of chronic risks to chemical exposure that only in recent times could be measured because of advances in analytical chemistry. As the public demands ever-decreasing levels of risk from these uncertain but low-level industrial and pub-

TABLE 2-6
FACTORS AFFECTING THE ACCEPTABILITY
OF RISK BASED ON THE PERCEPTION OF
PEOPLE

Greater acceptability	Lower acceptability
Voluntary	Involuntary
Natural	Synthetic
Controllable	Uncontrollable
Delayed effect	Immediate effect
Essential	Nonessential
Major benefits	Minor benefits
Experienced	Inexperienced
Understandable	Not understandable
Known	Unknown
Common	Uncommon
Mundane	Dramatic
Routine	Special
Low media coverage	High media coverage
Low controversy	Controversial

lic hazards, the cost to consumers and taxpayers escalates dramatically with each level of risk reduction. The risks associated with our safety, health, and environmental policies should be managed to obtain goods and services within the context of how we want the private and government sectors to be now and in the future.

Communities and the public fear certain types of industrial facilities more than is warranted because of lack of effective communication in dealing with risk perception. Table 2-7 shows the basis for much of the negative risk perception of communities toward the siting of industry facilities. This is particularly true for certain facilities like hazardous waste management, nuclear power, and chemical process plants; oil refineries; and other operations with hazardous chemicals. Communications must be open, logical, understandable, and empathetic to the public and the employees or local negative risk perception will threaten the facility. For example, the Kanawha River Valley in West Virginia, with its large concentration of chemical industries, has discussed worst-case accident scenarios with the public. [26]

The government sometimes intervenes with regulations and warnings on products and exposures to bring risk into better focus for industry and the public. Warning labels are required on cigarette packages and alcoholic beverages, certain industrial equipment, flammable containers, and seat-belt use. Nutrition labels are required on packaged food products. Some classes of chemicals have been banned from manufacture or distribution. Workplace exposure to toxic chemicals is regulated, as is their discharge into the environment. The EPA risk management program for chemi-

TABLE 2-7
THE BASIS FOR NEGATIVE RISK PERCEPTION BY COMMUNITIES FOR INDUSTRIAL
FACILITIES

Risks are unfamiliar.

Involuntary risks.

Risks are controlled by outsiders.

Undetectable risks.

Risks are unfair.

Individual protective action not permitted.

Dramatic and memorable risks.

Uncertain risks.

Unrelated hazard comparisons.

Risk estimation, not reduction, emphasized.

Source: P. M. Sandman, *Seton Hall Legislative Journal,* vol. 9, 1985.

cal accidental release prevention is integrated with the OSHA process safety management. [30] Vehicles must obey posted speed limits on roadways. These regulations and warnings influence public and private perception of risk.

People react to the risk of hazards as they are perceived. If the perception is inaccurate, the response will also be inappropriate. Such an inappropriate response may be harmful, or even disastrous; hence the need for risk assessment to identify the hazards and measure the frequency of occurrence and the magnitude of the consequences.

RISK ASSESSMENT

Since no activity or technology can be absolutely safe, the question arises, "How safe is safe enough?" The answer to this question does not have a simple or easy answer. First, a safety risk is defined as possible consequences for human death, disease, injury, and for property destruction or damage to the environment.

Risk equals the probability of the occurrence times the severity of the harmful effects. It may be difficult to precisely quantify these parameters, so they must be estimated. [20,24]

Take the problem of developing a nuclear waste disposal plan. South Carolina has been the final resting place for the disposal of about two-thirds of the low-level radioactive waste in the United States. Radioactive medical equipment, nuclear plant materials and tools, and other types of irradiated waste had been sent from all states to the Barnwell County landfill for 23 years. As this waste disposal business grew, so did the public concerns about risk and the general local view that other states' waste was not wanted in South Carolina.

The federal government's 1980 response to this age-old problem was to tell all states to either dispose of their own radioactive waste or to form compacts for re-

TABLE 2-8
INTERDEPENDENT STEPS IN DETERMINING AN ACCEPTABLE RISK

Specify the objectives and measures of effectiveness to be achieved.

Define the possible alternatives that could achieve the objectives and their associated risks.

Identify all possible consequences of each alternative.

Quantify the various consequences, using consistent assumptions.

Analyze the results and prioritize the alternatives.

Select and implement the best choice for an acceptable risk.

Obtain feedback and iterate the process as necessary.

---

gional landfills. Five years of studies and political posturing by the states produced no results. In 1985 the Congress informed all of the states that South Carolina could restrict state access to the Barnwell landfill in 1993. After this 8-year advance warning, South Carolina shut its doors to all but 8 states. The only other low-level waste site, in Hanford, Washington, has already been restricted to accommodate 11 states. This leaves 31 states without a permanent disposal site.

Until these 31 states find a permanent disposal site, their low-level wastes will pile up at the locations where the wastes are being generated. Radioactive materials are used extensively in the private and public sectors: by nuclear power plants, hospitals, medical research centers, and national laboratories, and for defense-related activities. Most states have yet to settle on their own disposal site, much less construct and operate a suitable disposal facility. In the meantime, interim on-site storage at generator sites may continue to create greater risks than that of a well-conceived central disposal facility for low-level radioactive waste.

A standardized analysis is used to determine an acceptable risk (see Table 2-8). Based upon the objectives and measures of effectiveness, the best alternative is selected. This approach can be used to manage the risk once it has been identified and quantified.

The process of risk assessment is best achieved through a systematic approach, shown in Table 2-9. [11,15] In this process the hazard and its accompanying risk are characterized so that they may be effectively managed.

TABLE 2-9
THE PROCESS OF RISK ASSESSMENT

Identification of the potentially harmful hazards.

Measurements to estimate the consequences of the hazards.

Estimation of the probability of the occurrence of each hazard consequence.

Quantitative calculation of risks and comparison with potentially acceptable hazard levels.

Characterization of the hazard risks to be managed, along with the assumptions and uncertainties.

Ranking of the risk hazards for management decision making.

The understanding of risk involves three specific questions: What can go wrong? What are the consequences? How likely is it to happen? [7]

The quality and quantity of information that can be obtained about these three questions are critical in accurately understanding the risk situation. This necessary information should be based upon the best available combination of knowledge, experience, analysis, and judgment.

In industrial operations employees, materials, and equipment come together in the work environment to produce products or perform services. Productivity is best when the operation at the facility runs smoothly, thereby allowing time and resources to be used efficiently and effectively. Accidents, illnesses, injuries, and fatalities interrupt operations, disrupt the normal routine, and result in additional costs and other losses. Accidents increase the time to accomplish a job, reduce productivity, and raise operation costs. Lost-time injuries and illnesses, wasted materials, damaged equipment, and property loss can also increase operational and hidden costs and further decrease effectiveness. [2]

Risk assessment and management in many respects have goals similar to productivity improvement. Both of these endeavors must be integrated into daily activities and management systems. An organization must manage well all of its resources and use the most cost-effective technology. This is best achieved with trained, educated, and skilled employees. [5] Losses from interruptions can prevent production schedules from being met, generate off-specification products, and reduce employee morale, all of which result in business losses.

In safety, health, and environmental protection an operation should first set realistic goals and objectives. Then the significant risk hazards to achieving these goals and objectives should be identified, characterized, and managed to an acceptable level. Monitoring systems should be installed to follow the effectiveness of control measures, which may include management and engineering controls as well as personal protective equipment.

As tasks are performed in the workplace, there are risks associated with employees, equipment, materials, and the work environment that may cause accidents and other disruptions. Workers monitor or gather information, use that information to make decisions, and control the situation within acceptable limits. Materials and equipment must be properly selected, installed, or distributed and handled, maintained, and used within defined levels of risk. The work environment should not detract from the safety and health of the employee and must be designed to function accordingly. The work environment should also be managed to achieve desirable interpersonal relationships to properly coordinate information, materials, and human behavior.

There are risks associated with human, situational, and environmental factors. People can cause an accident, injury, or illness by their actions (commission) or by their failure to act (omission). This applies to both workers and management. Human errors, rather than engineering defects, are responsible for most workplace accidents. [19,37] Examples of human factor risks may include:

- Unauthorized use of equipment
- Operating equipment in an improper manner

- Removing safety devices
- Rendering safety devices inoperative
- Knowingly using defective equipment or tools
- Deviation from safety rules, regulations, policies, or procedures

Situational factors introduce risk that may be related to material hazards, unsafe equipment and tools, facility hazards, and unsafe operating procedures. Examples of situational risk factors include:

- Inappropriate process design
- Improper equipment selection
- Unsafe materials of construction
- Poor equipment installation
- Improper storage of hazardous materials
- Inadequate facility planning
- Poor handling of hazardous wastes

Environmental factors introduce risk through physical, chemical, biological, and ergonomic hazards that either cause or contribute to accidents, injuries, illnesses, or fatalities. Examples of environmental factor risks include:

- *Physical factors:* Noise, vibration, radiation, illumination, temperature, pressure, or humidity
- *Chemical factors:* Gases, vapors, fumes, mists, smoke, and dusts that may be toxic, flammable, explosive, corrosive, or reactive
- *Biological factors:* Bacteria, viruses, fungi, or parasites that may cause illness or death
- *Ergonomic factors:* Human and situational factors that are not compatible with the physical and behavioral limitations of the employees

Lead and asbestos are both major known chemical health hazards. [27] Unstable chemicals like polymerization initiators or catalysts are also examples of dangerous workplace chemicals. [6] About 40 percent of all natural gas transmission line incidents in the United States are caused by external force like earthquakes or heavy equipment. [30]

Computer control systems are used in many operations to improve the efficiency of manufacturing facilities. During hazard review hardware and software are often treated as black boxes. Considerable safety and operating improvements can be realized by including these computer systems in the hazard review process. [30]

Considerable risk reduction can be realized by hiring and properly training quality employees; purchasing only recommended and approved equipment and materials; having strong management commitment to safety, health, and environmental protection; and generally being attentive to the risks of hazards.

The first step in risk assessment is to identify and evaluate the hazards that exist in the workplace. The hazards may be related to any of the human, situational, or environmental factors previously discussed. There are many examples of major losses of industrial property because of mechanical failure of equipment, such as piping, storage tanks, reactors, drums, pumps, and compressors. A risk-based inspection of equipment should be used to optimize the use of resources to achieve an acceptable level of safety. [35] Probably the best place to begin to obtain this information is to interview the employees, who are working in operations. Insurance companies, professional organizations like ASSE, NSC, AIHA, ACGIH, AIChE, manufacturers and product suppliers, labor representatives, safety and health staff personnel, and peers in the safety and health field could be helpful in locating hazards. Inspections by OSHA and other government regulatory agencies can provide helpful hazard information. Historical accident reports or other records about injuries, illnesses, or fatalities should provide guidance. OSHA incident rates help to identify potential problem areas.

Another technique that can be used to identify hazards is hazard analysis, which studies operational and management systems. [42] This systematic procedure is used to find hazards that have been overlooked in process and equipment design or facility layout, unforeseen operating hazards, and hazards from modifications or changes in the original system. As each step in the system is analyzed, the relationships between employees and the work environment, materials, equipment, and the process are studied. This analysis identifies hazardous conditions and potential accidents; helps develop control measures; points out the knowledge, skill, and physical qualifications required of employees for specific tasks; and uncovers unsafe procedures and actions in the workplace.

Workers' compensation information is useful in identifying risk of injury and illness in the workplace. The primary purpose of the workers' compensation reporting system is to provide facts to determine eligibility and amount of compensation. Other uses for these reports include injury frequency counts from specific employers that can identify the most common injury patterns and sources of injury-related costs. Industrywide trends and common areas of concern can be identified by pooling employer-specific information. [39]

Risk assessment in safety, health, and environmental protection often involves exposure to potential chemical hazards. The potential chemical exposure locations and pathways to humans must be carefully planned (see Table 2-10). The health risks should begin by identifying the chemicals and their exposure levels for the various employee activities and behavior patterns. [9] Care must be taken in obtaining accurate, reliable data to ensure the validity of the risk assessment. [17] Once these data have been obtained, they should be evaluated to characterize the

TABLE 2-10
POTENTIAL CHEMICAL EXPOSURE PATHWAYS FOR HEALTH RISK
ASSESSMENT IN HUMANS

Pathway	Route of body entry
Air	Inhalation of gases, vapors, and particulates
Food	Ingestion of meat, fish, dairy products, grains, and produce
Surface water	Ingestion, inhalation, dermal contact
Groundwater	Ingestion, dermal contact

workplace and define the magnitude of the potential chemical hazards. After the potential exposure pathways have been identified, the exposure concentrations and chemical entry levels into the human body may be estimated. The degree of toxicity of the chemicals and the dose-response relationship and risk must be characterized. This methodology provides the degree of risk at the facility or workplace and an estimate of acceptable levels of chemical exposure to the employees and the general public. [3] Environmental, safety, and health risks have played a major role in the shift from organic solvents to aqueous formulations. [22]

Once the hazards have beeen identified and evaluated, they should be ranked according to degree of risk. Probability, or frequency, and the severity of the consequences must be taken into account. [34] This ranking process should place the most unsafe or worst hazard first.

It is usually advisable to begin ranking hazards according to potential destructive consequences, followed by an estimate of the probability of the hazard actually resulting in an occurrence. [39]

Hazards that may cause death or destruction of a facility have the greatest severity. Hazards would be minor or negligible if they probably would not result in a lost workday because of injury or illness. Any and all quantitative and qualitative data should be used to determine the probability estimates.

The reliability of a process component may be determined as follows:

$$R(t) = e^{-\mu t}$$

where $R(t)$ = component reliability over the time interval
t = time interval
μ = component constant failure rate

The failure probability then becomes

$$P(t) = 1 - R(t)$$
$$= 1 - e^{-\mu t}$$

Qualitative hazard probabilities could range from immediate occurrence of the hazard exposure to an unlikely exposure situation. The number of persons who might be exposed to the potential hazard should also be included in this ranking. If a large number of people could be exposed to the hazard, this is obviously a greater danger than only one or two employees being exposed. The risks are then ranked relative to each other based upon the consequence and probability. This methodology should generally follow the risk assessment process in Table 2-11. Risk may then be defined mathematically as

$$\text{Risk} = \text{probability} \times \text{consequences}$$

The consequences may happen to individuals (individual risk) or to groups. Economic risk and societal risk would be examples of group risks. High-risk and low-risk situations are usually readily apparent and result in straightforward decision making. Intermediate risk is more difficult for decision making because con-

TABLE 2-11
POTENTIAL RISK FACTORS IN THE IMPACT OF HAZARDS ON
PEOPLE, FACILITIES, AND THE SURROUNDING COMMUNITY

Type and length of hazard exposure
Number of people exposed inside and outside the facility
Demographics of the exposed people
Effectiveness of emergency response inside and outside the facility
Lost time of employees and outside people
Reduction in employee morale
Damage to public image
Property damage inside and outside the facility
Cost of cleanup, repairs, and lost production inside and outside the facility
Personal injury and damage lawsuits
Backlash legislation and additional regulatory constraints

siderable judgment is needed from management. All of this information then goes to the next step: management decision making.

Managers can only make intelligent, informed decisions for risk assessment if they have complete and accurate information. All of the alternatives that were considered should be presented in the risk ranking, along with the recommended course of action. These recommendations may include improved procedures or technology; equipment maintenance, repair, or replacement; process redesign; employee training; or even maintaining the status quo. The recommendations must include all of these assumptions and uncertainties.

The people in the decision process will want to know how reliable your numbers and quantitative results are. Data from laboratory analysis will have a basis for accuracy and precision. Accuracy is the difference between a given answer and the true value. Precision is the measure of identical samples, analysis, apparatus, and procedures to replicate themselves. The true value then becomes a consensus of experts, often as recognized by some official agency. Determinate or systematic errors are consistent and reproducible: for example, poor equipment calibration, errors in the methodology, or faulty apparatus. Indeterminate errors are not reproducible and are random: for example, poor analytical technique, sloppy housekeeping, calculation errors, or interpretation errors. Acceptance or rejection of data should be done with the aid of statistical mean and standard deviation criteria.

Financial and other practical considerations also enter into the management decision. These constraints may include money, employee limitations, or customer demands. Throughout all of this assessment and management process, the principal objective of safety should guide the decisions of management. [2]

Most people seek to avoid risk while gaining a desired result. There can hardly ever be a gain without at least some risk. Industrial operations involve some of that risk. An acceptable state of safety has sufficient gain to make the risk of operation desirable. We travel in aircraft for gain even though airplane accidents can kill people, for example. The high risk of a potentially lifesaving drug might be acceptable to many people. [23]

Catastrophic industrial accidents and other safety, health, and environmental problems have made the general public wary of most industrial process facilities. Protection of human safety and health and the environment has caused the public to oppose the siting of many plant facilities in their locale. While a new or expanded facility may bring numerous economic benefits to a community, the same facilities also are often associated with declining property values, health risks, and a perceived reduction in the local quality of life. Political legislation and subsequent regulations stemming from these fears have made siting process facilities an unpopular, complex, and expensive process.

The concept of risk assessment and management is crucial in the siting of process facilities. Many government agencies have accepted the use of risk assessment for evaluating concerns for human safety and health. The "not in my back yard" (NIMBY) sentiment has made the siting of many facilities difficult. There are numerous examples of technologically sound and economically attractive industrial plants that have been unable to obtain construction and operating permits because of the NIMBY sentiment.

The general public cannot be expected to voluntarily accept a risk to human safety, health, and the environment when there are no corresponding benefits. As a result, decisions involving risk should include the benefits associated with the siting of the process facility. The public has not uniformly accepted the product of frequency times consequence as a valid measure of true risk. Industry and government need to recognize this public perception and include it in their evaluation of siting the process facility. Failure to recognize this misconception on the part of process and plant designers has encouraged the growth of the NIMBY sentiment. Industry and government can no longer ignore the fear, lack of understanding, and hostility that can result in a community when trying to site an otherwise attractive process facility.

The design and operation of facilities usually involve multiple processes and auxiliary systems with a wide variety of raw materials, chemical reactions, and physical transformations. Considerable emphasis should be placed upon the risk of facilities to prevent safety hazards and pollution by elimination or reduction. However, even with the best of prevention there will always remain the need for risk management at industrial process activities. [18,20]

There will never be unanimous approval and acceptance by local citizens for siting a new or expanded process facility in their locale. The NIMBY syndrome generally disregards the benefits of a facility and concentrates mainly on the associated risks. When made aware of the benefits and need for the process facility, the general public will usually accept reasonable risks. With acceptable controls, the economic benefits of a new facility may make the facility acceptable.

RISK ANALYSIS AND MANAGEMENT

The subject of risk management is a broad field that includes the areas previously discussed in this chapter and elsewhere in this textbook. The elements in an overall risk management program should be integrated into all aspects of the workplace

TABLE 2-12
ELEMENTS OF A RISK-MANAGEMENT PROGRAM

Hazard identification
Risk assessment
Administrative controls
Engineering controls
Emergency response planning
Operation and emergency training
Accident and incident investigation
Near-miss review
Internal and external audits
Feedback and iteration

(Table 2-12). Safety, health, and environment protections require the continual commitment of everyone. [43]

Effective risk management ensures an objective, consistent response to the identified risks. This requires thorough planning, organizing, implementing, and controlling to achieve a successful risk management program. [28]

There are a number of analysis methods. These methods all rely on decision-making processes to provide recommendations regarding degree of risk from exposure to hazards. The evaluations can also provide insight into control measures for risk reduction.

The regognition of potential hazards requires considerable familiarity with the operations and related process systems, storage and distribution of materials and products, and relevant job activities. The workplace is a dynamic situation; there are continual changes in operations, processes, and other work requirements. There are potential hazards that relate to human, situational, and environmental factors. These hazards need to be recognized, quantified, and controlled. For example, operations and process systems use raw materials to produce products, by-products, and wastes. [34] OSHA requires a chemical inventory (Hazard Communication Standard: 29 CFR 1910.1200). Suppliers must furnish a material safety data sheet (MSDS) with every chemical.

In another example, an oilwell hazard identified during a routine workover has led to safer operating practices. [4]

The examination of a process, operation, and facility for potential hazards is a complex task. Generally, it is best to begin with process flow diagrams and an outline of the facility layout. A more detailed piping and instrument diagram (P&ID) better identifies all of the potential hazards. These items, along with standard operating procedures, should provide a valuable description of the operations for study and review.

Many techniques or procedures can be used to identify and evaluate potential hazards to safety, health, and environmental protection. In 29 CFR 1910.119, OSHA requires one or more of the following methodologies to determine and evaluate process safety hazards:

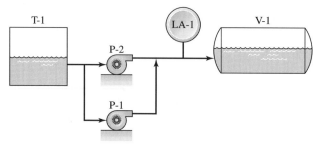

FIGURE 2-4
FLUID FLOW EXAMPLE FOR HAZARD ANALYSIS.

- What-if checklist
- Hazard and operability study (HAZOP) [29]
- Failure mode and effects analysis (FMEA)
- Fault tree analysis
- An appropriate equivalent methodology

FAULT TREE ANALYSIS

Let us take a simple example. Sulfuric acid is pumped from an atmospheric storage tank, T-1, to a pressurized process vessel, V-1 (Fig. 2-4). Two pumps can perform this service: P-1 is driven by a diesel engine; P-2 has an electric motor. During normal operations the system uses P-2. In the event of an emergency, P-1 can be switched into service. If P-2 fails, the operator has 4 minutes to switch to P-1 before a potential hazard occurs. There is a low-flow alarm, LA-1, to the process vessel, V-1.

The first step in fault tree analysis, a deductive reasoning process, is to clearly define the top event. In Figure 2-5, the top event is "system fails." There are two

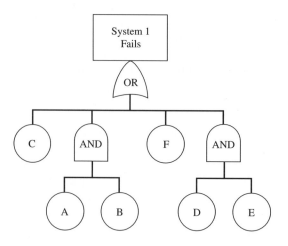

FIGURE 2-5
TYPICAL FAULT TREE ANALYSIS.

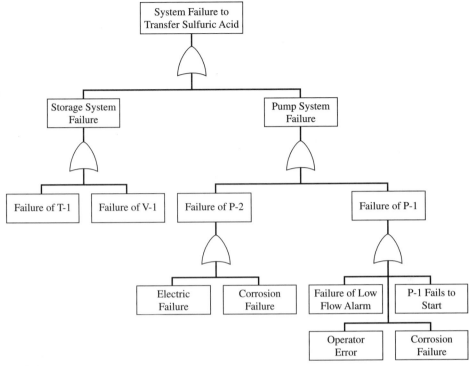

FIGURE 2-6
FAULT TREE ANALYSIS FOR THE FLUID FLOW EXAMPLE.

AND gates, one OR gate, and the A–F basic events. In Figure 2-6, the top event becomes "System fails to transfer sulfuric acid" from T-1 to V-1 for an uninterrupted period ≥ 4 minutes because of equipment failure or operator error.

The next step is to develop two or more intermediate events connected to the top event with a logic gate. Two intermediate events, either storage system failure or pump system failure, could result in the top event through an OR gate. Storage system failure could occur if either T-1 or V-1 fails; this would be the basic event through an OR gate.

Pump system failure could occur if both P-1 and P-2 fail; these are basic events through an AND gate. The failure of P-2 could be caused by an electric power failure or pump corrosion, which are basic events through an OR gate. Causes of P-1 failure could be: operator error, P-1 fails to start, the low-flow alarm fails, or pump corrosion; these are basic events through an OR gate (see Fig. 2-6).

The fault tree analysis method can be summarized into the following steps:

- Define the top event.
- Define the intermediate events.
- Identify all gates and basic events.
- Resolve all duplication and conflict.

EVENT TREE ANALYSIS

Event tree analysis (ETA) is an inductive logic model that identifies possible outcomes from a given initiating event. An initiating event will usually initiate an accident or incident. An ETA considers the responses of operators and safety systems to the initiating event. This technique is best suited for analyzing complex processes involving several layers of safety systems or emergency procedures.

The first step in the development of an event tree is to define an initiating event that could lead to failure of the system: equipment failure, human error, utility failure, or natural disasters (see Fig. 2-7).

The next step is to identify intermediate actions to eliminate or reduce the effects of the initiating event. The event tree develops two branches for each intermediate event, one for a successful and the other for an unsuccessful operation. The top path represents success, and the bottom path failure.

For the simplified example an initiating event becomes the failure of P-2. There are several response steps to the initiating event, including the low-flow alarm, the operator response, and the failure of P-1 to respond.

This event tree analysis, as shown in Figure 2-8, uses the following.

1 P-2 fails and becomes the initiating event. The probability of this event has been defined as 1.0.

2 The low-flow alarm to V-1 either works or fails. If it works, the upper arm is followed. If it does not work, the lower arm applies. The alarm has a 0.998 probability of success.

3 The operator either responds or does not respond to the alarm. The probability of responding is 0.952.

4 The last response is the operator getting P-1 to start. The probability of this occurring is 0.995.

Based on Figure 2-8, the probability of any branch of the event tree occurring is the product of the event probabilities on the branch. The top branch has a success probability of 1.0 + 0.998 + 0.952 + 0.995 = 0.945. The total additive value for the probabilities of all of the branches is 0.945 + 0.005 + 0.048 + 0.002 = 1.0.

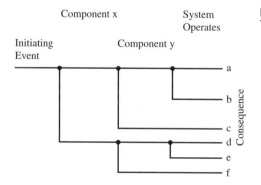

Component x   System Operates

Initiating Event   Component y

a
b
c
d
e
f

Consequence

FIGURE 2-7
TYPICAL EVENT TREE ANALYSIS.

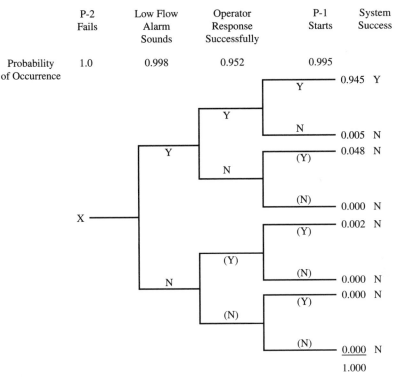

FIGURE 2-8
EVENT TREE ANALYSIS FOR THE FLUID FLOW EXAMPLE.

Figure 2-9 on page 64 shows a simplified event tree analysis; unsuccessful paths are ignored and terminated immediately. But all probabilities are not available to check the sum to unity.

Event tree analysis is best for analyzing initiating events that could lead to the final effect of the event. The event tree branches each represent a separate sequence of relationships between the safety functions for the initiating event.

The event tree can be summarized as follows:

- Identify initiating events that could result in an accident.
- Identify the safety functions to mitigate the initiating event.
- Construct the event tree.
- Describe accident sequence outcomes and their probabilities.

For the same system and probability assumptions, identical results should occur for fault tree and event tree analysis. Fault trees are more comprehensive than event trees because the event tree is based upon a single failure event. Most people tend to think logically about safety systems, using the event tree approach. The fault tree has a backward logic approach to many people. A combination of both

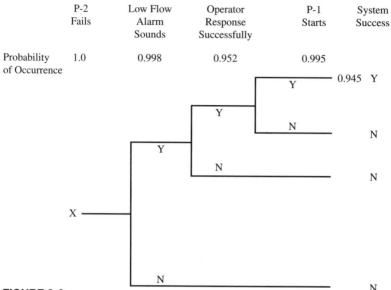

P-2 Fails	Low Flow Alarm Sounds	Operator Response Successfully	P-1 Starts	System Success

Probability of Occurrence	1.0	0.998	0.952	0.995

FIGURE 2-9
SIMPLIFIED EVENT TREE ANALYSIS FOR THE FLUID FLOW EXAMPLE.

fault tree and event tree analysis may take longer to perform, but both approaches have advantages.

TIMING OF RISK MANAGEMENT

Safety should be considered at all stages in the life cycle of a process system or a facility. [16] Safety should be considered early in the process because there is a greater opportunity to make the design inherently safe at a lower cost. Elimination or reduction of safety hazards is the simplest way to solve safety problems.

The need for safe R&D facilities may differ from the needs of safe commercial plants. [33] The life of an operating facility begins with research and development, followed by engineering design, plant construction, start-up, and ongoing operations. As this scenario progresses to achieve operations, it becomes more and more difficult and costly to modify or change equipment, processes, or procedures to assure safe operations. As construction is completed, the only alternative is to add controls, redundancy, alarms, and interlocks and to provide additional training to operating personnel.

Since there is never zero risk, risk reduction becomes key. The remaining or residual risk must be managed in an acceptable manner. [8] This can only be done if management has a good working knowledge of the processes, facility layout, and potential community impact. This working knowledge must be kept current through all changes and modifications at the facility. Management should conduct periodic reviews to confirm the risk of hazards. [21] All of these risk management activities should be openly communicated to employees and the public in a timely manner.

BIBLIOGRAPHY

1 *Accident Facts,* National Safety Council, Itasca, Ill., 1993.
2 *Accident Prevention Manual for Business and Industry: Administration and Programs,* 10th ed., National Safety Council, Itasca, Ill., 1992.
3 *Accident Prevention Manual for Business and Industry: Environmental Management,* National Safety Council, Itasca, Ill., 1995.
4 Bailey, J. M., "Well Control Problem Leads to Better Workover Procedures," *Oil and Gas Journal,* January 16, 1995.
5 Baldwin, D. A., *Safety and Environmental Training,* Van Nostrand Reinhold, New York, 1992.
6 Bravo, F., and D. Contreras, "Safely Handle Polymerization Initiators," *Chemical Engineering Progress,* July 1994.
7 Center for Chemical Process Safety, *Guidelines for Hazard Evaluation Procedures,* 2d ed., AIChE, New York, 1992.
8 Center for Chemical Process Safety, *Plant Guidelines for Technical Management of Chemical Process Safety,* AIChE, New York, 1992.
9 Chao, S., and A. Attari, "Methods Developed for Detecting Hazardous Elements in Producing Gas," *Oil and Gas Journal,* January 16, 1995.
10 Cropper, C. M., "Skier's Knee," *Forbes,* February 14, 1994.
11 Crowl, D. A., and J. F. Louvar, *Chemical Process Safety: Fundamentals with Applications,* Prentice-Hall, Englewood Cliffs, N.J., 1990.
12 Ember, L. R., "The Nicotine Connection," *C&E News,* November 28, 1994.
13 "The Facts about Fats," *Consumer Reports,* May 1994.
14 Finucane, M. T., and J. N. Edmundson, Risk Assessment and the Major Hazard Industries, with Particular Reference to Offshore Installations. Proceedings of the 2d International Conference on Fire Engineering, Gulf Publishing Company, Houston, 1989.
15 Glickman, T. S., and M. Gough, *Reading in Risk,* Resources for the Future, Washington, D.C., 1990.
16 Greenberg, H. R., and J. J. Cramer, *Risk Assessment and Risk Management for the Chemical Process Industry,* Van Nostrand Reinhold, New York, 1991.
17 Hajian, H. G., and R. L. Pecsok, *Working Safely in the Chemistry Laboratory,* American Chemical Society, Washington, D.C., 1994.
18 Hammer, W., *Occupational Safety Management and Engineering,* 4 ed., Prentice-Hall, Englewood Cliffs, N.J., 1989.
19 Kamp, J., "Worker Psychology," *Professional Safety,* May 1994.
20 Kavianian, H. R., and C. A. Wentz, *Occupational and Environmental Safety Engineering and Management,* Van Nostrand Reinhold, New York, 1990.
21 Kim, I., et al., "Risk and the CPI," *Chemical Engineering,* February 1995.
22 Kirschner, E. M., "Environment, Health Concerns Force Shift in Use of Organic Solvents," *C&E News,* June 20, 1994.
23 Klaassen, C. D., et al., *Casarett and Doull's Toxicology,* 3d ed., Macmillan, New York, 1986.
24 LaGrega, M. D., et al., *Hazardous Waste Management,* McGraw-Hill, New York, 1994.
25 Lepkowski, W., "Bhopal Ten Years Later," *C&E News,* December 19, 1994.

26 Lepkowski, W., "Chemical Companies Make Public Worst Case Accident Scenarios," *C&E News,* June 20, 1994.
27 Macenski, A. G., "Lead: Health Hazard for the 1990s," *Professional Safety,* January 1994.
28 Morgan, D. J., and G. H. Swett, "Simplify Remedial Decisions by Using Risk-Based Management," *Hydrocarbon Processing,* August 1994.
29 Mukesh, D., "Include HAZOP Analysis in Process Development," *Chemical Engineering Progress,* June 1994.
30 Murphy, M. R., "Prepare for EPA's Risk Management Program Rule," *Chemical Engineering Progress,* August 1994.
31 Nimmo, I., "Extend HAZOP to Computer Control Systems," *Chemical Engineering Progress,* October 1994.
32 Obradovitch, M. M., and S. E. Stephanow, *Project Management: Risks and Productivity,* Spencer, Bend, Ore., 1990.
33 Palluzi, R. P., "Develop and R&D Safety Standards Program," *Chemical Engineering Progress,* February 1995.
34 Plog, B. A., *Fundamentals of Industrial Hygiene,* 3d ed., National Safety Council, Itasca, Ill., 1988.
35 Reynolds, J. T., "Risk-Based Inspection Improves Safety of Pressure Equipment," *Oil and Gas Journal,* January 16, 1995.
36 Roland, H. E., and B. Moriarty, *System Safety Engineering and Management,* 2d ed., Wiley, New York, 1990.
37 Schurman, D. L., and S. A. Fleger, "Human Factors in Hazops," *Professional Safety,* December 1995.
38 Sharn, L., "Earlier Damage a Suspect in Blast," *USA Today,* March 29, 1994.
39 Slote, L., *Handbook of Occupational Safety and Health,* Wiley, New York, 1987.
40 Sutton, I. S., *Process Reliability and Risk Management,* Van Nostrand Reinhold, New York, 1992.
41 Tolpa, G., "Simplify Process Hazard Reviews with 3-D Models," *Hydrocarbon Processing,* October 1994.
42 Wakeling, H., Advances in Hazard Identification. Proceedings of the 2d International Conference on Fire Engineering, Gulf Publishing Company, Houston, 1989.
43 Wentz, C. A., *Hazardous Waste Management,* 2d ed., McGraw-Hill, New York, 1995.
44 U.S. Department of Labor, *Work Injuries and Illnesses by Selected Characteristics, 1992,* Washington, D.C., 1994.

PROBLEMS

1 A half-a-pack-a-day cigarette smoker has a 0.15 average lifetime risk of dying of lung cancer. Assuming a normal life expectancy of 70 years, what would be the risk of the smoker dying of cancer from one cigarette?
2 You have just won the grand prize in a contest, which is an expense-paid week in Paris for you and your companion. You will be flown by jet airplane, be wined and dined in Michelin three-star restaurants, and stay in a magnificent French chateau near a nuclear power plant. You must take the entire trip package as planned with no exceptions or substitutions. Discuss the personal risks that are associated with the trip, based on the types of activities in Table 2-5. Given all of the risks involved, would you take the trip?
3 You have an opportunity to have unsafe sex with a companion who has the HIV virus. Assume the probability of contracting the HIV virus is 50 percent. Also assume that the probability of dying from the AIDS disease caused by the HIV virus is 95 percent.

What is the probability of you dying from this single unsafe sex act? Would you be willing to take this risk?

4 Compare the average risk per person for these two situations: (a) 1 chance in 10,000 chance of 10,000 people dying; (b) 1 chance in 10 of 10 people dying. Which situation is likely to be less acceptable to the public?

5 The following workers' compensation data has been compiled for a specific industry.

Work activity	Type of injury	Percent of total cost
Handling materials	Strained back,	45
	broken toes	10
Operating machinery	Cut hands,	17
	eye injury	8
Construction	Head injury	11
Maintenance	Steam burns	9

As the safety director, where should you concentrate your resources to reduce safety risks?

6 The amount of ascorbic acid or vitamin C in vitamin supplement pills has been analyzed from production samples.

Sample	Vitamin C (mg)
1	62
2	58
3	60
4	63
5	59
6	58
7	62
8	61
9	60
10	59

What is the mean and standard deviation of the vitamin C content of the 10 samples? Discuss the significance of these results.

7 Using the simplified example for fluid flow in Figure 2-4, perform
   A what-if hazard analysis

8 A checklist hazard analysis

9 A HAZOP hazard analysis

10 A catalytic chemical reaction has two raw-material feed streams and a solvent. A HAZOP analysis is to be performed on the process. Discuss the kinds of questions that might be generated from the HAZOP guidewords.

11 A serious hazard and incident scenario and an offshore oil and gas platform could involve major fires and explosion. Develop a fault tree for this incident. Develop an event tree for a flammable release that may become either a controlled or an uncontrolled fire.

12 The failure rate of a control valve is estimated to be 0.75 faults per year. What is the failure probability for the control valve?

13 For each of the following component failure rates, determine their failure probabilities.
   a Thermocouple temperature, 0.45 faults per year
   b Liquid level measurement, 2.16 faults per year
   c Liquid flow measurement, 1.34 faults per year

14 You have been named the new environmental and safety manager for a large oil-refining facility that processes hydrocarbons. From your past experience you know there have been hydrocarbons entering the air, soil, surface water, and groundwater. Which of these hydrocarbons pose the greater hazards?

15 As the manager of an oil refinery you are aware of the potential for hydrocarbon releases to the environment. How would you assess and prioritize both the past and present hazards for a comprehensive risk management plan?

16 A plant operator, using computer control, attempted to fill a liquid storage vessel that was already full. The operator mistakenly confused the readout on the computer screen, and a large spill resulted. What caused the incident?

17 An operator with an automated control system added water to a tank that contained reactive chemical residues. The reaction of the water caused increased tank pressure, which burst the rupture disk, releasing a flammable toxic gas into the atmosphere. The control panel pressure gauge for the tank had a high-pressure warning light, which was missed or ignored because of the confusing arrangement on the display panel. What type of error caused this incident?

18 A computer programmer inadvertently typed the code for pressure instead of temperature into a control software program. The intent of the program was to control at 125°C and 2 atm pressure. Instead 125 atm of pressure became the control point. The process could not be controlled and a large quantity of off-specification was produced. How could this situation have been avoided?

3

TOXICOLOGY IN
THE WORKPLACE

Toxicology is the study of the adverse effects of chemicals on living organisms. The science of toxicology is concerned mainly with the toxic or poisonous properties of chemical substances. Although the news media provide daily warnings about the presence of chemicals in our water, air, and food, most chemicals do not represent a hazard under normal circumstances. [12] But at sufficiently high concentrations and levels of exposure, *all* chemicals have the potential of being a hazard. In an analogous manner, at a sufficiently low concentration and level of exposure, all chemicals are safe and do not have the potential of being a hazard.

If a chemical has a noxious effect on the body as a result of contact with the respiratory tract, the mouth and digestive system, the skin, or the eye, it is considered toxic. The effect may be reversible or irreversible, result in benign or malignant tumors, have mutagenic or teratogenic effects, or cause bodily injury or death.

Medication, vaccines, and chemical exposure can result in side effects that are life-threatening. The benefit of medicines must be weighed against their adverse effects.

A chemical is considered acutely toxic when it causes harm after only short-term exposure or exposure in limited amounts. Chronic toxicity measures similar adverse health effects, including death, but over a much longer time frame, possibly many years or a lifetime. Even potential carcinogens and highly toxic substances may exhibit no effect at the moment of exposure, if the quantity is very small or the exposure time is short.

The Delaney amendment which was added to the Food, Drug and Cosmetics Act in 1952, assumes there is no safe level of chemical exposure that is really safe for food, if the chemical has been proven to be a carcinogen regardless of how low the level of exposure. This has become a major scientific controversy that ignores an effect level for a carcinogen.

In order to be effective in the workplace, toxicology must consider all types of chemical exposure and the resulting effects on living organisms. The main objective should be to define how much is unacceptable and to recommend precautionary measures and constraints to assure that under normal workplace conditions employees are not exposed to those unacceptable levels. [5, 6, 7, 13]

There are many chemicals that are essential to health. Some of these are toxic in quantities larger than the small quantities essential for human health. We could not live without small amounts of zinc, manganese, copper, potassium, chromium, nickel, tin, selenium, and molybdenum. Exposure to large quantities of these chemicals produces severe acute and chronic toxicity in humans.

Many factors contribute to toxicity: [14]

Route of entry
Dosage level
Physiological state of the receiver
Environmental conditions
Physical properties of the chemical
Chemical properties of the chemical

The route of entry affects the timing and ability for toxic chemicals to enter and be transported throughout the body. Some chemicals are highly selective in their action on tissue, blood cells, and organs. Other chemicals exert harmful effects on all living matter. Certain biological species may have built-in mechanisms to protect these plants and animals from specific chemical toxins. It then becomes important to designate the biological species and exposure conditions when comparing one chemical to another.

A toxic effect is the response to a chemical exposure that satisfies all of the criteria shown in Table 3-1. [14] It is impossible to generalize about the toxicity of a chemical based upon its composition or structural formula. There are some analo-

TABLE 3-1
CRITERIA USED TO DEFINE A TOXIC EFFECT FROM CHEMICAL EXPOSURE

1 There is an observable or measurable physiological deviation in an organ or organ system that catalyzes a normal physiological process or results in a specific biochemical change.
2 The physiological change reduces the ability of an organ or system to resist or adapt to other normal stimuli, or the protective mechanism is impaired in its defense against adverse stimuli.
3 The deviation is attenuated or removed on removal of the stimulus.
4 The observable deviation can be duplicated in more than one animal species, but not necessarily at the identical dose-response relationships.
5 The observable deviation is confirmed by other research investigators.

Source: National Safety Council.

gies between chemical composition, structure, and toxicity, but each chemical must be evaluated on a case-by-case basis.

The physical and chemical properties of the substance are a major factor in the associated hazard potential. Many liquid solvents are highly volatile and will vaporize to form high concentrations in air at ambient conditions. The vapor pressure of a liquid is important hazard information because it contributes greatly to the liquid hazard potential from vapor inhalation. Liquids with low boiling points are much greater inhalation hazards than those with high boiling points, all other things being equal.

Injuries can be local or systemic. A local injury is the result of direct contact of a chemical with living tissue. The skin can be severely burned or vision can be impaired. The respiratory system can be injured from the inhalation of toxic vapors, fumes, mists, and dusts that injure the lining of the trachea and the lungs. A chemical must gain entrance to the bloodstream to produce a systemic injury.

ROUTE OF BODY ENTRY

The route of entry into the body plays an important role in chemical toxicity. The toxic effects of a substance are dependent upon how it gains extrance into the body and, further, into the bloodstream. The most common routes of entry into the body are inhalation, absorption through the skin, ingestion, and injection.

A substance can enter via more than one route at a time, depending upon the chemical properties and surrounding conditions. For example, a toxic vapor in air can enter the body by both inhalation and skin absorption. Once the chemical has entered the bloodstream, the toxic effect may be general or specific to certain organs or tissue.

The most important route of entry for industrial exposure to chemicals is by inhalation. When air and its contaminates are inhaled, they first pass through the upper respiratory tract: the nose, throat, trachea, and bronchial tubes (Fig. 3-1 on page 72). The air is then transported to the alveoli, where the gases are diffused across thin membrane cell walls (Fig. 3-2 on page 72). [14] As the air passes through the respiratory system the airborne particle flow and size becomes more restricted. The successively smaller branches of the respiratory system and the winding passageways are the basis for physical removal of particles. The upper respiratory system removes about half of the particles that are 2 micrometers (µm) in diameter, and the alveolar or pulmonary air spaces remove the rest.

By the time airborne particles reach the alveolar sac, the airborne particles are generally less than 5 µm in diameter because nearly all particles larger than 4 to 5 µm are deposited in the upper respiratory tract. This highly efficient particulate collection removes about half of the 1 µm particles from the inhaled air; the other half are exhaled.

The concentration of oxygen in the alveolar air is greater than the oxygen content of the blood coming to the lungs from the right ventricle of the heart, causing oxygen diffusion from the alveoli through the permeable membrane into the blood. This gas diffusion model is based mainly upon the differential partial pressures of oxygen and carbon dioxide in the respiratory system (Fig. 3-3 on page 73). In an analogous situation the oxygen concentration in the bloodstream is greater than in the tissue cells, causing oxygen to permeate the capillary walls to increase the level

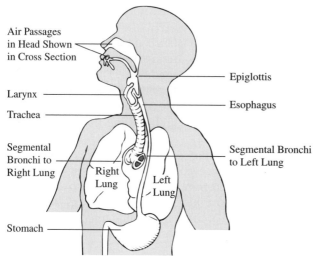

FIGURE 3-1
THE MAJOR PARTS OF THE HUMAN RESPIRATORY
SYSTEM. (*Source:* Used by permission of National Safety
Council, Itasca, Illinois.)

of oxygen in the tissue cells (Fig. 3-4). The flow of waste carbon dioxide is the reverse of the oxygen flow. Carbon dioxide diffuses from the tissue cells through
vein walls where it is carried in the blood to the alveolar sacs for expulsion from
the respiratory system.

Oxygen is carried by the blood plasma and is also combined with hemoglobin
in the red blood cells to form oxyhemoglobin. Whenever the free oxygen partial
pressure in blood plasma is reduced by tissue cell transfer, the combined oxygen is

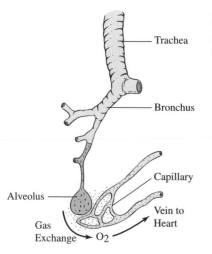

FIGURE 3-2
THE GAS EXCHANGE OF AIR IN AN ALVEOLUS.
(*Source:* Used by permission of National Safety
Council, Itasca, Illinois.)

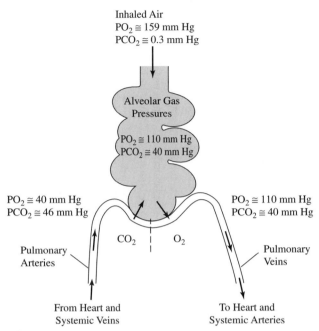

Inhaled Air
$PO_2 \cong 159$ mm Hg
$PCO_2 \cong 0.3$ mm Hg

Alveolar Gas
Pressures

$PO_2 \cong 110$ mm Hg
$PCO_2 \cong 40$ mm Hg

$PO_2 \cong 40$ mm Hg
$PCO_2 \cong 46$ mm Hg

$PO_2 \cong 110$ mm Hg
$PCO_2 \cong 40$ mm Hg

CO_2 O_2

Pulmonary
Arteries

Pulmonary
Veins

From Heart and
Systemic Veins

To Heart and
Systemic Arteries

FIGURE 3-3
GAS DIFFUSION MODEL FOR THE RESPIRATORY
SYSTEM AND THE BLOODSTREAM. (*Source:* Used by
permission of National Safety Council, Itasca, Illinois.)

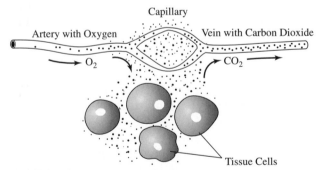

Capillary

Artery with Oxygen

Vein with Carbon Dioxide

O_2

CO_2

Tissue Cells

FIGURE 3-4
GAS DIFFUSION MODEL FOR THE BLOODSTREAM AND
TISSUE CELLS. (*Source:* Used by permission of National
Safety Council, Itasca, Illinois.)

released by the oxyhemoglobin to maintain free oxygen in the plasma. The reduced hemoglobin is later replenished by oxygen exchange with alveolar air. The oxygen content of the lungs is based upon the content of the incoming air, the flushing of the lungs to regulate carbon dioxide, and the rate of oxygen diffusion into the blood as it passes through the lungs.

Healthy adults should not experience immediate effects of oxygen deficiency in air when the normal volume of oxygen in air declines from 21 percent to as low as 15 percent. At 10 percent oxygen, healthy adults experience dizziness and shortness of breath. Stupor and life-threatening situations occur at 5 to 7 percent oxygen. At less than 5 percent oxygen, death in humans may occur in a matter of minutes. [9]

While the respiratory system is continually bringing in oxygen and discharging carbon dioxide, the constituents of various other gases and vapors, as well as particulate matter, are also entering the system. Some of these other constituents may be toxic substances. The type and quantity of a toxic chemical entering the respiratory system, and deposited, absorbed, or otherwise remaining in the respiratory pathways depend upon the concentration in the air, the exposure duration, and the pulmonary ventilation volumes. Gases and vapors pass through the alveolar membranes into the blood, where they are then distributed to various organ sites. The gases and vapors with lower water solubility and higher fat solubility are more likely to enter the blood via inhalation. As long as the airborne contaminate concentration remains constant, the level of the toxic substance in the blood will reach equilibrium between the absorption, metabolism, and elimination processes of the body.

Another important route of entry into the body is absorption through the skin. When a chemical comes in contact with the skin, there are a number of possible outcomes.

- The skin may block entry into the body.
- The chemical may cause skin irritation.
- The chemical may produce skin sensitization.
- The chemical may penetrate the skin and enter the bloodstream.

There are three layers of skin tissues; epidermis, dermis, and subcutaneous tissue. The total thickness of the skin ranges from 0.5 mm on the eyelid to about 4 mm on the soles of the feet and palms of the hand.

The top skin layer is the *epidermis,* which consists of dead cells. This layer is the rate-limiting barrier against absorption of water and various aqueous solutions. It offers little protection against absorption of organic liquids or gases in general. The layer flakes off over time. When this layer becomes damp or wet, it flakes off even faster. As the layer is depleted, it is replaced by newly formed cells from below. This regeneration and sloughing process also protects against chemical exposure. The epidermis contains no blood, but it is bathed in lymph, a blood-derived fluid. The nerve endings in the dermis layer supply the sense of touch through the epidermis layer.

The *dermis* is 15 to 40 times thicker than the epidermis. It is tough and resilient, making it the main protection against trauma. When the dermis is injured, it forms new tissue and repairs itself. Blood vessels, nerve fibers, hair follicles, and sweat glands are contained in the dermis.

The *subcutaneous layer* contains fatty and resilient elements that cushion and insulate the skin layers above it. The subcutaneous layer forms an important link between the dermis and tissue that covers muscles and bones.

The skin serves to regulate body heat through the production of sweat. The temperature of the environmental surroundings and physical activity usually controls the generation of sweat. The loss of body heat by evaporation of sweat is much more effective than by heat radiation. Sweat is more apparent and visible when the humidity is high, because the evaporation rate slows in high humidity. Low humidity results in high evaporation rates and drier skin. The skin absorption rate for some chemicals increases as the temperature and perspiration rate increases. Chemicals that are water-reactive or generate heat upon contact with moisture can severely damage skin when it has perspired excessively.

The absorption of chemicals by the skin may result from surface contamination of the skin or clothing. Other chemicals may be absorbed from the vapor phase in direct proportion to their concentration in air. Once the chemicals are absorbed into the skin, they can then migrate through the skin layers to eventually enter the bloodstream. Absorption of chemicals through the skin increases dramatically when the skin continuity is interrupted by abrasion, laceration, or puncture. Several physical and chemical properties are important in determining the ability of a chemical to be absorbed through the skin, including aqueous and lipid solubilities, extent of ionization in water, size of the molecule, pH of the skin, and concentration of hair follicles in the skin.

Ingestion of toxic chemicals into the body is less of a problem in the workplace than either inhalation or skin absorption. Occasionally, a hazardous amount of a chemical may be swallowed accidentally via hand-to-mouth contact. Eating lunch or snacks in contaminated work areas should be avoided. Previously inhaled chemicals may be ingested when contaminated saliva or respiratory tract mucus is swallowed.

Injection of chemicals directly into the bloodstream rarely occurs in the workplace. Injection is sometimes used in laboratory animal studies to determine chemical toxicity. This technique is far less costly than animal inhalation studies, but it does have some drawbacks. For example, injection bypasses the normal body protective mechanisms that may prevent the chemical from entering the body.

DOSE-RESPONSE RELATIONSHIP

The dose-response relationship is the basis for toxicological considerations. A dose of a chemical is administered to a test animal species and the outcome is observed. This methodology is continued by trial and error until a range is identified where all or most of the test animals die at one end and all or most of the test animals survive at the other end of the range. The data from the specified test conditions are then

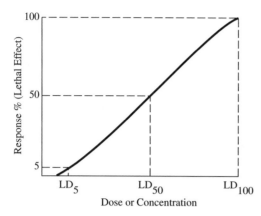

FIGURE 3-5
TYPICAL DOSE-RESPONSE
RELATIONSHIP. (*Source:* National Safety
Council.)

plotted to form a typical relationship, as shown in Figure 3-5. This dose-response relationship is a fundamental observation for all chemicals in the workplace.

Dosage levels are reported as

- Quantity per unit of body weight
- Quantity per unit of air volume respired
- Quantity per unit of exposed skin surface

The length of time of exposure is also critical to the dose-response relationship. When the chemical concentration and the exposure time are considered together, their product becomes an important criterion because high concentrations for short time periods produce similar effects as low concentrations over longer time frames. While this relationship will probably not be valid at either end of the dose-response curve, it can be used with caution in anticipating toxic effects of chemicals in the midrange of the curve. This technique can be used to predict potentially safe exposure limits in the absence of actual experimental data.

There are several basic assumptions to establishing a dose-response relationship. [11]

- The response is due to the chemical administered.
- The response is related to the dosage level.
- The dose and response can be measured quantitatively.

Since low-level exposures to most chemicals are not harmful, it follows that there is a threshold level below which there is either no effect or a potentially beneficial effect. As shown in Figure 3-6, as the dose is increased, there is a point that begins to produce a measurable adverse effect. This initial toxicity observation and the rate of increasingly adverse effects are used to define the degree of toxicity of a substance. This toxicity is also based upon the response produced in a biological system or target organ.

Multiple doses of low levels of a chemical are less harmful to the body than the same total amount when it is administered in a single dose. The body can detoxify

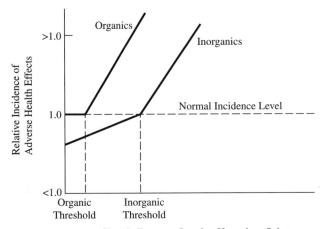

FIGURE 3-6
THRESHOLD EFFECT OF CHEMICAL EXPOSURE.

or excrete smaller doses fairly quickly. A toxic substance becomes a poison when the rate of intake is greater than the rate of detoxification or excretion. The accumulation of the chemical will be observed with increasing levels in the blood, urine, or exhaled air.

The dose-response curve can be used to study the lethal dose (LD) in the test animal population. For example, LD_{50} is the dose of a substance that will be fatal to 50 percent of a defined test animal population by any route of body entry other than inhalation. LD_0 and LD_{100} refer to zero or 100 percent fatalities, respectively. The LD units are usually expressed in milligrams per kilogram body weight. The experimental conditions of the LD_{50} rate should be specified. These include test animal species, route of chemical administration, chemical purity, and time period of administration. The acute LD_{50} values for some selected chemical substances are shown in Table 3-2 on page 78.

The slope of the dose-response curve can provide a safety margin in going from a nonlethal to a lethal dose. If the slope is steep, the safety margin is low. Also, the ability to compare relative toxicity between chemicals becomes more difficult with a wide variation in the slope of the curve.

Inhalation exposures of airborne chemicals are expressed as lethal concentration (LC), that is, the concentration of the substance in air that is likely to cause death to a certain percentage of the test animals in a known length of time. LC values should give the animal species studied, the exposure concentration in parts per million (ppm), and exposure time.

Many types of adverse responses may be evaluated. The action of the toxic chemical can be determined by examining the animal organs, particularly the liver and kidney. Variations in the animal growth rate may be observed even though the chemical does not cause death. The ratio of the lungs to body weight may define

TABLE 3-2
ACUTE LD_{50} VALUES FOR SELECTED CHEMICALS

Chemical	LD_{50}*
Ethyl alcohol	10,000
Sodium chloride	4,000
Ferrous sulfate	1,500
Strychnine sulfate	2
Nicotine	1
2,3,7,8 tetrachlorodibenzo-p-dioxin	0.001
Botulinus toxin	0.00001

*In milligrams per kilogram of body weight.

pulmonary edema from irritants like ozone or NO_x. Chronic exposure to chemicals may be studied with physiological function tests.

The liver and kidneys have considerable binding capacity for chemicals and concentrate more toxins than any other organs. The liver and kidneys are the most important organs in the metabolic elimination of toxic chemicals from the body. When ingested chemicals enter the body they are passed to the liver from the small intestine. The chemicals are processed, transformed, or stored in the liver and then passed into the bloodstream. Chemicals that enter the body from inhalation, skin absorption, and injection pass into the bloodstream directly, thus going to the kidneys prior to the liver. The kidneys have the capacity to excrete chemicals, while the liver has the capacity to biotransform them. [11]

Bones can also store toxic chemicals like lead and strontium obtained from blood by ion exchange with calcium in bones. [11] These toxins can later reenter the bloodstream by reverse ion exchange.

Short-term, acute LD_{50} tests are the most common toxicity measurements in animal experiments. Once an LD_{50} value is obtained for a chemical exposure, it can be compared to the level of toxicity for other chemicals, as shown in Table 3-3.

TABLE 3-3
RELATIVE TOXICITY CLASSIFICATIONS

Classification	Rat oral LD_{50} (g/kg)	Rat 4-h LC_{50} (ppm)	Rabbit skin LD_{50} (g/kg)	Equivalent human oral LD_{50}
Extremely toxic	≤0.001	≤10	≤0.005	taste (1 grain)
Highly toxic	0.001–0.05	10–100	0.005–0.043	1 tsp
Moderately toxic	0.05–0.5	100–1000	0.044–0.34	1 oz
Slightly toxic	0.5–5	1000–10,000	0.35–2.81	1 pt
Practically nontoxic	5–15	10,000–100,000	2.82–22.6	1 qt
Relatively harmless	>15	>100,000	>22.6	>1 q

Source: National Safety Council.

[14] Those chemicals that are extremely, highly, or moderately toxic should be handled and treated as such in the workplace. Caution should be used in interpreting laboratory animal data and applying the data to potential human exposure. Even so, the data for laboratory animals are useful as guides in estimating likely ranges of toxicity in the workplace. LD and LC values for industrial chemicals are available from manufacturers' material safety data sheets. [1]

ACUTE AND CHRONIC EFFECTS

Chemical toxicity is divided into acute and chronic effects, which are interrelated with acute and chronic exposure. Acute exposures and effects are short-term at high concentrations with almost immediate results. Acute exposures usually last less than 24 hours and are often related to an accident. Leaking valves and storage vessels, piping ruptures, and equipment damage are typical workplace sources of acute exposure. These exposures happen suddenly and usually are severe. The chemical enters the body rapidly by inhalation or skin absorption. The critical survival period is shortened and occurs suddenly because of a single exposure that damages vital organs. The acute effect of such an exposure is apparent within minutes or hours.

Chronic effects have symptoms of illness of long duration or frequent recurrence. The chronic effects develop slowly over a long period of time. The continued chronic exposure in the workplace may last throughout an entire working career. The symptoms of chronic poisoning occur at low levels of contaminant and are usually not apparent for a long time period—years or even decades of exposure.

AIR-CONTAMINATE EXPOSURE

Inhalation of air contaminates is the most common exposure route for toxic chemicals in the workplace. Based upon physiological action, most air contaminants can be classified into irritants, asphyxiants, and narcotics. The response to the physiological action depends upon the properties of the chemical, the concentration, and the duration of exposure.

An *irritant* causes aggravation upon contact with tissue. When chemical irritants come in contact with the face and upper respiratory system, the eyes, nose, and mouth are affected. The solubility of an irritant gas greatly influences where the respiratory tract is affected. A highly water-soluble gas like ammonia or hydrogen chloride will irritate the nose and throat as surface moisture absorbs the irritant.

A less soluble gas like nitrogen dioxide will be transported much deeper into the lungs before significant absorption occurs. A comparison of chemicals that exhibit this type of behavior is shown in Table 3-4. [14] These irritants can be inhaled as a gas, a mist, or coatings on particles.

The toxic action of irritants can vary widely. Primary irritants usually exhibit significant irritant action that is far in excess of the systemic toxic action. Also the absorption products formed on the tissues of the respiratory tract are not of serious concern. On the other hand, secondary irritants produce systemic effects that are much greater than any irritant action on mucous membranes. Many aromatic and

TABLE 3-4
IRRITANTS IN THE RESPIRATORY TRACT

	Chemical formula	Toxic effect
Upper respiratory tract irritants		
Formaldehyde	CH_2O	Suspected human carcinogen; severe breathing difficulty and cough
Acrolein	C_3H_4O	Eye and nose irritant
Ammonia	NH_3	Respiratory tract and eye irritant
Sulfur dioxide	SO_2	Eye and respiratory tract irritant; may cause suffocation
Entire respiratory tract irritants		
Chlorine	Cl_2	Intense coughing
Ozone	O_3	Pulmonary congestion
Lower respiratory tract irritants		
Nitrogen dioxide	NO_2	Pulmonary edema
Phosgene	$COCl_2$	Throat irritant; lethal

polyaromatic hydrocarbons exhibit this secondary irritant behavior. These irritant effects are usually completely reversible after short-term exposures. Some irreversible damage may be experienced after long-term exposure.

Asphyxiants block or otherwise interfere with the transfer of oxygen to body tissues. This normally will result in suffocation. Some asphyxiant gases are inert physiologically, but dilute oxygen levels in the respiratory tract and bloodstream, which in turn reduces oxygen levels below the normal requirements for body tissues. Nitrogen and carbon dioxide are examples of such inert gases; they deprive the body of the oxygen being transported from the lungs in the blood. Brain cells can perish in a few minutes, if they are cut off from oxygen. Partial reduction of oxygen in the blood for an extended period of time can result in brain damage or even death. Normal ambient air contains about 21 percent oxygen and 79 volume percent nitrogen. As the oxygen content of the air is reduced, the life process is slowed and threatened. The interrelationship between blood oxygen content and time combines to produce these physiological effects on body tissues.

Another type of asphyxiant gas reacts chemically to prevent the transport of oxygen from the respiratory tract to the blood or blocks the transfer of blood oxygen to the tissues. Carbon monoxide, hydrogen cyanide, and hydrogen sulfide are examples of chemical asphyxiants. Carbon monoxide blocks oxygen transport by preferentially combining with hemoglobin in the blood. Hydrogen cyanide inhibits the ability of body tissues to use oxygen. Hydrogen sulfide will paralyze the brain respiratory center and the sense of smell. Any of these chemical asphyxiant gases can cause collapse and unconsciousness at the following approximate levels in air:

Carbon monoxide	1000 ppm
Hydrogen cyanide	150 ppm
Hydrogen sulfide	500 ppm

Carbon monoxide is a highly poisonous gas that will cause death if inhaled for even a few minutes. It has no smell or taste. More accidents are caused by carbon monoxide than by all other toxic gases combined. [8]

Hemoglobin will combine 300 times more readily with carbon monoxide than with oxygen, forming carboxyhemoglobin. In the normal process of metabolism, carbon monoxide is always being formed in the body and creating low concentrations of carboxyhemoglobin. In smokers the effects from carbon monoxide far outweigh the effects from industrial operations. A typical pack-a-day smoker will have a 5 to 10 percent carboxyhemoglobin blood level, which is comparable to 50 ppm CO in air.

Oxygen uptake by cell tissues will be blocked by hydrogen cyanide. Cyanide poisons the enzyme systems of the body upon contact. The life process is dependent upon these enzyme systems and interference with their catalytic action results in toxic effects.

Narcotics interfere with the central nervous system by causing anesthesia, the loss of sensation, ranging from mild symptoms to the loss of consciousness or even death if the dose is large enough. These narcotic substances and anesthetics include short-chain aliphatic alcohols and ketones, alkyne hydrocarbons, and ethers. Ethyl alcohol, acetone, acetylene, and ethyl ether are among the more common examples of narcotics or anesthetics. Many industrial organic solvents have at least some narcotic effect. Anesthetics can produce desirable results, as evidenced by the loss of sensation or consciousness prior to medical procedures with ethyl ether or chloroform.

The nervous system is made up of the central nervous system (brain, cranial nerves, spinal cord, and spinal nerves) as well as the peripheral nervous system (autonomic and somatic nervous system). The nervous system regulates and coordinates body activities and responses to stimuli. The senses of sight, hearing, taste, smell, and touch are these stimuli and involve the organs of the eye, ear, mouth, and nose and receptors like the skin, joints, and muscles. The central nervous system is composed of the brain, the spinal cord, and their nerves. The central nervous system allows sensations to be received and recognized, and voluntary appropriate action to be taken by muscles under the direct control of the brain.

The autonomic nervous system, a branch of the peripheral nervous system, contains the network of nerves lying outside the brain and spinal cord that regulate body functions by automatically controlling a complex network of involuntary muscles. The autonomic nervous system involuntarily acts continuously whether the person is awake or asleep. The somatic nervous system controls and regulates voluntary activities of the body.

A nerve is a bundle of nerve fibers, made up of neurons, which are nerve cells, that have body cells to initiate and conduct impulses. The concentration gradient of sodium and potassium ions inside and outside the cell body is regulated by the permeability of the nerve membrane of the cell body. When the nerve is at rest, the potassium concentration in the cell is greatest. When the nerve is stimulated, the membrane permits a large flow of sodium ions to enter the nerve cell body and potassium ions to exit, inducing a local current that propagates the impulse. The impulse reaches the end of a nerve fiber and requires a chemical messenger to cross a synapse or point of contact with another nearby neuron. Acetylcholine is the chemical messenger that plays an important part in the transmission of nerve impulses. An imbalance of acetylcholine, which is quickly destroyed by the acetylcholinesterase enzyme, may result in neuromuscular blockage. The synapses are polarized and allow impulses to pass in only one direction. Synapses are susceptible to fatigue, oxygen deficiency, anesthetics, therapeutic drugs, and toxic chemicals.

Neurotic chemicals produce their main toxic effect on the central nervous system. Acetylcholine ($C_7H_{17}NO_3$) must be maintained at sufficient levels to assure transmission of the nerve impulse from the end of one nerve to receptors of the receiving nerve. This chemical level is controlled by acetylcholinesterase enzyme action. Acetylcholinesterase breaks down acetylcholine to form acetic acid and choline. When the enzyme action is inhibited by the neurotoxin, the acetylcholine level increases and eventually reaches the level that does not permit efficient transfer of nervous energy. This causes the nervous system to collapse and cease to function. The neurotoxins prevent or stimulate the continuation of the electric impulse, based upon the relationships in Table 3-5.

TABLE 3-5
CHEMICAL SUBSTANCES THAT AFFECT THE NERVOUS SYSTEM

Chemical substances	Effect on nerves
Inhibitors	
Butane	Decreases neuronal activity
Methylene chloride	Decreases neuronal activity
Botulinus toxin	Prevents release of acetylcholine
Ethyl alcohol	Decreases conductivity of sodium and potassium
Tetrodotoxin	Blocks sodium membrane permeability
Tellurium, thallium	Damages nerve insulators
Cyanate, cyanide (chronic)	Damages nerve insulators
Carbon monoxide (chronic)	Damages nerve insulators
Arsenic, methyl mercury	Damages peripheral nerves
Tetraethyl lead	Damages peripheral nerves
Ethylene glycol	Damages peripheral nerves
Methyl butyl ketone	Damages peripheral nerves
Curare	Blocks neuromuscular activity
Stimulants	
Caffeine, theobromine	Increases neuronal activity
Organophosphates	Inhibits acetylcholinesterase
DDT, pyrethrin	Increases neuronal activity
Batrachotoxin	Increases sodium membrane permeability

The kidneys play a central role in excreting waste in the form of urine, which regulates the fluid balance and environment of the body. The rate of blood flow through the kidneys is greater than with other organs, causing the kidneys to be exposed to relatively high concentrations of foreign chemicals. The kidneys filter blood and remove organic and inorganic substances such as urea, uric acid, creatinine, sodium salts, and potassium salts. The kidneys maintain the equilibrium between the blood and the body fluids by removing these substances from the blood. Urine is transported from the kidneys to the bladder, which is periodically emptied and discharged to the outside. The formation of urine is a continuous process and depends upon the blood pressure, the daily fluid intake, and the velocity of the blood flow. The volume of urine excreted by a normal human is 1000 to 2000 mL per day. Diet, liquid intake, ambient temperature, body temperature, age, blood pressure, and the health of the individual will affect the volume and properties of the urine.

Nephrotoxic agents produce toxic chemical effects on the kidneys. Edema, which causes excessive accumulation of watery serous fluids in tissues or body cavities, is an observable result. Proteinuria, the presence of protein in the urine, can also result from exposure to nephrotoxins.

Heavy metals like cadmium, mercury, lead, arsenic, bismuth, chromium, platinum, and uranium can damage the kidneys. Cadmium and mercury reduce the ability of the kidneys to conserve protein and cause high protein levels in urine. Long-term lead exposure can cause kidney failure and death.

When hydrogen ions from acids contact arsenic compounds, arsine (AsH_3) is produced. Arsine has a garlic odor and is poisonous. It destroys red blood cells and causes the kidneys to fail.

Carbon tetrachloride and chloroform are nephrotoxic agents that cause kidney damage and death. Both of these substances are converted to a toxic metabolite that destroys kidney tissue.

The liver receives blood from the intestines and hepatic artery. Any chemicals that are ingested and pass into the blood become exposed to the liver. As the first organ to receive blood from the intestines, the liver is where the major biotransformation processes take place. The liver is a mass of cells packed around a bundle of arteries and veins. It is a chemical-processing plant that synthesizes important body chemicals and converts other chemicals into forms for disposal by the kidneys. It is the largest organ of the body, producing bile, that is stored in the gallbladder for subsequent use in the lower intestine for digestive purposes. A normal adult will produce 800 to 1000 mL of bile per day, with the majority of the bile secreted during digestion.

The liver receives blood that contains the final products of digestion and decomposition in the gastrointestinal tract. The liver removes glucose from the blood and synthesizes glycogen, a polysaccharide $(C_6H_{10}O_5)_x$ that is a storage form of carbohydrates, the main fuel for the body. Urea and uric acid are produced in the liver, as are blood-clotting agents. The liver stores vitamins A, D, and B_{12}. The biotransformation processes of the liver convert many toxic substances to less hazardous chemicals.

Many factors affect biotransformation processes of the liver. Low-protein diets and diets with deficiencies of calcium, copper, zinc, iron, vitamin C, or vitamin E reduce the rate of biotransformation. Cirrhosis of the liver is often caused by chronic alcoholism or infectious hepatitis. Failure to properly wash hands following a bowel movement is the most common cause of infectious hepatitis. Liver cancer is usually caused by the spread of cancer from some other body source by dissemination through the bloodstream. Chemical injury to the liver may be caused by a variety of hydrocarbons, for example, carbon tetrachloride, chloroform, trichloroethylene, and dinitrobenzene. The liver may be injured by metals like arsenic, manganese, phosphorus, and beryllium.

Hepatotoxic chemicals can cause damage or otherwise injure the liver. Carbon tetrachloride can adversely affect the cellular membranes of the liver and is the most common example of the hepatotoxin in the workplace. Accumulation of fat, mostly triglycerides, in the parenchymal cells is the result of an imbalance between the synthesis rate and the release rate of the triglycerides into circulation. Hepatotoxins block the secretion of the triglycerides into the blood plasma, reducing the concentration of plasma lipids and plasma lipoproteins. This fat accumulation is a key factor in liver injury. Phosphorus and tannic acid are other examples of hepatotoxins.

The heart can undergo cardiac sensitization from the inhalation of certain hydrocarbons like chloroform and cyclopropane. This can lead to cardiac arrhythmias, which are alterations in the heartbeat. Glue sniffing or exposure to solvent aerosols can lead to cardiac sensitization.

Inhaled dust and other solid particulate matter can have adverse health effects, especially if the particles are small enough to enter the alveoli. These small particles, typically 1 to 2 μm in diameter, can be inert dust that is merely deposited deep in the lung without an adverse tissue response unless inhaled in excessive amounts. Examples of inert dust include calcium carbonate and calcium sulfate. So long as these particles do not alter the structure of the lung air spaces, cause reversible tissue reaction, or produce significant scar formation, their dust is generally considered to be only a nuisance.

Other types of inhaled dust can cause acute or chronic health effects. The acute reactions to inhaled dust can be described as irritant, toxic, or allergenic. Chronic dust exposure can result in serious types of pneumoconiosis, which are diseases of the lung.

Insoluble materials like silica and asbestos are more difficult for the body to remove once they have entered the lungs. These materials cause fibrotic changes that result in scarlike tissue and other lung tissue damage, as well as cancerous changes in the lung. Airborne asbestos is a known carcinogen in the respiratory system, causing lung cancer. Smokers who have had occupational exposure to asbestos increase their risk of lung cancer by a factor of about 100. This synergistic effect between smoking and airborne asbestos is a classic example of the need for accurate and complete information before reaching conclusions on toxicity hazards in the workplace.

Inhaled dust that is soluble in body fluids will dissolve in the tissue fluids that line the lung. The resulting solution could damage the lung or be absorbed into the bloodstream to be distributed throughout the body to target organs. These target organs could include the kidney, the liver, or the nervous system. Heavy metals such as lead, mercury, and manganese, and their compounds, once absorbed into the bloodstream usually result in slow, cumulative, chronic toxic symptoms. Excessive inhalation exposure to inorganic lead dust or fumes may cause anemia, headaches, loss of appetite, and weight loss. Chronic high-level exposures can cause nerve damage, colic, and bone marrow changes. On the other hand, organic lead compounds can concentrate in the brain and damage the functions of that organ or possibly be fatal.

Inorganic mercury compounds tend to concentrate in the kidneys and brain. Organic mercury concentrates mostly in the brain. Manganese poisoning is a chronic disease that affects the nervous system and causes psychomotor instability.

NEOPLASMS

A neoplasm is a new growth of tissue that serves no physiological function. This abnormal tissue growth is often used to describe cancerous or potentially cancerous tissue. Cancerous or malignant neoplasms are able to invade tissues directly or can spread to new distant locations in the body, in a process known as metastasis.

A *carcinogen* is an agent that produces or causes cancer. There are 26 known human carcinogens associated with occupational exposure and medical treatment. [15] These include common chemicals, such as asbestos, benzene, chromium, radon, and vinyl chloride. An additional 157 substances associated with occupational exposures and medical treatment are potential carcinogens. [15] Caution should be used in classifying a substance as a human carcinogen, if it has only caused cancer in a single laboratory animal experiment. NIOSH usually labels a substance a suspected human carcinogen if it has caused cancer in at least two different laboratory animal species.

The major causes of human cancer are tobacco smoke and improper diet. The way in which industrial chemicals cause cancer is not fully understood. Chemical composition and structure, route of entry to the body, length of exposure, synergism with lifestyle and diet, heredity, and other possible factors are all potentially involved in the risk associated with cancer. The general public has tremendous fear of cancer, making any industrial chemical that may be a carcinogen a target for elimination. Whenever the cancer deaths of a few people can be related to industrial chemical exposure, the public and environmental groups demonstrate for the removal of that chemical from commerce. Tobacco smoke, which kills more than 400,000 Americans a year from lung cancer and heart disease, seems to be ignored and tolerated because of its general familiarity and historic place in commerce and our lifestyle. The nicotine in tobacco is addictive and has been cited as the reason why smokers continue to smoke. [4]

Coal tars were among the first industrial chemicals associated with cancer. The polynuclear aromatic hydrocarbons in coal tar are suspected carcinogens and are also found in the exhaust of combustion engines and tobacco smoke.

Many carcinogens are formed in nature. Aflatoxins in tobacco and doxorubicin in mushrooms are examples of naturally occurring carcinogens. [11]

The classic Ames test, using salmonella bacteria, is an inexpensive test used to detect mutagenesis and carcinogenesis for chemical substances. The Ames test has a high (90 percent) degree of sensitivity and specificity in determining these important criteria in potentially toxic chemicals.

Mutagens affect the genetic system of exposed humans by altering hereditary material. Future generations could be damaged by mutagens. The time lag between exposure and effect makes it difficult to identify the connection between the exposure and actual effect of the hazard. Mutagens can cause changes in chromosomes or a biochemical change in genes. Ionizing radiation and ultraviolet light can cause genetic mutations. As an example, leukemia begins when a mutagen causes such rapid growth and widespread proliferation of white blood cells that they cannot be removed from the blood. This excessive level of immature white blood cells interferes with normal body functions, making the mutagen that caused the phenomenon a carcinogen. All carcinogens are also mutagens, but all mutagens are not necessarily carcinogens.

Teratogens are agents in pregnant women that cause the development of congenital malformations in the fetus. They interfere with normal human embryonic development. Unlike mutagens, malformations caused by teratogens are not hereditary. Examples of teratogens include thalidomide, rubella, and ionizing radiation. Teratogens require a developing fetus to be present, while mutagens do not. Since the workplace contains more and more women, some of whom are pregnant, the problem of teratogens in the work environment is much greater than in the historic past. Damage to the fetus is more likely to occur during the first trimester of pregnancy, when many women are only becoming aware of their pregnancy.

PERMISSIBLE EXPOSURE LIMITS

When OSHA was formed in 1970 it adopted permissible exposure limits (PELs) for industrial chemical exposure based upon the threshold limit values (TLVs) of the American Conference of Governmental Industrial Hygienists (ACGIH). The ACGIH is a professional society, not a government agency. ACGIH data for establishing TLVs for chemicals is based mainly on industrial experience and chronic laboratory animal inhalation studies. It is regularly reviewed and updated as appropriate by this committee of qualified members.

If the TLV for a substance is exceeded in the workplace, a person could be harmed. In the anticipation of such a health problem, OSHA requires the employer to initiate certain safety provisions if the action level (one-half of the PEL) is exceeded, for example, employee exposure measurement, employee training, and medical surveillance. The action level concept helps protect employees from overexposure with a minimum expense to the employer.

Laboratory animal studies are important in determining chemical toxicity before a new chemical is introduced into commerce. Based upon the results of such tests, preventive measures can be designed to protect workers, consumers, and the public. It is difficult to extrapolate a laboratory animal LD_{50} or LC_{50} to a TLV that is acceptable in the workplace. Acute toxicity data is not very useful in predicting chronic results. Certain animal species are more sensitive to a chemical than others. Also the animal response will not necessarily parallel the human response as the dosage level is varied.

We are developing better information on the effects of industrial chemical exposure with humans every day in the workplace. Many common chemicals, such as lead, benzene, vinyl chloride, asbestos, and carbon monoxide, have sufficient historic human exposure data to permit intelligent toxicity decisions. The committee of the ACGIH uses all of the toxicity information sources at its disposal in setting TLVs.

OSHA responds quickly to TLV changes by the ACGIH and adjusts the OSHA PEL to reflect the TLV change. Threshold limit values are based on the best available information from industrial experience and from human and animal experimental studies.

THRESHOLD LIMIT VALUES

There are three TLV categories, which are expressed in ppm or milligrams per cubic meter (mg/m³). [2] TLVs for gases or vapors are usually detemined as ppm and mg/m³. TLVs for particulates are usually expressed as mg/m³. The asbestos TLV is expressed in fibers per cubic centimeter of air.

The volume occupied by a given weight of a gas depends upon the gas's molecular weight. The following relationship can be used to convert ppm to mg/m³ at 25°C.

$$mg/m^3 = \frac{ppm(mol\ wt)}{24.5}$$

1 *Threshold limit value, time-weighted average (TLV-TWA)* Time-weighted average concentration for a normal 8-hour workday and a 40-hour work week, to which nearly all workers may be repeatedly exposed, day after day, without adverse effect.

$$TLV - TWA = \frac{\sum_{i=1}^{n} t_i C_i}{\sum_{i=1}^{n} t_i}$$

where TLV-TWA = threshold limit value, time-weighted average (ppm)
$\quad\quad$ t_i = exposure time at concentration i (h)
$\quad\quad$ C_i = concentration of chemical (ppm)
$\quad\quad$ n = total time periods

2 *Threshold limit value, short-term-exposure limit (TLV-STEL)* Concentration to which workers can be exposed continuously for a short period of time without suffering irritation, chronic or irreversible tissue damage, or narcosis that could increase the likelihood of accidental injury, impair self-rescue, materially reduce work efficiency, and without exceeding the daily TLV-TWA. The STEL is defined as a 15-minute TWA exposure that should not be exceeded at anytime during a workday even if the TLV-TWA is not exceeded. Exposures between the TLV-TWA and TLV-STEL should not be longer than 15 minutes or occur more than four times per day.

3 *Threshold limit value, ceiling (TLV-C)* Concentration that should not be exceeded during any part of the working exposure.

$$\text{TLV} - \text{STEL} = \frac{\sum_{i=1}^{n} t_i C_i}{0.25}$$

In practice TLV-TWA and TLV-C values are used more commonly than TLV-STEL because of the difficulty in monitoring TLV-STEL values in the workplace.

The TLVs are meant to be used as guidelines to assist in the control of health hazards. They should not be used as exact measurements between safe and potentially dangerous workplace conditions. They should not be used as precise legal standards, but can be useful in developing a well-conceived health and safety program. OSHA has come to place considerable reliance on the TLVs because of the reliability and reputation of the ACGIH as an accurate unbiased organization. The ACGIH lists planned TLV changes in their Notice of Intended Changes, with sufficient time to solicit comments from concerned parties.

The TWA concept allows concentrations higher than the recommended value provided the elevated levels are offset by periods of lower concentrations. Risk assessment becomes an important part of the TWA concept. The type of harm that may be experienced from exposure to these various contaminate levels needs to be assessed by health and safety professionals.

When TLVs are measured in the workplace, they should be determined in the breathing zone of the worker to allow a TWA to be calculated. There is sometimes difficulty in obtaining a representative breathing-zone sample and uncertainties about the extent of absorption of the amount inhaled. The changing conditions of the work environment can make uniform, predictable air sampling difficult. Also, contamination of the air samples must be avoided. [15]

Analysis is needed to measure the airborne concentrations of chemicals in the workplace because in many instances the odor threshold of toxic substances is well above the TLV for the chemical, as shown in Table 3-6. [2,3]

When more than one hazardous substance that could affect the same organ system is present the combined effect must be considered. The principle of additive effects should be used for these mixtures. It is based on the concept that each hazardous substance will contribute to the overall hazard as though the substance was singularly present.

$$\text{TLV}_{\text{mix}} = \sum_{i=1}^{n} \frac{C_i}{\text{TLV}_i}$$

TABLE 3-6
ODOR THRESHOLDS FOR CHEMICAL SUBSTANCES THAT ARE GREATER THAN THE CHEMICAL TLV

Chemical	Odor threshold (ppm)	Inhalation TLV-TWA (ppm)
Benzene	100	10*
Carbon monoxide	—	25
Carbon tetrachloride	79	5 skin
Chlorine	5	0.5
Chlorobromomethane	400	200
Chloroform	200	10
Epichlorohydrin	10	2* skin
Ethylene oxide	300	1
Isopropyl amine	10	5
Methanol	2000	200 skin
Methylene chloride	300	50
Nitromethane	200	20
Propylene oxide	200	20
Toluene-2,4-disocyanate	0.4	0.005
Turpentine	200	100
Vinyl chloride	4100	5

*ACGIH intends to reduce further.

where C_i = observed ambient concentration, ppm
    TLV_i = corresponding threshold limit, ppm

If the sum of the mixture, TLV_{mix}, exceeds unity, then the threshold limit of the mixture has been exceeded. [2,10]

There are assumed to be no synergistic or antagonistic effects of the combination of substances present. If such a combination of effects are present, they need to be factored into the system of concern. Normally such combined effects are difficult to quantify without long-term studies that target the situation.

Heat, ultraviolet light, ionizing radiation, elevated pressure, and high altitudes may cause additional body stress and could alter the effects of chemical exposure. Large deviations from normal are usually necessary before an observable effect can be experienced.

Analysis of biological samples from humans who have been exposed to toxic agents is a useful method of assessing occupational exposure to hazardous substances. The amount of a toxin circulating in the blood and being excreted or exhaled can be determined from blood, urine, and breath samples. Other samples that might be helpful in monitoring toxins are hair, fingernails, and feces. Urine samples should be used for heavy metals and their compounds, mineral acids, benzene, and acrylonitrile. Blood samples can be useful for heavy metals and carbon dioxide. Breath samples can be used to measure organics like alcohols, aliphatics, alkenes, ketones, and chlorinated hydrocarbons. The observed effects in urine and blood samples sometimes do not occur for extended periods of time, for example, several weeks for heavy metals to a few days for organics. Substances may be stored in target organs for extended periods of time before an equilibrium is established within body systems. The liver and kidney are common target organs. The body converts many toxic organic compounds to less harmful metabolites, for example, benzene to phenol, methyl alcohol to formic acid, and parathion to *p*-nitrophenol. The detection of the related metabolites in urine is indicative of the toxin precursor in the body. When inhaled gases and vapors enter the respiratory tract and are not dissolved or metabolized, they are exhaled from the body. These exhaled substances will continue to clear from the body for several hours after exposure with a progressive decreasing rate of excretion as their internal body concentration is diminished.

BIOLOGICAL EXPOSURE INDICES

As employees are exposed to atmospheric TLVs of a chemical and equilibrium is reached, there should be a corresponding biological or excretory threshold limit in the employee. The ACGIH has adopted biological limit values called biological exposure indices (BEIs) for certain chemicals. The BEIs are reference values intended as guidelines for evaluation of potential health hazards in the workplace. The BEIs are levels of determinants that are likely to be observed in specimens form health workers who have been exposed to chemicals at the TLV exposure level. [2]

There is no fine line between hazardous and nonhazardous exposures based upon BEIs, which are based on 8-hour exposures during a 5-day work week. This biological monitoring should be considered complementary to workplace air monitoring because it can confirm and verify the results of the air monitoring. The BEIs are based on human exposure data, which show a relationship between intensity of exposure and the determinant biological levels or between the determinant biological levels and human health effects. Animal studies are not considered for BEI values because they do not usually provide suitable data for establishing the BEI limit.

The BEI values are usually expressed in units that are based on the type of sample being measured. The BEIs from urine samples are based upon milligrams or micrometers per gram of creatinine ($C_4H_7ON_3$), the end product of creatine metabolism. About 0.02 g/kg of body weight of creatinine is excreted by the kidneys per day. The BEIs from blood samples are based upon a percent of hemoglobin. The BEIs from breath samples are in ppm of exhaled air. Whenever the ACGIH revised TLVs, the corresponding BEIs also come under reassessment. The BEI concept, like the TLV, is based upon each chemical having a reasonably safe level of exposure below which significant illness, injury, or discomfort will seldom happen. Once these safe limits of exposure have been determined for a specific chemical, the environmental workplace conditions should be controlled so that the exposure levels do not exceed that limit.

BIBLIOGRAPHY

1 Altvater, T. S., "Material Safety Data Sheets," *Professional Safety,* October 1990.
2 American Conference of Governmental Industrial Hygienists, *Threshold Limit Values for Chemical Substances and Physical Agents and Biological Exposure Indices,* Cincinnati, 1994.
3 Dreisbach, R. H., and W. O. Robertson, *Handbook of Poisoning,* Appleton & Lange, Norwalk, Conn., 1987.
4 Ember, L. R., "The Nicotine Connection," *C&E News,* November 28, 1994.
5 Fawcett, H. H., *Hazardous and Toxic Materials,* Wiley, New York, 1984.
6 Gallant, R. W., "Highly Toxic Liquids: Teaching Operators to Handle Them," *Chemical Engineering,* April 1990, pp. 116–120.
7 Grossel, S. S., "Highly Toxic Liquids: Moving Them Around the Plant," *Chemical Engineering,* April 1990, pp. 110–115.
8 Hajian, H. G., and R. L. Pecsok, *Working Safely in the Chemistry Laboratory,* American Chemical Society, Washington, D.C., 1994.
9 Hammer, W., *Occupational Safety Management and Engineering,* Prentice-Hall, Englewood Cliffs, N.J., 1989.
10 Kavianian, H. R., and C. A. Wentz, *Occupational and Environmental Safety Engineering and Management,* Van Nostrand Reinhold, New York, 1990.
11 Klassen, C. D., et al., *Casarett and Doull's Toxicology,* 3d ed., Macmillan, New York, 1986.
12 Ottoboni, M. A., *The Dose Makes the Poison,* Vincente Books, Berkeley, Calif., 1984.
13 OSHA, Occupational Exposures to Hazardous Chemicals in Laboratories, *Federal Register,* 55, no. 21 (30 January, 1990): 3300–3335. Microfiche.
14 Plog, B. A., et al., *Fundamentals of Industrial Hygiene,* National Safety Council, Itasca, Ill., 1988.
15 U.S. Department of Health and Human Services, *Seventh Annual Report on Carcinogens,* Research Triangle Park, NC, 1994.

PROBLEMS

1 Define LD_{50} for a chemical substance.
2 In the sixteenth century, Paracelsus, a Swiss alchemist and physician who introduced lead, sulfur, iron, and arsenic into pharmaceutical chemistry said, "All substances are poisons, there is none which is not a poison, the right dose differentiates a poison and remedy." How does his statement relate to the present-day science of toxicology?
3 Which is more important in industrial toxicology—the toxicity of the substance or the risk or hazard associated with its use?
4 What constitutes an acceptable risk for exposure to a toxic chemical?
5 Name some factors that could be considered in determining an acceptable toxicology risk.
6 What constitutes safe and highly toxic oral chemical-dose levels for humans?
7 What are the major routes by which toxic chemicals gain access to the body?
8 What route to the body would produce the greatest toxic effect and the most rapid response for a dose of a toxic chemical?
9 A toxic chemical is known to be detoxified in the liver. Would it be more or less toxic if ingested rather than inhaled?
10 Which of the major routes to the body relate more frequently to industrial exposure?
11 What is the difference between acute and chronic exposure to toxic chemicals?

12 Describe the effects of a single benzene exposure as compared to repeated benzene exposures at the same dosage level.

13 In attempting to characterize the toxicity for a specific chemical, would you need acute or chronic exposure information?

14 How does fractionation of the chemical dose relate to the effect of the chemical exposure?

15 Name some known human carcinogens and their adverse effects on humans.

16 What lifestyle factors are known to cause cancer?

17 Why is the term *chemical* looked upon unfavorably by society?

18 How does the public perception of risk to chemical exposure differ from that of scientists and engineers?

19 What are carcinogens?

20 Define TLV-TWA, TLV-STEL, and TLV-C for hazardous chemicals.

21 Based on the acute LD_{50} values in Table 3-2, would you expect nicotine or strychnine sulfate to be more toxic?

22 Name some lifestyle factors that cause cancer.

23 Where can the list of known carcinogens be found?

24 Where can the list of suspected carcinogens be found?

25 Where can the TLV for chemicals be found?

26 What is the concentration in mg/m^3 of 5 ppm of carbon tetrachloride?

27 What is the concentration in mg/m^3 of 25 ppm of ammonia?

28 The 8-hour TWA of airborne acetic acid concentration in the workplace measures 15 ppm. The TLV-TWA measures 10 ppm. Is this location in compliance?

29 The 8-hour TWA of airborne methyl ethyl ketone (MEK) concentration in the workplace measures 40 ppm. The TLV-TWA measures 200 ppm. Is this location in compliance?

30 Workplace air contains 100 ppm butane (TLV, 800 ppm), 50 ppm pentane (TLV, 600 ppm) and 10 ppm hexane (TLV, 50 ppm). Are the combined effects of this mixture in compliance?

31 Workplace air contains 120 ppm isopropyl alcohol (TLV, 400 ppm), 0.3 ppm chlorine (TLV, 0.5 ppm), and 45 ppm nitroethane (TLV, 100 ppm). Are the combined effects of this mixture in compliance?

32 A liquid contains 40 wt % heptane (TLV, 400 ppm), 35 wt % methyl chloroform (TLV, 350 ppm) and 25 wt % perchloroethylene (TLV, 25 ppm). Assume the atmospheric composition to be the same as the liquid. What is the TLV of this mixture in mg/m^3?

33 What are the main chemical exposure routes in the workplace?

34 What is the key gas or vapor property that determines the rate of uptake from the alveoli into the bloodstream?

35 What is the target organ for the toxic effects of a narcotic substance?

36 What is the target organ for the toxic effects of chronic mercury exposure?

37 During an 8-hour workday the ammonia levels at a work location were 10 ppm (5 h), 20 ppm (1 h), and 30 ppm (2 h). Is the air level in compliance if the NH_3 PEL is 25 ppm?

38 The isopropyl alcohol vapor levels at a workstation during an 8-h workday were 450 ppm (2 h), 375 ppm (3 h), and 385 ppm (3 h). Is this exposure level in compliance if the isopropyl alcohol PEL is 400 ppm?

39 What is the partial pressure of oxygen PO_2 in air at standard conditions?

40 What adverse skin conditions could be caused by exposure to coal tar chemicals?

41 In lower molecular weight chlorinated aliphatic compounds, what is the TLV effect of increasing the chlorine content?

42 What is the chemical classification of benzene, toluene, and xylene?

43 What is the distinguishing property that differentiates lead fumes from lead dust?

44 At 25°C and 760 mmHg a solvent with a molecular weight of 105 has a vapor pressure of 390 mmHg. If a 55-gallon drum of this solvent at these conditions is spilled in a confined space, what is the vapor concentration of the solvent in ppm and mg/m³?

45 How many mL of liquid CCl_4 would be evaporated at 25°C and 760 mmHg into a 200-L drum to result in a 20 ppm CCl_4 concentration in air? Assume a liquid density of 1.5 g/mL.

4

ENVIRONMENTAL
PROTECTION

All industrial processes and plants generate contaminants that range from common
municipal trash to hazardous waste. The treatment and disposal of these wastes result in the release of some of these contaminates into the environment in gas, liquid, or solid form.

Gas releases usually result from stack gas or fugitive process emissions. These
may include combustion products like water and carbon dioxide, products of incomplete combustion like carbon monoxide and complex organics, and vaporized
hydrocarbons.

Liquid releases may come from wastewater, wastewater treatment discharges,
spills, and rainfall runoff. These releases may be from point sources, like end-of-
the-pipe discharges, or from nonpoint sources, like runoff.

Solid releases might include bulk trash, drummed wastes, suspended solids in
wastewater, and airborne dust and particulates. The solid releases are either hauled
away or released in combination with gases or liquids.

Once these contaminates have been released into the environment, they will
move in a variety of pathways and mechanisms. Chemistry, physics, geology, hy-
drogeology, engineering, and biology all play a role in the transport and fate of the
contaminates in the environment.

AIR POLLUTION

Point sources of air emissions are emitted from well-defined locations, such as exhaust stacks and vents. Fugitive emissions may come from surface impoundments,
storage tanks, process pumps, and roadways. Often these air pollution releases involve both gas and solid phases of contaminates.

The most common mechanism for fugitive emissions at industrial operations is the volatilization of liquids to the gaseous phase. The rate of liquid volatilization is dependent on the temperature land vapor pressure of the substance and the concentration differential between the gas and liquid phases.

The rate of diffusion of gases is inversely proportional to the square root of the density or molecular weight of the gas. The diffusion coefficient may be estimated from the relationship.

$$\frac{D_1}{D_2} = \left(\frac{p_2}{p_1}\right)^{1/2} = \left(\frac{M_2}{M_1}\right)^{1/2} \qquad (4\text{-}1)$$

where D = diffusion coefficient, cm^2/sec
   p = gas density, gm/cm^3
   M = gas mole wt

Example 4-1

What are the relative diffusion rates of hydrogen and oxygen gases through a pinhole leak in a pipe fitting?

Solution

The molecular weight for hydrogen is 1; for oxygen, 16.

$$\frac{D \text{ hydrogen}}{D \text{ oxygen}} = \left(\frac{32}{2}\right)^{1/2} = 4$$

The diffusion rate for hydrogen is about four times faster than oxygen from the pinhole leak.

Industrial facilities that process volatile organic liquids and gases have the potential to emit large quantities of these substances into the atmosphere. Valves, pipe fittings, pumps, open containers and vats, routine sample taps, cross-contamination of process streams, and spills commonly emit these volatile organics. Once airborne, the volatiles diffuse into the workplace and the area surrounding the facility. There will be a large dilution factor associated with these airborne contaminates as they are dispersed in the environment. Nevertheless these contaminates may cause long-term, chronic health problems to humans and gradually degrade the environment. Plant operations and maintenance programs should do everything practical to avoid or minimize these fugitive emissions. The EPA-required reporting of releases on Form R has identified numerous such industrial contaminate releases.

Volatile organics may also cause odor problems at industrial facilities by becoming an annoyance and potential health hazard in the workplace. [2] The surrounding community is usually highly sensitive to plant odors, particularly as the intensity or type of odor changes.

The odor intensity is related to the odorant concentration by the following relationship:

$$I = kC^n \tag{4-2}$$

where I = perceived odor intensity
 k = constant
 C = odorant concentration
 n = exponent

The exponent may range from 0.2 to 0.8 depending upon the odor intensity of the substance at various concentration levels, as shown in Figure 4-1. [2] The slope of this log-log plot is used to determine the exponent, n.

Example 4-2

Find the relative odor intensity level for odorant A ($n = 0.2$) and odorant B ($n = 0.8$) when the concentration of each odorant decreases by a factor of 10.

Solution

$$I = kC^n$$
$$\log I = \log k + n \log C$$

For odorant A
$$I \sim 10^n = 10^{0.2} = 1.58 \text{ reduction}$$

FIGURE 4-1
THE RELATIONSHIP BETWEEN PERCEIVED ODOR SENSATION AND ODORANT
CONCENTRATION. Individual points are shown for 8, 11, and 12 to illustrate the linearity.
Legend: 1. Thiophenol; 2. Ethyl selenomercaptan; 3. Ethyl sulfide; 4. Phenyl isocyanide; 5.
Ethyl selenide; 6. Methyl sulfide; 7. Coumarin; 8. Methylmercaptan; 9. Hydrogen sulfide; 10.
Pyridine; 11. Allyl alcohol; 12. Nitrobenzene. (*Source:* Buonicore and Davis: *Air Pollution
Engineering Manual; Air and Waste Management Association.*)

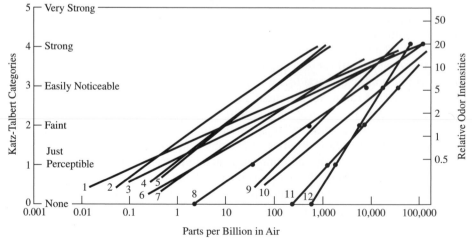

For odorant B

$$I \sim 10^{0.8} = 6.31 \text{ reduction}$$

The relative effect is a decrease in the perceived odor intensity of odorant A by a factor of 1.58, and of odorant B by a factor of 6.31. This is an important concept when using engineering controls like air dilution to control industrial odors.

The greater the slope in Figure 4-1, the greater the effect of dilution on the perceived odor intensity of the odorant. For example, the effect of air dilution of ethyl sulfide is less than the effect of air dilution for hydrogen sulfide.

The detection threshold of an odor is the minimum concentration that produces a perceived odor sensation. Odor panels are used to determine detection thresholds when at least half of the panel members correctly detect the odor sensation. Such techniques involve subjective determinations by the people on the panel and should be evaluated within the context of the test parameters.

The odor recognition threshold is the minimum concentration that produces an odor sensation that can be correctly identified by the specific odorant present. The recognition threshold always requires a higher concentration level than required for the detection threshold. The detection and recognition thresholds for a variety of odorous industrial substances are shown in Table 4-1 on page 98 [1, 7, 10, 13]

Industrial odors may be caused by many substances. Nearly everyone is familiar with the smell of ammonia gas and hydrogen sulfide gas. These inorganic gases are objectionable either in the workplace or the local community. Sulfur compounds like mercaptans and organosulfides have low odor threshold concentration levels. Organoamines are also problem odorants for industrial operations.

The petroleum industry produces enormous quantities of crude oil and natural gas, which are vital to the economy. During the exploration and production of these hydrocarbons, volatile organic compounds (VOCs) are emitted, including benzene, toluene, xylenes, and butadiene.

Much of the oil and gas is stored in various types of storage tanks prior to and after further processing into products for the marketplace. Emissions from these VOC and petroleum-products sources are a significant contribution to air quality. Gasoline is the largest petroleum product that must be transferred from petroleum refineries to the consumer. During the transfer and storage of gasoline, VOCs escape into the atmosphere. Gasoline is a complex mixture of hydrocarbons. It includes paraffins, naphthenes, olefins, and aromatics, among them benzene, xylenes, toluene, hexane, ethyl benzene, cumene, methyltertiary-butyl ether, and naphthalene. When these VOCs are emitted to the air, they contribute to air pollution.

The dry-cleaning industry uses huge quantities of solvents to clean fabrics. These include petroleum-based solvents and chlorinated solvents. Perchloroethylene and trichloroethane are the most common chlorinated solvents (Table 4-2 on page 98). Stoddard solvents and higher-boiling petroleum mixtures are the most common petroleum-based solvents used in dry cleaning. The dry-cleaning process includes a washer, extractor, and dryer with additional tanks and pumps. Most of the hydrocarbon air emissions from dry clearning occur during the washer-dryer

TABLE 4-1
DETECTION AND RECOGNITION OF INDUSTRIAL ODORS

Compound name	Formula	Detection threshold (ppm)	Recognition threshold (ppm)	Description
Acetaldehyde	CH_3CHO	0.067	0.21	Pungent, fruity
Allyl mercaptan	CH_2CHCH_2SH	0.0001	0.0015	Disagreeable, garlic
Ammonia	NH_3	17.0	37.0	Pungent, irritating
Amyl mercaptan	$CH_3(CH_2)_4SH$	0.0003	—	Unpleasant, putrid
Benzyl mercaptan	$C_6H_5CH_2SH$	0.0002	0.0026	Unpleasant, strong
n-Butyl amine	$CH_3(CH_2)NH_2$	0.080	1.8	Sour, ammonia
Chlorine	Cl_2	0.080	0.31	Pungent, suffocating
Dibutyl amine	$(C_4H_9)_2NH$	0.016	—	Fishy
Diisopropyl amine	$(C_3H_7)_2NH$	0.13	0.38	Fishy
Dimethyl amine	$(CH_3)_2NH$	0.34	—	Putrid, fishy
Dimethyl sulfide	$(CH_3)_2S$	0.001	0.001	Decayed cabbage
Diphenyl sulfide	$(C_6H_5)_2S$	0.0001	0.0021	Unpleasant
Ethyl amine	$C_2H_5NH_2$	0.27	1.7	Ammonia
Ethyl mercaptan	C_2H_5SH	0.0003	0.001	Decayed cabbage
Hydrogen sulfide	H_2S	0.0005	0.0047	Rotten eggs
Methyl amine	CH_3NH_2	4.7	—	Putrid, fishy
Methyl mercaptan	CH_3SH	0.0005	0.0010	Rotten cabbage
Ozone	O_3	0.5	—	Pungent, irritating
Phenyl mercaptan	C_6H_5SH	0.0003	0.0015	Putrid, garlic
Propyl mercaptan	C_3H_7SH	0.0005	0.020	Unpleasant
Pyridine	C_5H_5N	0.66	0.74	Pungent, irritating
Sulfur dioxide	SO_2	2.7	4.4	Pungent, irritating
Thiocresol	$CH_3C_6H_4SH$	0.0001	—	Skunk, rancid
Trimethyl amine	$(CH_3)_3N$	0.0004	—	Pungent, fishy

TABLE 4-2
PROPERTIES OF SELECTED CHLORINATED AND PETROLEUM SOLVENTS

	Perchloroethylene	1,1,1-Trichloroethane	Stoddard solvent
Flash point, °F	None	None	105
Boiling point, °F	250	164	350
Density, lb/gal	13.55	11.07	6.5
Heat of vaporization, (Btu/lb)	90	102	115
Molecular weight	165.8	133.4	145

operation and the filter disposal. Collectively, these dry-cleaning solvents are a substantial contributor to air pollution in metropolitan areas.

Organic solvents are widely used for parts cleaning and other degreasing operations. The organic solvents may be chlorinated, such as methylene chloride, perchloroethylene, and trichloroethane. Petroleum-based solvents like Stoddard solvent and some water-based solvents are emitted into the air upon exposure of the solvent vapor-air interface and during the transfer of parts to or from the operation. These vapors contribute to air pollution wherever solvent parts cleaning occurs.

Surface coatings and corrosion protection for consumer products like automobiles, furniture, appliances, and containers contribute VOCs and related air pollution. Typically, these protective coatings use organic solvents in the range of 50 to 90 percent volatiles. There are considerable air pollutants emitted from such surface-coating operations, which include the application of the coating, drying, and baking.

Industrial operations like combustion, drying, and solids handling generate particulate air emissions. Most of the point-source particulates are emitted from stacks and vents.

Combustion of fuels allows the heat content of the fuel to be released and converted into energy like electric power, steam generation, and heating. The ash from combustion processes is the noncombustible, nonvolatile part of the fuel. Ash is a solid residue from the combustion process. Fly ash is smaller, airborne particles. Bottom ash is larger, heavier particles that are withdrawn from the bottom of the combustion process.

Coal and lignite are the most common examples of solid fuels that contain carbon, hydrogen, sulfur, oxygen, and nitrogen. Various minerals and mineral oxides make up the noncombustibles. The major air pollutants from coal combustion are inorganic mineral particulates, sulfur oxides, and nitrogen oxides. Unburned combustibles like organics and carbon monoxide are also emitted.

Sulfur dioxide and, to a lesser extent, sulfur trioxide compose the sulfur oxide emissions. Almost all of the sulfur in coal will be converted to sulfur oxides during the combustion process. If the ash is alkaline, various sulfate salts will also be present.

Nitrogen oxides are generated from molecular nitrogen in the combustion air and the nitrogen present in the fuel. The nitrogen in the fuel is much more reactive than the nitrogen in the air. NO_x formation in coal combustion is increased by higher combustion temperatures, longer residence times, and the presence of nearly stoichiometric air in the combustion chamber.

Fuel oil is the most widely used liquid fuel in electric utilities and industrial operations. The residual carbon content of the fuel oil is indicative of the particulate emissions expected from the combustion process. SO_x emissions are dependent upon the sulfur content of the fuel oil. SO_2, SO_3, and mineral sulfates are all formed from fuel oil combustion, with SO_2 being at least 95 percent of the sulfur oxide emissions. About half of the nitrogen present in the fuel is converted to NO_x. Only minor amounts of VOCs are produced from fuel-oil-fired utility plants.

Natural gas is used for power and steam generation. It is considered a clean fuel because it is primarily methane, with only minor amounts of ethane and other natural gas liquids, nitrogen, helium, and carbon dioxide. Typical emissions from natural gas combustion contain no significant particulates or sulfur oxides.

Wood-fired fuel combustors may generate considerable particulate matter, depending on the wood composition. Particulate emissions from wood-fired boilers range from 0.5 to 4.0 percent of the wood fuel.

Fugitive particulate emissions are usually either from industrial processes or open-dust surface areas. Process areas that feed solid raw materials often generate fugitive emissions, which are exhausted through ventilation systems. Open-dust areas include roadways, parking areas, storage piles, and barren ground. Solid particles are entrained into the atmosphere by wind or equipment acting on the exposed materials. This may occur when materials are transported, stored, or transferred at an industrial facility. It also occurs at construction sites and paved and unpaved roads. Open-dust areas usually experience large fluctuations in airborne particulates because of variations in the process cycle, the sporadic nature of materials-handling operations, and the effects of ambient conditions, for example, temperature, wind, and precipitation on air emissions. The agriculture industry is an excellent example of this type of particulate generation.

The generation of fugitive dust by surface vehicles on unpaved roads can be obtained from the following relationship. [12]

$$e = 2.1\left(\frac{s}{12}\right)\left(\frac{v}{30}\right)\left(\frac{m}{3}\right)^{0.7}\left(\frac{w}{4}\right)^{0.5}\left(\frac{365-p}{365}\right)$$

(4-3)

where e = PM$_{10}$ (particulate matter $\leq$µm diameter) emission factor, pounds per vehicle miles traveled

s = silt content of road surface material, %

v = mean vehicle velocity, mph

m = mean vehicle weight, tons

w = mean number of wheels

p = number of precipitation days ($\geq$0.01 in rainfall) per year

Example 4-3

An 18-wheel truck weighing 18 tons is traveling on an unpaved road at 50 mph. Determine the emissions factor, if the road surface silt content is 4% $\leq$10µm diameter particles. There are 30 precipitation days per year at this location.

Solution

Using equation 4-3

$$e = 2.1\left(\frac{s}{12}\right)\left(\frac{v}{30}\right)\left(\frac{m}{3}\right)^{0.7}\left(\frac{w}{4}\right)^{0.5}\left(\frac{365-p}{365}\right)$$

$$= 2.1\left(\frac{0.04}{12}\right)\left(\frac{50}{30}\right)\left(\frac{18}{3}\right)^{0.7}\left(\frac{16}{4}\right)^{0.5}\left(\frac{365-30}{365}\right)$$

$$= 0.075 \text{ lb particulates per vehicle mile}$$

Bulk handling of solid materials is a common industrial practice in transfer operations. The particulate emissions from materials handling can be estimated from the following relationship. [12]

$$e = 0.0032k\frac{(U/5)^{1.3}}{(M/2)^{1.4}} \tag{4-4}$$

where e = emission factor
$\quad k$ = particle size multiplier (see table below)
$\quad U$ = mean wind speed, mph
$\quad M$ = material moisture content, wt %

Particle diameter (μm)	k (dimensionless)
<30	0.74
<15	0.48
<10	0.35
<5	0.20
<2.5	0.11

Example 4-4

Finely divided solid raw materials are being continuously dropped onto an industrial conveyor. The average particle diameter is 12 μm and the average moisture content of the solids is 25 wt %. The mean wind speed is 4 mph. What is the daily particulate emission rate from this operation if 200 tons per day are transferred on the conveyor.

Solution

Using equation 4-4,

$$e = 0.0032k\frac{(U/5)^{1.3}}{(M/2)^{1.4}}$$

$$= 0.0032(0.48)\frac{(4/5)^{1.3}}{(0.25/2)^{1.4}} = 0.021$$

$$e = 0.021 \text{ (200 tons/day)} = 4.2 \text{ tons/day}$$

This estimate is sensitive to both wind speed and moisture content of the solids.

WATER POLLUTION

Surface water is found in many forms, including lakes, rivers, streams, the ocean, and lagoons. We are dependent on the quality of this water to support our lifestyle, ranging from drinking water to getting rid of wastes. Marine life is also dependent on surface water for supporting its lifestyle.

From the beginning of time pollutants have been degrading water quality. For many centuries these pollutants were not great enough to adversely effect the use of surface water. As populations and industrial growth increased, these same pollutants reached a level to make some waters no longer suitable for fishing, swimming, shipping, and waste disposal.

Point sources of water pollutants are mainly domestic sewage and industrial wastes. This type of pollution can be greatly reduced or possibly eliminated with wastewater treatment followed by effluent discharge to the receiving body of water.

Nonpoint sources of water pollutants include agricultural and urban runoff from rainfall. Treatment of these pollutant sources is much more difficult. Reduction in nonpoint source pollutants usually occurs through changes in land use.

Water pollutants include dissolved oxygen, nitrogen, phosphorus, bacteria, solid particles, heavy metals, organic substances, and thermal energy. Fish require oxygen and nutrients to survive, making some of these pollutants desirable in moderation.

When organic or inorganic wastes are introduced into water they usually demand oxygen, which may be depleted to a level below a critical point necessary to support marine life. The amount of oxygen required to degrade the waste is related to the amount of waste introduced into the water. Oxygen is also being added to the water from the atmosphere, with the net effect of these competing processes determining the dissolved oxygen concentration in the water. [4]

The quantity of organic oxygen-demanding material, called biochemical oxygen demand (BOD), is measured by the change in oxygen concentration in water caused by the oxygen demands of microorganisms, degrading the organic matter present. BOD is determined by the following relationship.

$$BOD_t = BOD_u \, (1 - e^{-kt}) \qquad\qquad (4\text{-}5)$$

where BOD_t = at time t (days)
  BOD_u = when all organics have been degraded
   k = rate constant, per day

Example 4-5

If the BOD_5 of a waste is 45 mg/L and the rate constant is 0.125 per day, what is the estimated ultimate BOD?

Solution

$$BOD_t = BOD_u\,(1 - e^{-kt})$$

$$45 = BOD_u[1 - e^{-0.125(5)}] = BOD_u\left(1 - \frac{1}{1.87}\right)$$

$$BOD_u = 96.8 \text{ mg/L}$$

For regulatory purposes and wastewater analysis, the most commonly used BOD values are based upon the 5-day test, BOD_5. The ultimate BOD is probably more indicative of total oxygen requirements for the wastewater.

Nutrients cause excessive plant growth, resulting in severely depleted dissolved oxygen, when the plants die and decay. Often the largest source of these nutrients is nonpoint-source runoff from the agricultural industry. Municipal and industrial wastewater treatment plants and home septic tanks are also significant nutrient sources.

The hydrologic cycle involves the continuous circulation of water between surface water, the atmosphere, and the subsurface. This cycle begins when rain or snow precipitation onto the surface of the earth becomes surface water runoff or groundwater. Eventually this water is returned to the atmosphere by evaporation or by plants through evapotranspiration and the cycle is repeated. Throughout the entire cycle, water is coming in contact with gases, liquids, and solids, and is thereby contaminated with pollutants, which may or may not be harmful, depending upon their toxicity and quantity. The resulting pollutant concentration in the water from the hydrologic cycle can significantly alter the water quality.

LAND POLLUTION

Releases of contaminates to the soil may occur from many sources. Transportation accidents involving hazardous materials are fairly common occurrences that pollute the land. When these same materials are transferred in and out of storage, the potential for spills exists. Storage tanks have finite lives, even with sound preventive maintenance programs. Eventually they will leak or develop structural failures. The transportation and storage of hazardous materials and wastes is a major source of potential pollutants to the land, as well as to the air and water, depending on how well a spill is contained and cleaned up.

The effluent from waste treatment facilities usually consists of treated liquids, to be discharged to bodies of water, and treated sludges or solids, to be disposed of in or on the land. Even low concentrations of pollutants in wastewater treatment plant effluent will eventually become land pollutants. The sludges and solids from these operations usually contain little, if any, organic contaminants because they have been consumed or destroyed in the treatment process. The remaining inorganics in the sludges and solids, particularly toxic heavy metals, hve become concentrated in the treatment process. These high pollutant concentrations may be harmful to human health and the environment when they are discharged to the land.

Abandoned landfills are a liability because of the historic mismanagement of wastes. They have created tremendous land pollution in the United States and their cleanup will be costly for both present and future generations.

POLLUTION PREVENTION

Pollution prevention and waste reduction have become the commonsense solution to the prevention of future environmental problems. We need to limit the historic practice of using environmental treatment processes, which merely transfer the waste from one medium to another. By using materials more efficiently, industry can reduce the generation of waste and achieve the desirable protection of human health and the environment. At the same time, the costs of waste management and regulatory compliance can be lowered and long-term liabilities and risks can be minimized. A properly managed pollution prevention program in the early stages of facility design can be a genuine asset for successful facility siting in a local community. If you eliminate the waste generation, no permits will be needed to manage it, since the permit process is usually the critical path to facility construction and operation.

Ideally, pollution should be prevented at the source through product reformulation, product substitution, process modification, equipment redesign, improved housekeeping, and segregation of incompatible toxic wastes. Wastes can also be reclaimed and recycled either at the source or off-site at another facility. [6]

Initially, industrial pollution prevention efforts should concentrate on those solutions that are obvious, simple, and low-cost, such as housekeeping improvements and direct recycling of process materials. Equipment redesign and process modifications are generally more complex and expensive and are generally implemented later in the pollution prevention program. The reformulation or substitution of products to prevent pollution must overcome many marketing and operational obstacles before becoming successful. Process and plant designers need to work closely with research and development, operations, and marketing personnel in order to achieve an optimal process and plant design that addresses these environmental options. [9]

The Resource Conservation and Recovery Act (RCRA) is a clear statement of national priorities in hazardous waste management. HSWA, in emphasizing waste reduction, recognizes the need to reduce the volume of hazardous waste and to recover and recycle valuable resources in the process. This is the most reasonable way to conserve raw materials and at the same time to reduce the amount of hazardous pollutants.

Under this scenario, the first priority should be to eliminate or reduce the generation of waste at the source (see Fig. 4-2). The ideal situation would be to completely eliminate the production of waste. Since it is normally impractical to completely eliminate waste from an industrial process, the reduction of waste volume should be a realistic goal. Waste reduction will lessen the environmental impact,

FIGURE 4-2
PRIORITIES IN POLLUTION PREVENTION.

lower the operating cost, decrease the complexity of waste management, and reduce the potential liability of the waste.

An inventory audit should be an important part of the management information
system for industrial waste materials. The inventory should list waste that would
be produced from operations. With such inventories, wastes should be readily iden-
tifiable and strategies more easily devised to mitigate their environmental impacts.
For example, some wastes may be produced only because of occasional malfunctions beyond the normal control of operations personnel, but the repetitive production of undesirable wastes may reveal processing problems. Correction of these
problems can often be accomplished with minimal investment.

A *waste-flow diagram* should produce a quantitative mass balance for the inputs
and outputs of the process operations and is necessary to establish the basis for any
meaningful pollution prevention program. The development of the waste-flow diagram requires a thorough understanding of the process engineering, the chemical
and physical properties of production materials, and the flow characteristics of the
system. [14]

Because it is highly unlikely that all of the necessary data regarding the process
system will be available initially, a sequenced approach is recommended for developing the waste-flow diagram. This approach, shown in Figure 4-3 on page 106,
begins with research operations, maintenance, environmental, procurement, and
other personnel who would manage the facility.

Achieving quantitative closure of the diagram within the system may be so difficult and time-consuming that its attainment would not be cost-effective. Therefore,
practical judgment must be exercised to maintain a reasoned approach to closure.

Once a reliable quantitative flow diagram—showing all incoming, intermediate,
and outgoing streams at the facility—has been achieved, it forms the basis for developing the relationships within the pollution prevention system shown in Figure

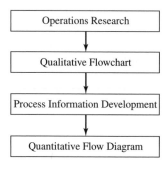

FIGURE 4-3
SEQUENCE OF DEVELOPMENT OF A WASTE-FLOW DIAGRAM.

4-4. Incoming materials are obtained according to predetermined specifications and competitive economics. The delivered materials are generally placed in storage, depending upon their physical and chemical properties. The materials are then distributed from storage to be utilized in plant processes.

As these incoming materials undergo physical and chemical process changes, products are created, streams are recycled, and wastes are generated. The process waste that is generated is accumulated through a collection system and placed in

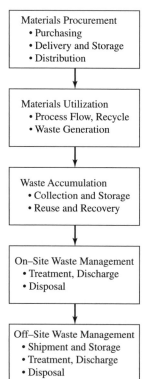

FIGURE 4-4
RELATIONSHIPS WITHIN THE POLLUTION PREVENTION SYSTEM.

storage, pending further disposition. Some of the wastes may be reusable in the process following physical or chemical treatment. Others may be recovered as valuable by-products.

At this point in the evaluation, the process potential for pollution prevention has been completed. The remaining steps focus on waste treatment to reduce the volume and hazard of waste generated and minimize its discharge and disposal costs, whether the treatment takes place on-site or off-site.

With the establishment of both the origin and quantities of the wastes and other process substances, the pollution prevention model can now be designed, as shown in Figure 4-5. This model should result from mass balances based upon the following relationship:

$$\text{Input materials} = \text{products} + \text{materials recovery} + \text{waste discharge} + \text{waste disposal}$$

Mass balance relationships should be developed for each process step in the pollution prevention model. Using process relationships, the system will serve as an important tool for collecting data necessary for the development of subsequent alternatives and their selection for implementation, as shown in Figure 4-6 on page 108. Design engineers should focus on the selection criteria for pollution prevention and be certain these criteria have the concurrence of the management of the facility.

The selection of the actual process should take into account all factors that fit into the overall program goals. There are numerous ways to reduce the generation of wastes. The ease of treating these wastes, the raw material costs, and the amount

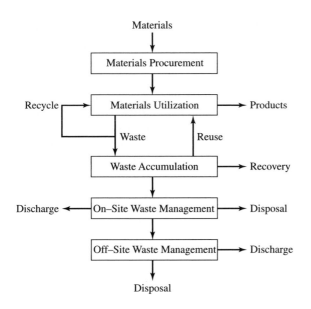

FIGURE 4-5
CONCEPTUAL DESIGN OF THE POLLUTION PREVENTION MODEL.

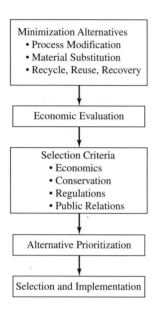

FIGURE 4-6
SELECTION PROCESS IN POLLUTION PREVENTION.

of wastes generated are all important factors that could influence on-site reduction and treatment by the generator. Also to be considered is possible conversion of wastes to end-use products.

It is important to recognize all potential alternatives and to accurately define the associated variables so that a realistic economic evaluation can be performed on each alternative. Selection criteria should be determined to permit an objective decision. In addition to economics, the selection criteria might include such diverse items as ease of procurement, conservation, environmental regulations, and the need to maintain desirable public relations. The alternatives should then be prioritized and reviewed by all concerned parties before the selection of the best alternative. Facility personnel must then commit to implementation of the alternative selected in order for it to be successful.

Identification of problem wastes being produced in significant volumes can encourage research and development efforts and can foster cooperation to seek solutions through coordinated efforts. Waste minimization considerations should be given priority in market development activities. A new chemical should be assessed early in its development to minimize or possibly eliminate waste generation. Substituting a raw material or modifying the end product may also eliminate a waste or possibly change a waste to a less hazardous form. This process substitution necessitates careful evaluation of the impact on the production process, safety, and the marketplace.

By improving the plant housekeeping and maintenance of equipment, considerable reduction of wastes can be achieved. The adjustment of operating procedures

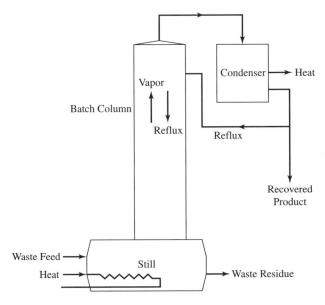

FIGURE 4-7
BATCH DISTILLATION FOR
SOLVENT RECOVERY.

may also reduce wastes. A management information system for cost analysis will normally be an effective tool in determining the feasibility of process changes.

As the final disposal of wastes becomes more strictly regulated and expensive, the recovery of waste energy and valuable materials will become more common. A batch distillation column is used for solvent recovery (Fig. 4-7). Wastes themselves may yield materials that might be useful to the generator. For instance, many wastes that contain hydrocarbons have considerable fuel value that could be recovered with modest additional investment. But hydrocarbon wastes may contain other contaminants that could cause operating problems during their thermal destruction. It is not unusual for severe corrosion, erosion, and fouling problems to increase maintenance costs when a waste stream is being used to generate and recover energy. Wastes do not have the same quality assurance criteria of commercial feedstocks. This makes their quality and quantity subject to fluctuations, which may be difficult to manage in the operation of the facility.

Economics will always be the driving force in choosing between resource recovery of desirable materials from hazardous wastes and their disposal. Recovery of materials has often been unprofitable, a factor that has prevented many industries from adopting resource-recovery techniques. Lack of process technology to accomplish resource recoveries within budget constraints has also hindered development.

For waste reduction and recycling to succeed, management must be strongly committed to supporting such a program. It should be recognized that many waste reduction projects will involve up-front investment of capital resources. In addition to the initial investment, operating and maintenance costs could make the return on these projects unattractive when compared to historic economic evaluation parameters. By working closely with research, operations, and marketing, engineers may be able to redesign or reformulate products to remain cost-effective with reduced impact on the environment.

Although prevention of pollution is certainly desirable, there are constraints—mainly economics, process technology, creativity, and innovativeness—that guarantee that some waste treatment and ultimate disposal will be necessary.

Once all practical alternatives have been examined to reduce the waste generation and to maximize the recycling and reuse of wastes, the next priority should be to treat the waste to render it harmless and reduce its volume. Treatment may cover a wide range of alternatives from physical, chemical, and biological technologies to thermal destruction processes. The treatment process ideally should convert the wastes into substances that can be recycled and reused, properly converted, and safely returned to the environment.

The final priority should provide secure land disposal for those wastes remaining from the waste elimination, reduction, treatment, and recycling processes. Virtually every waste treatment process will result in some residuals that will be subject to final disposal. It is impossible to completely eliminate all waste generation and disposal. There will always be a need for secure landfill capacity to accommodate these residuals.

Large quantities of hazardous wastes have been disposed of in surface impoundments, landfills, and injection wells. As more stringent land disposal regulations come into effect, emphasis should shift toward pollution prevention and resource recovery. Generators will no longer be using short-term solutions to waste management that merely shift the costs and liability risks into the distant future. Corporate responsibility and the pressure of public opinion will help force the solution of this issue toward the protection of human health and environment.

PLANT PROCESSES

In the design of plant processes it is desirable to critically examine the potential for waste reduction and pollution prevention. By critiquing each segment of plant process, considerable progress in pollution prevention can be made. Examples of environmental concerns for plant processes particularly related to the chemical and petroleum industries will now be discussed. Each of these processes should be analyzed in light of pollution prevention opportunities shown in Figures 4-5 and 4-6. [3, 4] Suggested environmental issues for design consideration are listed with each of these examples.

The synthesis of ammonia by the Haber-Bosch process involves a hydrogen-rich "synthesis gas" produced from natural gas that is purified and reacted with nitrogen. In most modern ammonia plants, methane from natural gas is the feedstock for synthesis gas production, as shown in Figure 4-8. In the steam hydrocarbon reforming process, gas and steam are fed to a primary reformer where the high temperature and nickel catalyst cause them to react to form carbon monoxide, carbon

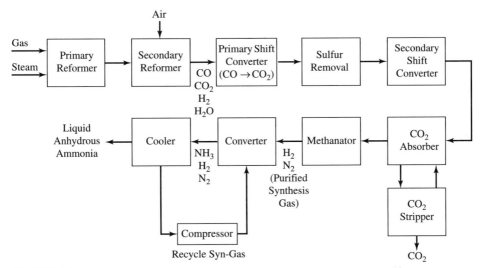

FIGURE 4-8
PROCESS FOR AMMONIA PRODUCTION. (*Source:* Cralley: *In Plant Practices for Job-Related Health Hazards Control.* Copyright © 1989. Reprinted by permission of John Wiley & Sons, Inc.)

dioxide, and hydrogen. The effluent gas, plus air, flow to a secondary reformer, where additional reactions yield carbon monoxide, water, and hydrogen.

Carbon monoxide in the raw synthesis gas is reacted with water in the two-stage shift converter to form carbon dioxide and hydrogen. This reaction proceeds in the presence of catalysts such as iron oxide and may be promoted by compounds such as chromium oxide and copper. Passing the stream through a zinc oxide bed removes sulfur compounds by conversion to hydrogen sulfide, which is then adsorbed. Carbon dioxide and any remaining sulfide are scrubbed out with an amine absorbent. Next, the methanator converts the remaining carbon dioxide and carbon monoxide to methane and water by reacting them with hydrogen over a nickel catalyst.

Ammonia is synthesized by the reaction of the hydrogen and nitrogen in the purified synthesis gas. The reaction, which occurs at elevated temperature and pressure, requires an iron catalyst promoted with Al_2O_3, K_2O, or CaO. The purified synthesis gas is compressed and fed into the top of the catalyst-containing converter, where up to 30 percent of the synthesis gas is converted. The ammonia in the stream leaving the converter is liquified by refrigeration and then removed for storage. The unreacted synthesis gas is returned to the converter via the compressor.

Ammonia production may produce the following potential wastes:

Spent nickel catalyst from the primary reformer
Spent nickel catalyst from the secondary reformer
Spent iron oxide, chromium oxide, and copper catalyst from the primary and secondary shift converters

Spent zinc oxide catalyst from sulfur conversions
Spent amine absorbent
Carbon dioxide and hydrogen sulfide by-products
Spent nickel catalyst from the methanator
Spent iron conversion catalyst with Al_2O_3, K_2O, or CaO
Unreacted synthesis gas
Anhydrous ammonia storage
Fugitive ammonia emissions
Empty catalyst containers
Cooling water treatment

Benzene, toluene, xylenes, and related aromatics are industrial chemicals that are major feedstocks in the production of plastics, synthetic rubber, and fibers. The catalytic reforming of petroleum naphtha and the steam cracking of petroleum feedstocks for the production of olefins are the major sources of aromatics, as shown in Figure 4-9.

Benzene, toluene, and xylenes result from catalytic naphtha reforming and steam cracking. The liquid product from catalytic naphtha reforming is especially suitable for the production of toluene and xylenes. On the other hand, steam cracking for the production of ethylene, propylene, and other olefins yields a liquid stream relatively high in benzene.

The production of benzene, toluene, and xylenes from aromatic feedstocks is accomplished by separating the aromatics from the nonaromatics and then fractionating the individual aromatics. Distillation cannot be used to separate the aromatics from the nonaromatics because cyclohexane, *n*-heptane, and the other alkanes produce azeotropes with benzene or toluene that are not separable by distillation. The most widely used method of separating the aromatics from the nonaromatics is liquid-liquid extraction. Typically, the feed stream is extracted with a solvent—usually sulfolane, which extracts the aromatics—leaving a nonaromatic raffinate that can be used as a motor fuel blendstock. The solvent is fed to the top of the column and the aromatic-containing feedstock enters in the middle of the column. The nonaromatic raffinate then leaves the extraction column at the top, while the aromatic-containing solvent is removed from the lower part of the column. The raffinate from the extractor is water-washed to recover dissolved sulfolane.

The stream containing the solvent and the aromatics is sent to an extract recovery column, and the aromatics are distilled off under vacuum. The aromatic-containing stream then proceeds to further separation processes yielding the individual aromatics. The lean solvent is returned to solvent purification. A slip stream of solvent is vacuum-distilled to remove any high-boiling-point contaminants that may accumulate in the solvent.

The aromatics stripped from the extract are separated from each other by fractionation. Benzene and toluene are recovered from the stream of mixed aromatics by distillation. A C_{8+} aromatics stream from this first distillation is then fractionated further to produce a stream of mixed xylenes in the overhead and orthoxylene

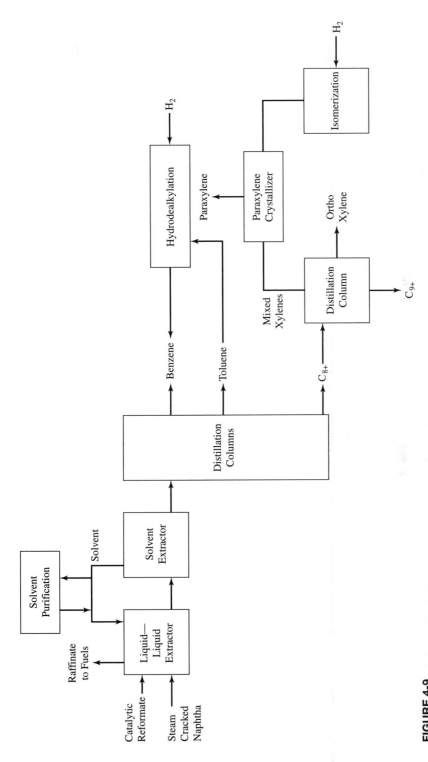

FIGURE 4-9
PROCESS FLOW DIAGRAM FOR AROMATICS PRODUCTION. (*Source:* Cralley: *In Plant Practices for Job-Related Health Hazards Control.* Copyright © 1989. Reprinted by permission of John Wiley & Sons, Inc.)

with the C_{9+} fraction in the bottom. The orthoxylene in this bottom stream is then separated from the heavier C_{9+} fraction in another distillation column.

Para-xylene can be recovered from the mixed xylenes by crystallization. The overhead stream containing the mixed xylenes and ethylbenzene is dried and cooled to as low as −105°F in stages using a refrigerant. High-purity *para*-xylene crystals form on the cold walls of the heat exchanger and are scraped off and separated from the mother liquor by centrifuging. To purify the crude *para*-xylene, it is partially remelted, recrystallized, and recentrifuged in one to three stages to achieve the desired product purity.

If maximization of *para*-xylene is desired, the filtrate from the *para*-xylene recovery unit is sent to a catalytic isomerization process that restores the equilibrium between xylene isomers to about 20 percent *para*-xylene for recycling to the crystallization unit.

Toluene may be converted to benzene by a dealkylation process. Hydrogen is reacted with the toluene at high temperatures and pressures to produce benzene and methane. This process also produces small amounts of miscellaneous by-product hydrocarbons including biphenyls.

Aromatics production may produce the following potential wastes:

Hydrocarbon storage
Fugitive hydrocarbon emissions
Wash water from sulfolane recovery
Hydrocarbon light ends from solvent distillation
High-boiling-point contaminates in the sulfolane solvent
Heavy hydrocarbons in the C_{9+} stream from the distillation column
Xylenes remaining after isomerization
Light ends from hydrodealkylation
Hydrocarbon by-products from hydrodealkylation
Cooling-water treatment

ENVIRONMENTAL PROTECTION CONTROL PROCESSES

During the design of a plant process and facility, every effort should be made to prevent pollution at the source of generation. Since it is usually impossible to completely eliminate all of the waste streams, the optimal plant design should be surveyed to quantify the various waste streams that will be generated and managed by the facility. The most cost-effective and environmentally safe manner for managing these wastes should then be determined.

The selection of waste control technologies is generally based upon economics. Government regulations and the adaptability of process technology are also considerations, as are public relations and geographic location. The final decision can be influenced by subjective political reasons. If after exploring the various alternatives for pollution prevention and assuming waste remains to be treated, then treatment alternatives should be systematically explored. Physical, chemical, biological, and thermal treatment processes should be considered. Many inorganic wastes can be treated by physical and chemical treatment techniques. These types of processes are desirable solutions because they are both cost-effective and well-known technologies that can be implemented on virtually any scale that the facility can accommodate on-site. It is also true that the proponents of the not-in-my-back-yard (NIMBY) position are more likely to be receptive to the siting of a facility based upon physical, chemical, and biological treatment because there are no exhausts or smokestacks that represent point sources of air emissions. [14]

Government regulations to protect air, water, and land play an important role in the design, construction, and operation of process facilities. The lead time required for the planning and construction of new facilities because of these environmental regulations can add as much as several years to the total time required for regulatory approval prior to construction and operation. Usually these environmental approvals require a public hearing, making the viewpoints of local communities and elected officials extremely important in the ultimate siting of the facility.

Air Quality

Air pollutants are either naturally occurring or manufactured. Naturally occurring air pollutants make up the background pollution in any given geographic location and may include the following natural sources:

Wind-blown dust
Ash and gas emissions from volcanic activity
Smoke and gases from forest fires
Pollen from plants
Hydrocarbons from vegetation
Ozone from lightning
Gases from natural decomposition
Natural radioactivity

Air pollutants from human sources are major contributors to urban air pollution and generally come from motor vehicles, municipal trash, electric power generation, and the production of commodities and consumer products.

Particulate matter in air is generally defined as particles <1000 μm in diameter. The relative size of various types of particles in air includes both natural and human sources of air pollutants (Fig. 4-10 on page 116). The removal of these particulate pollutants from air is generally based on the particle size. Some of the more common air pollution control technologies are shown in Table 4-3 on page 117 and include a number of options for particulate reduction.

Other control technologies for the pollutants listed in the national ambient air quality standards are also shown in Table 4-3. These common pollution control technologies should be considered for each pollutant identified under the NAAQS. These same pollution control technologies for air emissions should also be considered to reduce air emissions from industrial processes involving the hazardous air pollutants.

The ultimate selection in the design of equipment to control air pollution usu-

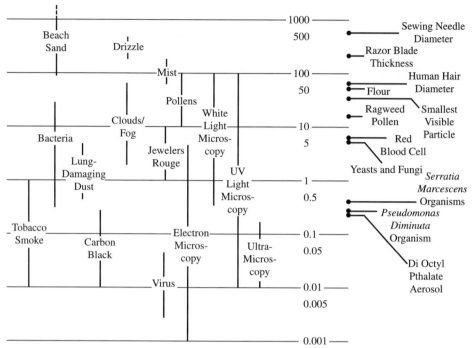

FIGURE 4-10
RELATIVE SIZE OF VARIOUS TYPES OF PARTICLES IN AIR. (*Source:* Davis and Cornwell, *Introduction to Environmental Engineering,* 2nd. ed. Copyright © 1991. Reprinted with permission of The McGraw-Hill Companies.

ally is based upon the most cost-effective solution for achieving compliance with environmental regulations. Because air pollution regulations may vary from state to state, air pollution technology should be selected on a case-by-case basis.

Cyclones are compact, low-cost, simple equipment that can be used to collect particulate matter >10 μm in size. Cyclones are not able to handle damp or tacky particulate matter.

Baghouse filters offer relatively simple operation with extremely high collection efficiency for particulates. At temperatures in excess of 550°F, baghouse filters require special refractory fabrics, which are expensive.

Scrubbers have high mass transfer efficiencies, low pressure drop, and low capital cost; are compact; and can collect particulates as well as gases. The waste from the scrubber often presents a waste disposal problem, and scrubbers have relatively high maintenance costs.

Electrostatic precipitators have high collection efficiencies, low pressure drops, and low operating costs. They generally are capital-intensive, have large space requirements, and do not adapt well to fluctuations in the gas stream operating conditions.

The design and operating advantages and disadvantages of absorption systems are similar to scrubbers. Adsorption systems can remove gaseous vapor or liquid

TABLE 4-3
COMMON POLLUTION CONTROL TECHNOLOGIES FOR AIR
EMISSIONS

Pollutant	Control technology
Particulates	Cyclone Baghouse filter Scrubber Electrostatic precipitator Wetting
Sulfur dioxide	Chemical reaction systems Scrubber Spray chamber
Carbon monoxide	Incinerator Flare Afterburner, catalytic converter
Nitrogen dioxide	Scrubber Catalytic converter
Ozone	Source reduction
Lead	Scrubber Cyclone Baghouse, precipitator
Hydrocarbons	Incinerator, afterburner Flare, adsorber, catalytic converter
Photochemical oxidants	Source reduction

contaminants from process streams at low concentrations of contaminants. The cycles of absorption operations are predictable and can be automated, while adsorption processes are cost-effective only at low concentrations of contaminants.

Combustion and incineration systems are capable of destroying organic contaminants, thereby generating heat and offering potential energy recovery. Combustion and incineration systems are usually capital-intensive. It is often very difficult to obtain operating permits for waste treatment processes.

Emissions stacks are the most common method of disposing of industrial waste gases. When the height of the emission stack is increased, the concentrations of these waste gases can be significantly reduced at ground level, but there is no reduction in the amount of pollutants released into the atmosphere. Wind direction and speed will be major contributing factors in determining the ground-level concentration of the pollutant being exhausted from the stack. The concentration of air pollutants downwind from the emission source is generally inversely proportional to the wind speed. Buildings, trees, and other obstacles will cause mechanical turbulence in the atmosphere that will diffuse the plume from the emission.

Another important factor that will influence the degree of turbulence and speed of diffusion is the variation of the atmospheric temperature with the height above

Temperature Gradient	Observation	Description

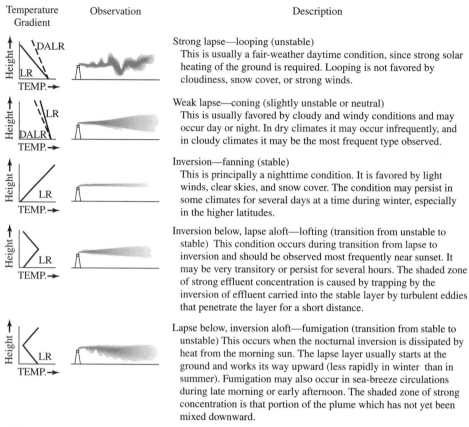

Strong lapse—looping (unstable)
This is usually a fair-weather daytime condition, since strong solar heating of the ground is required. Looping is not favored by cloudiness, snow cover, or strong winds.

Weak lapse—coning (slightly unstable or neutral)
This is usually favored by cloudy and windy conditions and may occur day or night. In dry climates it may occur infrequently, and in cloudy climates it may be the most frequent type observed.

Inversion—fanning (stable)
This is principally a nighttime condition. It is favored by light winds, clear skies, and snow cover. The condition may persist in some climates for several days at a time during winter, especially in the higher latitudes.

Inversion below, lapse aloft—lofting (transition from unstable to stable) This condition occurs during transition from lapse to inversion and should be observed most frequently near sunset. It may be very transitory or persist for several hours. The shaded zone of strong effluent concentration is caused by trapping by the inversion of effluent carried into the stable layer by turbulent eddies that penetrate the layer for a short distance.

Lapse below, inversion aloft—fumigation (transition from stable to unstable) This occurs when the nocturnal inversion is dissipated by heat from the morning sun. The lapse layer usually starts at the ground and works its way upward (less rapidly in winter than in summer). Fumigation may also occur in sea-breeze circulations during late morning or early afternoon. The shaded zone of strong concentration is that portion of the plume which has not yet been mixed downward.

FIGURE 4-11
DRY ADIABATIC LAPSE RATE FOR PLUME BEHAVIOR. (*Source:* Robert Perry and Don Green, *Perry's Chemical Engineers' Handbook,* 6th. ed. Copyright © 1984. Reprinted with permission of The McGraw-Hill Companies.)

the ground. This variation in temperature, or lapse rate, is shown in Figure 4-11 for differnt types of plume behavior. The dry adiabatic lapse rate (DALR) is the temperature change for rising dry air and has a value of approximately 1°C per 100 meters. The DALR is graphically depicted in Figure 4-11 as a reference standard for an atmosphere that is neutrally stable. If the atmospheric temperature falls at a rate greater than the DALR, the lapse rate is superadiabatic and the atmosphere is unstable. If the lapse rate is less than the DALR a subadiabatic condition exists and the atmosphere will be stable. The stack plumes that may be expected from these variations in the lapse rate profile are shown in Figure 4-11. Low-level inversions, which are caused by radiative cooling of the ground at night, and stable atmospheric layers are undesirable from an air pollution perspective because they minimize the rate of dilution of the contaminants in the atmosphere.

The design of an emission stack height must satisfy the maximum acceptable ground-level contaminant standards, which are determined by local air pollution regulations. The effective height of the air emission from the stack is the actual stack height modified by the plume rise, based on the exhaust emission velocity, and the buoyancy rise, based on the temperature of the emission gases and the atmospheric conditions. The maximum ground-level concentration is determined by an appropriate atmospheric diffusion equation.

Water Quality

Industrial process operations normally produce wastewater that must be managed before it can be returned to the environment. Domestic wastewater that is produced by the personnel at the operating facility is normally sent to the local public treatment works or other municipal wastewater treatment operation for treatment prior to discharge. If the operating facility is large or is in a remote location, domestic wastewater treatment operations are usually included in the plant design. [4]

Process wastewater streams from the facility may contain dilute quantities of hazardous materials and wastes that must be removed prior to discharge. Normally process wastewaters are treated on-site at the operating facility prior to discharge.

Solid Wastes

During the construction and operating life of the facility the generation of solid wastes will be considerable. Various types of trash and refuse will be generated that must be managed in an environmentally acceptable manner. Separation and sorting of various types of solid waste are the most desirable management techniques at the facility. The segregated solid waste components may then be recovered for reuse or recycling based upon economics. The solid waste management technique shown in Table 4-4 may be integrated to produce the most effective overall management technique for the operating facility. Each solid waste stream should be analyzed on a case-by-case basis to determine the appropriate overall management technique.

TABLE 4-4
SOLID WASTE MANAGEMENT TECHNIQUES

Desired performance	Management technique
Material separation	Manual, screening, trommel, pneumatic, sedimentation, flotation, magnetic, electromechanical
Material storage	Pits, bins, silos, warehouses
Material transfer	Trucks, railcars, belt, vibrating, pneumatic, and screw conveyors
Volume reduction	Compactors, crushers, incinerators
Particle size reduction	Shredders, crushers, grinders, chippers, hammer mills
Moisture removal	Dryers, filters, centrifuges
Material conservation	Resource recovery

Hazardous Waste

The selection of treatment processes for hazardous waste management is highly regulated by the Environmental Protection Agency. Often the EPA has mandatory processes for treatment of general categories and specific hazardous wastes. These hazardous waste treatment processes were selected by the EPA based on their ability to either destroy the waste or make it less hazardous to human health and the environment. The specific treatment technologies that are mandated by the EPA may be found in the 40 CFR 268.42–268.44.

Generally, organic wastes should be treated by incineration or biological treatment processes under these EPA regulations. Inorganic wastes should be treated by physical or chemical treatment processes to destroy the waste or render it less harmful. Toxic heavy metals and other treatment process residues that must ultimately be put into a landfill must first be stabilized, as shown in Figure 4-12.

In the selection and design of treatment processes for hazardous waste management, adherence to the EPA-mandated technologies is recommended. Otherwise, it becomes necessary to prove to the EPA that alternative treatment processes are at least as effective as the EPA-mandated treatment processes.

Many low-concentration effluents and other wastewaters can usually be treated in a cost-effective way prior to discharge into municipal sewers. Strong acids and alkalis can be neutralized. Heavy metals are virtually impossible to destroy, so they must be fixated so that they are nonleachable; then they can be landfilled. Reactive wastes must be carefully handled prior to undergoing physical and chemical process treatment to make them environmentally acceptable. Organic wastes offer considerable recovery potential and can be used for energy or recycled back into a chemical process.

Wastes that cannot be eliminated or recovered may be incinerated or undergo

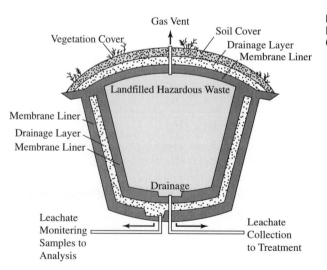

FIGURE 4-12
LANDFILL AFTER CLOSURE.

biological treatment. Incineration is preferred because it eliminates potential problems involved in landfill disposal and other interim hazardous waste management processes. Biological treatment also offers the potential of complete destruction of a waste. The development of specialized microbes for highly efficient destruction systems may eliminate the need for subsequent landfill disposal. Future hazardous waste management disposal strategies are expected to emphasize the more desirable alternatives of biological treatment and incineration in the selection of process technology. The EPA has added further emphasis in this area by specifying certain treatment technologies for hazardous waste disposal.

WATER TREATMENT

Treatment and disposal of wastes in an environmentally acceptable manner should generally be tailored to the type of waste and its medium: wastewater, solids, or air emissions. The use of chemical, physical, biological, and thermal waste treatment processes will now be discussed for these types of waste streams. [8, 14]

Aqueous waste streams vary widely with respect to volume, level, and type of contaminants and level of solids. The major sources of aqueous wastes may include the following:

- Contaminated process wastewater
- Contaminated runoff collected in impoundments or basins
- Contaminated water generated from equipment cleanup
- Aqueous waste generated from sediment or sludge dewatering
- Highly concentrated wastewater streams generated from certain aqueous waste treatment processes, for example, backwash from filtration, concentrate from reverse osmosis

Because these waste streams are so diverse in volume, type, and concentration of contaminants, a wide variety of treatment processes will have application. Rarely will any one unit treatment process be optimal for aqueous waste treatment. Therefore, treatment processes are frequently used in various combinations, and pretreatment requirements may be a prerequisite for effective use of a treatment process. The unit treatment processes include:

- Activated carbon
- Activated sludge
- Filtration
- Precipitation-flocculation
- Sedimentation
- Ion exchange
- Reverse osmosis
- Neutralization
- Gravity separation
- Air stripping

- Chemical oxidation
- Chemical reduction

Activated Carbon

The process of adsorption onto activated carbon involves putting a waste stream in contact with carbon, usually by flow through a series of packed-bed reactors. The activated carbon selectively adsorbs hazardous constituents by a surface attraction phenomenon in which molecules are attracted to the internal pores of the carbon granules.

Adsorption depends on the strength of the molecular attraction between adsorbent and the waste, molecular weight, type and characteristic of adsorbent, electrokinetic charge, pH, and carbon surface area.

Once the micropore surfaces of the carbon are saturated with organics, the carbon is spent and must either be replaced with virgin carbon and removed or thermally regenerated to be returned. The time to reach breakthrough or exhaustion is the most critical operating parameter, as shown in Figure 4-13. The carbon longevity balanced against influent and effluent concentration governs operating economics.

Activated Sludge

Microorganisms are capable of breaking down many organic compounds considered to be environmental and health hazards. Laboratory, pilot, and field studies have demonstrated that it is feasible to use this capability of microorganisms to control hazards. The function of biological treatment is to remove organic matter from the waste stream through microbial degradation by either aerobic processes, in which oxygen is required, or anaerobic processes, in which sulfate or nitrate serves as a terminal electron acceptor.

The method that has been most developed and is most feasible for treatment relies on aerobic microbial processes by providing an oxygen source and nutrients to

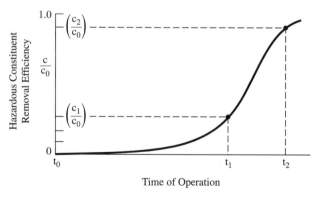

FIGURE 4-13
TYPICAL
BREAKTHROUGH CURVE.

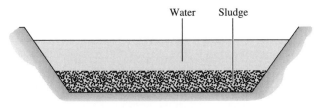

Holding Lagoon

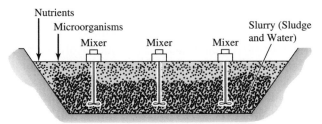

Lagoon Modified for Slurry Phase Treatment

FIGURE 4-14
LAGOON TREATMENT
SYSTEM.

enhance microbial activity. The microorganisms can generally be relied upon to de-grade a wide range of compounds given proper nutrients and sufficient oxygen. Such biological treatment processes may be applicable to treatment of aqueous haz-ardous wastes in lagoons, as shown in Figure 4-14, including activated sludge, and fixed film systems like rotating biological contactor disks and trickling bed filters.

In the conventional activated-sludge process, shown in Figure 4-15, aqueous waste flows into an aeration basin, where it is aerated for several hours. During this time, a suspended active microbial population, which has been maintained by recy-

FIGURE 4-15
ACTIVATED SLUDGE PLANT. (*Source:* Davis and Cornwell, *Introduction to Environmental Engineering,* 2nd. ed. Copyright © 1991. Reprinted with permission of The McGraw-Hill Companies.)

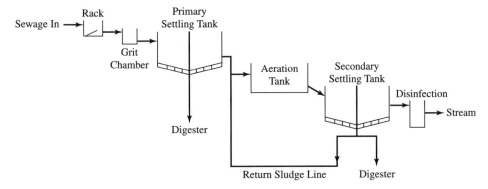

cling sludge, aerobically degrades organic matter in the stream while producing new cells. A simplified equation for this process is:

$$\text{Organics} + O_2 \rightarrow CO_2 + H_2O + \text{new cells} \tag{4-7}$$

The new cells produced during aeration form a sludge, which is settled out in a clarifier. A portion of the settled sludge is recycled to the aeration basin to maintain the microbial population, while the remaining sludge is disposed of; that is, it undergoes volume reduction and disposal. The clarified water flows to disposal or further processing.

In the oxygen-activated-sludge process, oxygen or oxygen-enriched air is used instead of air to increase the transfer of oxygen. Extended aeration involves longer detention times than conventional activated sludge and relies on a higher population of microorganisms to degrade wastes. Contact stabilization involves only short contact of the aqueous wastes and suspended microbial solids, with subsequent settling and treatment of the sludge to remove sorbed organics. Fixed film systems involve contact of the aqueous waste stream with microorganisms attached to some inert medium, such as rocks or specially designed plastic materials.

The activated-sludge process shown in Figure 4-15 is most widely used for biological treatment of aqueous wastes with relatively high organic loads. However, the high-purity oxygen system has advantages for hazardous waste site remediation. In addition, a number of other parameters may influence the performance of the biological treatment system, such as concentration of suspended solids, oil and grease, organic load variations, and temperature. Design of the system for a particular application can be achieved best by determining the necessary parameters with laboratory or pilot tests.

The trickling filter shown in Figure 4-16 consists of a bed of rocks over which the contaminated water was sprayed. The microbes form a slime layer on the rocks and metabolize the organics, while the oxygen provided as air moves countercurrent through the water.

A rotating biological contactor (RBC) consists of a series of rotating disks connected by a shaft, set in a basin or trough, as shown in Figure 4-17. The contaminated water passes through the basin, where the microorganisms, attached to the disks, metabolize the organics present in the water. Approximately 40 percent of the disk surface area is submerged to allow the slime layer to alternately come in contact with the contaminated water and the air where the oxygen is provided to the microorganisms.

Although a number of compounds are considered to be relatively resistant to biological treatment, it is recommended in practice that the treatability of waste be determined through laboratory biochemical oxygen demand (BOD) tests on a case-by-case basis.

Rotating biological contactors also have advantages for hazardous waste treatment. The units are compact. They can handle large flow variations and high organic shock loads, and they do not require use of aeration equipment. Sludge pro-

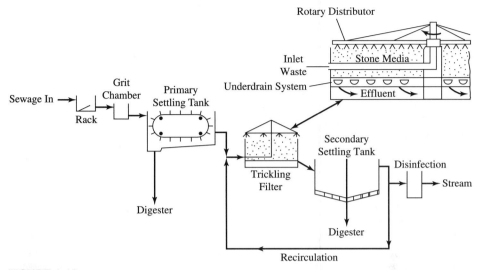

FIGURE 4-16
TRICKLING FILTER PLANT (*Source:* Davis and Cornwell, *Introduction to Environmental Engineering,* 2nd. ed. Copyright © 1991. Reprinted with permission of The McGraw-Hill Companies.)

duced in biological waste treatment may be a hazardous waste itself due to the sorption and concentration of toxic and hazardous compounds contained in the wastewater. If the sludge is hazardous, it must be disposed of in the RCRA-approved manner. If the sludge is not hazardous, disposal should conform with government sludge disposal guidelines.

Anaerobic microorganisms are also capable of degrading certain organic contaminants. Methanogenic consortiums, groups of anaerobes that function under very reducing conditions, are able to degrade halogenated aliphatics better than aerobic organisms. The potential for anaerobic degradation has been demonstrated in numerous laboratory studies and in industrial waste treatment processes that

FIGURE 4-17
RBC TREATMENT SYSTEM. (*Source:* Davis and Cornwell, *Introduction to Environmental Engineering,* 2nd. ed. Copyright © 1991. Reprinted with permission of The McGraw-Hill Companies.)

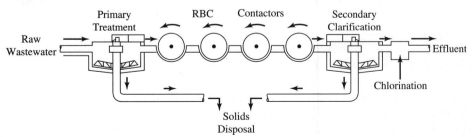

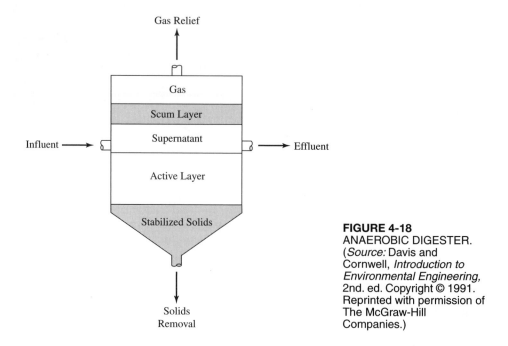

FIGURE 4-18
ANAEROBIC DIGESTER.
(*Source:* Davis and
Cornwell, *Introduction to
Environmental Engineering,*
2nd. ed. Copyright © 1991.
Reprinted with permission of
The McGraw-Hill
Companies.)

use anaerobic digesters (Fig. 4-18) or anaerobic waste lagoons as part of the treatment process.

Filtration

Filtration is a physical process whereby suspended solids are removed from solution by forcing the fluid through a porous medium. Granular media filtration is typically used for treating aqueous waste streams, as shown in Figure 4-19. The filter media consists of a bed of granular particles, typically sand or sand with anthracite coal. The bed is contained within a basin and is supported by an underdrain system that allows the filtered liquid to be drawn off while retaining the filter media in place. As water laden with suspended solids passes through the bed of filter medium, the particles become trapped on top of and within the bed. This either reduces the filtration rate at a constant pressure or increases the amount of pressure needed to force the water through the filter. In order to prevent plugging, the filter is periodically backflushed to dislodge the particles. The backwash water contains high concentrations of solids and requires treatment.

Often, granular media filters are preceded by sedimentation to reduce the suspended solids load on the filter. Granular media filtration is also frequently installed ahead of biological or activated carbon treatment units to reduce the suspended solids load and in the case of activated carbon to minimize plugging of the carbon columns.

FILTRATION CYCLE

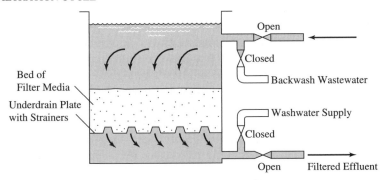

BACKWASH CYCLE

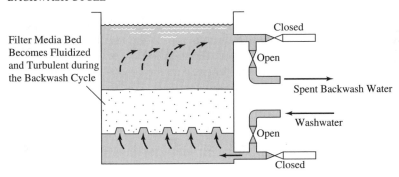

FIGURE 4-19
GRANULAR MEDIA FILTER CYCLE.

Filtration is a reliable and effective means of removing solids from wastes, provided the solids content does not vary greatly and the filter is backwashed at appropriate intervals. Filtration equipment is relatively simple, readily available in a wide range of sizes, and easy to operate and control. Filtration is also easily integrated with other treatment steps.

Precipitation-Flocculation

Precipitation is a physiochemical process whereby some or all of a substance in solution is transformed into a solid phase. It is based on alteration of the chemical equilibrium relationships affecting the solubility of contaminate species. Removal of metals as hydroxides or sulfides is a common precipitation application in wastewater treatment. Generally, lime or sodium sulfide is added to the wastewater in a rapid mixing tank along with flocculating agents. The wastewater flows to a flocculation chamber in which adequate mixing and retention time is provided for agglomeration of precipitated particles. Agglomerated particles are separated from the liquid phase by settling in a sedimentation chamber, or by other physical processes such as filtration.

Although precipitation of metals is governed by the solubility product of the metal species, in actual practice effluent concentrations comparable to the solubility product are rarely achieved. Usually, the amount of lime added is about three times the stoichiometric amount. The metal sulfides have significantly lower solubility than their hydroxide counterparts, and more complete sulfide precipitation is achieved, as shown in Figures 4-20 and 4-21. Metal sulfides are also stable over a broader pH range. Many metal hydroxides, on the other hand, are stable only over a narrow pH range. Metals reach a minimum solubility at a specific pH, but further addition of lime causes the metal to become soluble again. Therefore, dosages of lime need to be accurately controlled. This may be particularly challenging when working with aqueous wastes, where wide variations in flow rates and quantities of metals are to be expected. The stabilities of metal carbonates are also quite dependent on pH.

Flocculation is used to describe the process by which small, unsettleable particles suspended in a liquid medium are made to agglomerate into larger, more settleable particles. The mechanisms by which flocculation occurs involve surface chemistry and particle charge phenomena. In simple terms these various phenomena can be grouped into two sequential mechanisms:

• Chemically induced destabilization of the requisite surface-related forces, thus allowing particles to stick together when they touch

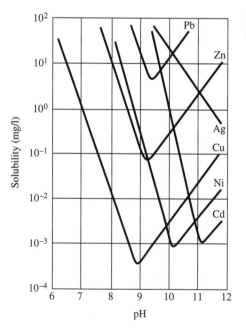

FIGURE 4-20
SOLUBILITY OF METAL HYDROXIDES IN WATER.

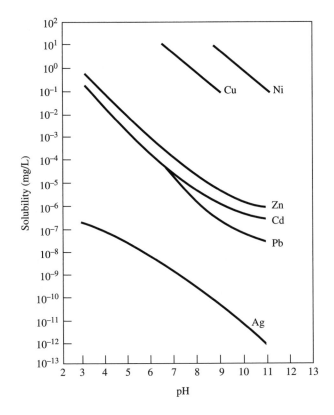

FIGURE 4-21
SOLUBILITY OF METAL
SULFIDES IN WATER.

• Chemical bridging and physical enmeshment between the attracting particles, allowing for the formation of large particles

Flocculation involves three basic steps:

• Addition of flocculating agents to the waste stream
• Rapid mixing to disperse the flocculating agent
• Slow and gentle mixing to allow for contact and bridging between small particles

Typically, chemicals used to cause flocculation include alum, lime, ferric chloride, ferrous sulfate, and organic flocculating agents, often referred to as polyelectrolytes. These materials generally consist of long-chain, water-soluble polymers, such as polyacrylamides. They are used either in conjunction with the inorganic flocculants, such as alum, or as the primary flocculating agent. A polyelectrolyte may be termed cationic or anionic, depending upon the type of ionizable groups, or nonionic if it contains no ionizable groups. The range of physical and chemical characteristics of available polymers for such applications is extremely broad.

Precipitation is applicable to the removal of most metals from wastewater, including zinc, cadmium, chromium, copper, lead, manganese, and mercury. Also, certain anionic species can be removed by precipitation, notably, phosphate, sulfate, and fluoride.

Selection and optimum doses of the most suitable precipitate or flocculant are determined through laboratory jar-test studies. Other important parameters that need to be determined as part of the overall design include:

- Most suitable chemical addition system
- Optimum pH requirement
- Rapid mix requirements
- Sludge production
- Sludge flocculation, settling, and dewatering characteristics

Precipitation is nonselective in that compounds other than those targeted may be removed. Both precipitation and flocculation are nondestructive and generate a large volume of sludge, which must be disposed of.

Precipitation and flocculation pose minimal safety and health hazards to field workers. The entire system is operated at near-ambient conditions, eliminating the danger of the high pressure or high temperatures associated with some other operations. While the chemicals employed are often skin irritants, they can be handled in a safe manner.

Sedimentation

Sedimentation is a process that relies upon gravity to remove suspended solids in an aqueous waste stream. The fundamentals of a sedimentation process include:

- A basin or container of sufficient size to maintain the liquid to be treated in a relatively quiescent state for a specified period of time
- A means of directing the liquid to be treated into the above basin in a manner conducive to settling.
- A means of physically removing the settled particles from the liquid

Sedimentation can be carried out either as a batch or as a continuous process in lined impoundments, conventional settling basins, clarifiers, and high-rate gravity settlers. Sedimentation provides a reliable means to remove suspended matter from a waste stream, provided the suspended matter is settleable and the treatment process, including the use of flocculants and coagulants, has been appropriately designed from laboratory settling tests. Most clarifiers are capable of removing 90 to 99 percent of the suspended solids in an aqueous waste.

Ion Exchange

Ion exchange is a process whereby the toxic ions are removed from the aqueous phase by being exchanged with relatively harmless ions held by the ion exchange material. Modern ion exchange resins are primarily synthetic organic materials containing ionic functional groups to which exchangeable ions are attached. These synthetic resins are structurally stable; that is, they can tolerate a range of temperature and pH conditions, exhibit a high exchange capacity, and can be tailored to show selectivity toward specific ions. Exchangers with negatively charged sites are cation exchangers because they take up positively charged ions. Anion exchangers have positively charged sites and, consequently, take up negative ions. The exchange reaction is reversible and concentration-dependent. It is possible to regenerate the exchange resins for reuse.

Ion exchange is a well-established technology for removal of heavy metals and hazardous anions from dilute solutions. Ion exchange can be expected to perform well for these applications with fed wastes of variable composition, provided the system's effluent is continually monitored to determine when resin-bed exhaustion has occurred. However, as mentioned previously, the reliability of ion exchange is markedly affected by the presence of suspended solids.

Reverse Osmosis

Osmosis is the spontaneous flow of solvent, for example, water, from a dilute solution through a semipermeable membrane to a more concentrated solution. Reverse osmosis (RO) is the application of sufficient pressure to the concentrated solution to overcome the osmotic pressure and force the net flow of water through the membrane toward the dilute phase (Fig. 4-22). This allows the concentration of solute impurities to be built up in a circulating system on one side of the membrane while relatively pure water is transported through the membrane.

The basic components of a reverse-osmosis unit are the membrane, a membrane support structure, a containing vessel, and a high-pressure pump. The membrane and membrane support structure are the most critical elements.

Reverse osmosis is used to reduce the concentrations of dissolved solids, both organic and inorganic. In treatment of hazardous waste–contaminated streams, use

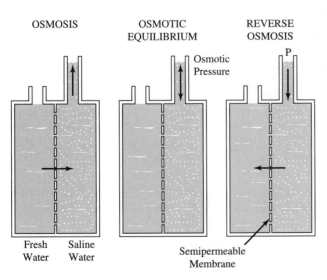

FIGURE 4-22
DIRECT AND REVERSE OSMOSIS. (*Source:* Davis and Cornwell, *Introduction to Environmental Engineering,* 2nd. ed. Copyright © 1991. Reprinted with permission of The McGraw-Hill Companies.)

of reverse osmosis would be primarily limited to polishing low-flow streams containing highly toxic contaminants. In general, good removal can be expected for high molecular weight organics and charged anions and cations. Multivalent ions are treated more effectively than are univalent ions. Recent advances in membrane technology have made it possible to remove such low-molecular-weight organics as alcohols, ketones, amines, and aldehydes.

Reverse osmosis is an effective treatment technology for removeal of dissolved solids presuming appropriate pretreatment has been performed for suspended solids removal, pH adjustments, and removal of oxidizers, oil, and grease. Because the process is so susceptible to fouling and plugging, on-line monitors may be required to monitor pH, suspended solids, and so on, on a continuous basis.

Neutralization

Neutralization consists of adding an acid or a base to a waste in order to adjust its pH. The most common system for neutralizing acidic or basic waste streams utilizes a multiple-compartment basin. This basin is lined with a material resistant to the expected environment.

In order to reduce the required volume of the neutralization basin, mixers are installed in each compartment to provide intimate contact between the waste and neutralizing reagents, thus speeding up reaction time (Fig. 4-23). Basin inlets are baffled to provide for flow distribution, while effluent baffles can help to prevent foam from being carried over into the receiving stream.

The choice of an acidic reagent for neutralization of an alkaline wastewater is generally between sulfuric acid and hydrochloric acid. Sulfuric acid is usually used due to its lower unit cost. Hydrochloric acid has the advantage of more soluble reaction end-products. The most economical acid should be determined on a case by case basis.

FIGURE 4-23
CHEMICAL NEUTRALIZATION TREATMENT SYSTEM.

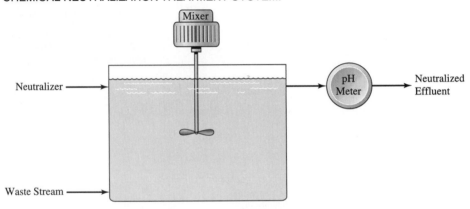

The caustic reagent is usually sodium hydroxide or lime, but ammonium hydroxide is occasionally used. The factors to be considered include purchase cost, neutralization capacity, reaction rate, storage and feeding requirements, and neutralization products.

Although sodium hydroxide costs much more than the other materials, it is frequently used because of its uniformity, ease of storage and feeding, rapid reaction rate, and soluble end-products. Lime is a low-cost material but savings are at least partially offset by increased capital and operating costs for the more complex feeding and reaction system required.

Neutralization of hazardous wastes may produce air emissions. Acidification of streams containing certain salts, such as sulfides, will produce toxic gases. Feed tanks should be totally enclosed to prevent escape of fumes. If the wastes being treated are concentrated, adequate mixing should be provided to disperse the heat of reaction. The process should be controlled from a remote location, if possible.

Gravity Separation

Gravity separators are primarily used to treat two-phased aqueous wastes. A typical application would be separation of gasoline or fuel oil from a hydrocarbon-contaminated wastewater. Gravity separation has also been used to separate oils from contaminated wastewater. For efficient separation, the nonaqueous phase should have a significantly different specific gravity than water and should be present as a nonemulsified substance. Emulsions between water and oil are common, and an emulsion-breaking chemical must frequently be added to the waste for efficient treatment.

Gravity separation offers a straightforward, effective means of phase separation provided the oil and water phases separate adequately within the residence time of the tank. Simple, readily available equipment can be used and operational requirements are minimal. If emulsion-breaking chemicals must be added to promote oil-water separation, laboratory tests should be conducted periodically to ensure adequate dosing.

Consideration must also be given to the disposal of the extracted waste constituents collected. For gravity separation processes, this material may consist of immiscible oil siphoned from the separator.

Air Stripping

Air stripping is a mass transfer process in which volatile contaminants in water or soil are transferred to gas.

Air stripping is frequently accomplished in a packed tower equipped with an air blower (Fig. 4-24 on page 134). The packed tower works on the principle of countercurrent flow. The water stream flows down through the packing while the air flows upward and is exhausted through the top. Volatile, soluble components have an affinity for the gas phase and tend to leave the aqueous stream for the gas phase. In the cross-flow tower, water flows down through the packing as in the counter-

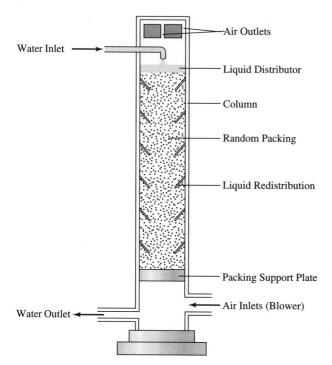

FIGURE 4-24
PACKED-TOWER AIR
STRIPPER.

current packed column; however, the air is pulled across the water-flow path by a fan. The coke tray aerator is a simple, low-maintenance process requiring no blower. The water being treated is allowed to trickle through several layers of trays. This produces a large surface area for gas transfer. Diffused aeration stripping and induced draft stripping use aeration basins similar to standard wastewater treatment aeration basins. Water flows through the basin from top to bottom or from one side to another, with the air dispersed through diffusers at the bottom of the basin. The air-to-water ratio is significantly lower than in either the packed-column or cross-flow tower.

Chemical Oxidation

In chemical oxidation, the valence state of the treated compounds is raised. For example, in the conversion of cyanide to cyanate under alkaline conditions using chlorine, the state of the cyanide ion is raised as it combines with an atom of oxygen to form cyanate. This reaction can be expressed as follows:

$$NaCN + Cl_2 + 2NaOH \rightarrow NaCNO + 2NaCl + H_2O$$
$$2NaCNO + 3Cl_2 + 4NaOH \rightarrow 2CO_2 + N_2 + 6NaCl + 2H_2O$$

Common commercially available oxidants include potassium permanganate, hydrogen peroxide, calcium or sodium hypochlorite, as well as chlorine gas.

Chemical oxidation is used primarily for detoxification of cyanide and treatment of dilute waste streams containing oxidizable organics. Among the organics for which oxidative treatment has been reported are aldehydes, mercaptans, phenols, unsaturated acids, and certain pesticides.

Chemical oxidation can be an effective way of pretreating wastes prior to biological treatment. Compounds that are refractory to biological treatment can be partially oxidized, making them more amenable to biological oxidation.

Chemical Reduction

Chemical reduction involves addition of a reducing agent in order to reduce toxicity or solubility or to transform the substance to a form that can be more easily handled. For example, in the reduction of hexavalent chromium to trivalent chromium using sulfur dioxide, the valence state of chromium changes from 6^+ to 3^+ (reduction) and the valence state of sulfur increases from 4^+ to 6^+ (oxidation). The reduction takes place simultaneously with oxidation in chemically equivalent ratios.

$$2H_2CrO_4 + 3SO_2 + 3H_2O \rightarrow Cr_2(SO_4)_3 + 5H_2O$$

Commonly used reducing agents include sulfite salts, for example, sodium bisulfite, sodium metabisulfite, sodium hydrosulfite, sulfur dioxide; and base metals, for example, iron, aluminum, and zinc. Chemical reduction is used primarily for reduction of hexavalent chromium and other heavy metals.

SOLIDS SEPARATION

The objective of separating solids from slurries is to attain two distinct waste streams: a liquid waste stream that can be subsequently treated for removal of dissolved and fine suspended contaminants, and a concentrated slurry of solids and liquid that can be dewatered and treated.

Classification by grain size is important in managing hazardous waste–contaminated soils and sediments because of the apparent tendency of contaminants to adsorb preferentially onto fine-grained materials such as clay and organic matter. The separation of solids by grain size and level of contamination could prove to be extremely beneficial to the overall management, treatment, transport, and disposal of contaminated soil material. Whereas relatively noncontaminated soils and sediments may be disposed of in ordinary sanitary landfills or discharged back into the stream, the highly contaminated solids must be disposed in a hazardous waste landfill, incinerated, or treated to render them nonhazardous.

Many of the methods presented in this section are discussed in terms of their ability to treat soil slurries or sediments containing particles within a specific size range. Table 4-5 summarizes the particle sizes that correspond to various soil types.

Solids separation methods addressed in this section include sieves and screens, hydraulic and spiral classifiers, cyclones, settling basins, and clarifiers. Sieves or screens consist of bars, woven wire, or perforated plate surfaces that retain parti-

TABLE 4-5
PARTICLE SIZES FOR VARIOUS TYPES OF SOILS

Classification	U.S. standard sieve size	Particle size (µm)
Gravel		
Coarse	>3/4 in	
Fine	3/4 in. to No. 4	
Sand		
Coarse	No. 4 to No. 10	
Medium	No. 10 to No. 40	
Fine	No. 40 to No. 200	
Silt		10–74
Coarse clay		1.0–10
Fine clay		<1.0

cles of a desired size range while allowing smaller particles and the carrying liquid to pass through the openings in the screening surface. Several types of screens and sieves have application for solids separation at hazardous waste sites.

Grizzlies consist of frame-mounted parallel bars set on an angle to promote materials flow and separation. Hoppers are provided beneath the grizzly to collect removed material. Bars are generally spaced 1 to 5 inches apart, depending upon the desired separation. Both fixed and vibrating grizzlies are available. Grizzlies are generally 6 to 9 feet wide by 12 to 18 feet long. They are used primarily for scalping, that is, removing a small amount of oversized material from a stream that is predominantly fines. They are generally limited to separating materials ≥2 inches in diameter.

Grizzlies offer a reliable method for removing coarse-grained material from slurries, thus significantly improving the reliability and performance of subsequent solids separation methods and also reducing maintenance costs by minimizing the amount of abrasive material that reaches the screen, cyclone, etc. Grizzlies contain no moving parts and are tough and abrasion-resistant. Maintenance and space requirements are minimal. They can easily be arranged in series or parallel to accommodate very high flows or achieve classification of coarse materials.

Sedimentation employing impoundment basins and conventional clarifiers is a well-established technology for removing particles ranging in size from gravel to fine silt (Fig. 4-25). However, proper flocculation is essential to ensure removal of silt-sized particles. Circular clarifiers are generally highly efficient in solids removal and are widely used in wastewater treatment. However, rectangular tanks are more suitable for barge mounting and where construction space is limited. [14]

High-rate clarifiers use multiple "stacked" plates or trays to increase the effective settling surface area of the clarifier and decrease the actual surface area needed to effect settling. High-rate clarifiers allow a higher flow or loading rate per unit of actual surface area than do conventional clarifiers. The trays or plates also induce optimum hydraulic characteristics for sedimentation by guiding the flow, reducing short circuiting and promoting better velocity distribution. High-rate clarifiers are

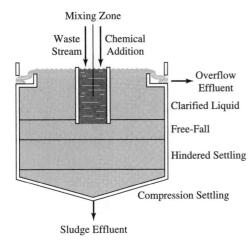

Mixing Zone

Waste Stream

Chemical Addition

Overflow Effluent

Clarified Liquid

Free-Fall

Hindered Settling

Compression Settling

Sludge Effluent

FIGURE 4-25
CENTER-FEED CLARIFIER.

able to handle 2 to 10 times the loading rate of conventional clarifiers and therefore require limited land use. Package units capable of handling 1000 to 2000 gal/min. are available and easily transportable by truck or barge. High-rate clarifiers are best suited to cleanup operations, where construction of earthen impoundments will not adequately protect groundwater, where land space is limited, or where barge mounting of clarifiers is required.

Impoundment basins have a high capital and operating cost. For this reason their use is generally limited to large-scale cleanup operations. Impoundment basins also pose potential for secondary impacts. Contaminants may leach into groundwater if the liner system is not properly designed, and the large surface area of the impoundment can result in volatilization of contaminants and localized air pollution problems.

Impoundment basins require a long setup time, and the need to obtain construction permits can further delay cleanup operations. Conventional and high-rate clarifiers eliminate some of the problems associated with impoundment basins. Clarifiers pose no threat to groundwater contamination. However, capital costs associated with the use of clarifiers can also be quite high for a large-scale waste treatment operation.

DEWATERING

Dewatering is used to reduce the moisture content of slurries or sludges in order to facilitate handling and prepare the materials for final treatment or disposal. Devices used to dewater slurries or sludges include gravity thickeners, centrifuges, filters, and belt presses. Selection of the most appropriate method depends on such factors as the volume of the slurry, solids content of the waste stream, land space availability, and the degree of dewatering required prior to treatment or disposal.

Although several of the dewatering methods are extremely effective in removing water, the solids are often not sufficiently dry to meet requirements for final disposal and require further treatment to fix or solidify the wastes. The contaminated water generated during dewatering generally contains hazardous constituents, as well as several hundred to several thousand mg/L suspended solids, and usually requires additional treatment.

Gravity thickening is generally accomplished in a circular tank, similar in design to a conventional clarifier. The slurry enters the thickener through a center feedwell designed to dissipate the velocity and stabilize the density currents of the incoming stream.

Centrifugal dewatering is a process that uses the force developed by fast rotation of a cylindrical drum or bowl to separate solids and liquids by density differences under the influence of centrifugal force. Dewatering is usually accomplished using solid-bowl or basket centrifuges.

The operation of the solid-bowl centrifuge is a continuous process. The unit consists of a long bowl, normally mounted horizontally and tapered at one end. Sludge is introduced into the unit continuously and the solids concentrate on the periphery. A helical scroll within the bowl, spinning at a slightly different speed, moves the accumulated sludge toward the tapered end, where additional solids concentration occurs prior to discharging the solids.

In the basket centrifuge, the flow enters the machine at the bottom and is directed toward the outer wall of the basket. Cake continually builds up within the basket until the centrate, which overflows a weir at the top of this unit, begins to increase in solids. At that point, feed to the unit is shut off, the machine decelerates, and a skimmer enters the bowl to remove the liquid layer remaining in the unit. A knife is then moved into the bowl to cut out the cake, which falls out the open bottom of the machine. The unit is a batch device with alternate charging of feed sludge and discharging of dewatered cake.

Centrifugation offers a simple, clean, and reliable method for dewatering sludges and other solids. It is less effective than filtration methods and dewatering lagoons, but more effective than gravity thickeners. Centrifuges are compact and well-suited to use in mobile treatment systems.

Filtration is a physical process whereby particles suspended in a fluid are separated from it by forcing the fluid through a porous medium. Three types of filtration are commonly used for dewatering: belt press, vacuum, and pressure filtration.

Belt filter presses employ single or double moving belts to continuously dewater sludges (Fig. 4-26). The belt-press filtration process includes three stages: chemical conditioning of the feed, gravity drainage to a nonfluid consistency, and dewatering. A flocculant is added prior to feeding the slurry to the belt press. In the next step, free water drains from the conditioned sludge. The sludge then enters a two-belt contact zone, where a second upper belt is gently set on the forming sludge cake. The belts, with the captured cake between them, pass through rollers. This stage subjects the sludge to continuously increasing pressures and shear forces. Progressively more and more water is expelled throughout the roller section to the end, where the cake is discharged. A scraper blade is often employed for the belt at the discharge point to remove the cake from the belt.

A vacuum filter consists of a horizontal cylindrical drum that rotates partially submerged in a vat of sludge (Fig. 4-27). The drum is covered with a continuous

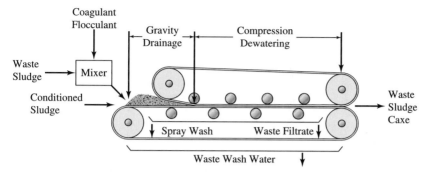

FIGURE 4-26
BELT PRESS FOR WASTE SLUDGE.

belt of fabric or wire mesh. A vacuum is applied to the inside of the drum. The vacuum causes liquid in the vat to be forced through the filter medium, leaving wet solids adhering to the outer surface. As the drum continues to rotate, it passes from the cake-forming zone to a drying zone, and finally to a cake-discharge zone, where the sludge cake is removed from the medium.

Pressure filtration is used to describe a category of filters in which rigid individual filtration chambers are operated in parallel under relatively high pressure. The filter press, the most common representative of the group, consists of vertical plates that are held rigidly in a frame and are pressed together (Fig. 4-28 on page 140). The liquid to be filtered enters the cavity formed by the frame. Pressed against this hollow frame are perforated metal plates covered with a fabric filter medium. The plate operates on a cycle that includes filling, pressing, cake removal, media washing, and press closing. As the liquid flows through the filter medium, solids are entrapped and build up within the cavity until the cavity is full. The slurry is dewatered until no filtrate is produced. The press is then opened, the de-

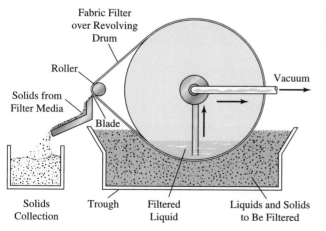

FIGURE 4-27
ROTARY-DRUM VACUUM FILTER.

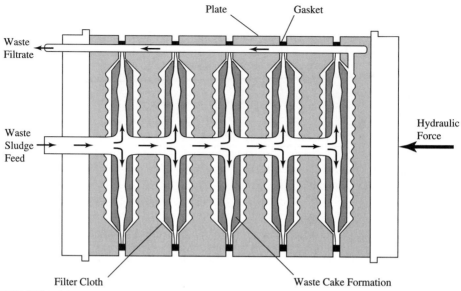

FIGURE 4-28
PLATE AND FRAME FILTER PRESS FOR WASTE SLUDGE.

watered slurry is removed, the plates are cleaned, and the cycle is repeated. In certain applications, the filter media is precoated with diatomaceous earth, fly ash, or other filter aids to improve performance.

Filtration can be used to dewater solids over a wide range of solids concentrations and particle sizes. Effectiveness for a particular application depends on the type of filter, the particle-size distribution, and the solids concentrations. For dewatering of municipal sludges, where considerable performance data are available, typical ranges for solids content and solids removal or capture are as follows:

	Solids content (%)	Solids capture (%)
Belt press filtration	15–45	80–95
Vacuum rotary filtration	15–40	88–95
Pressure filter	30–50	98

SOLIDIFICATION AND STABILIZATION

Solidification and *stabilization* are terms used to describe treatment systems that can

- Improve waste handling or other physical characteristics of the waste
- Decrease the surface area across which transfer or loss of contained pollutants can occur
- Limit the solubility or toxicity of hazardous waste constituents

The contaminants do not necessarily interact chemically with the solidification reagents but are mechanically locked within the solidified matrix. Contaminant loss is minimized by reducing the surface area. Stabilization methods usually involve the addition of materials that limit the solubility or mobility of waste constituents even though the physical handling characteristics of the waste may not be improved. Methods involving combinations of solidification and stabilization techniques are often used.

Cement-based solidification involves mixing the wastes directly with Portland cement, a very common construction material. The waste is incorporated into the rigid matrix of the hardened concrete. This method physically or chemically solidifies the wastes, depending upon waste characteristics. The end product may be a standing monolithic solid, or it may have a crumbly, soil-like consistency, depending upon the amount of cement added. Most hazardous wastes slurried in water can be mixed directly with cement, and the suspended solids will be incorporated into the rigid matrix.

Cement solidification is most suitable for immobilizing metals, because at the pH of the cement mixture most multivalent cations are converted into insoluble hydroxides or carbonates. However, metal hydroxides and carbonates are insoluble only over a narrow pH range and are subject to solubilization and leaching in the presence of even mildly acidic leaching solutions. Portland cement alone is also not highly effective in immobilizing organics.

Silicate-based processes refer to a broad range of solidification and stabilization methods that use a silicate material with lime, Portland cement, gypsum, and other suitable setting agents. Many of the available processes use proprietary additives and claim to stabilize a broad range of compounds from divalent metals to organic solvents. The basic reaction is between the silicate material and polyvalent metal ions. The silicate material that is added in the waste may be fly ash, bottom ash, or other readily available pozzolanic materials. Soluble silicates such as sodium silicate or potassium silicate are also used. The polyvalent metal ions that act as initiators of silicate precipitation may come from the waste solution. Portland cement and lime are most commonly used because of their availability and strength characteristics.

Vitrification at temperatures above 1300°C can be used to immobilize wastes in a glass monolith. Organics, heavy metals, and radioactive wastes have been managed with this process.

THERMAL DESTRUCTION OF HAZARDOUS WASTES

Thermal destruction is a treatment method that uses high-temperature oxidation under controlled conditions to degrade a substance into products that generally include CO_2, H_2O vapor, SO_2, NO_x, HCl gases, and ash. The hazardous products of the thermal destruction, such as particulates, SO_2 NO_x, HCl, and products of incomplete combustion, require air pollution control equipment to prevent release of undesirable species into the atmosphere. Thermal destruction methods can be used to destroy organic contaminants in liquid, gaseous, and solid waste streams. [11, 14]

The most common incineration technologies applicable to hazardous wastes include:

- Liquid injection
- Rotary kiln
- Fluidized bed
- Multiple hearth

The operating principles and general applications of these methods are summarized in Table 4-6.

A *liquid injection incinerator*, shown in Figure 4-29, consists of a refractory-lined combustion chamber and a series of atomizing nozzles. Two chamber systems are common. The primary chamber is usually a burner where combustible liquid and gaseous wastes are introduced. Noncombustible liquid and gaseous wastes are introduced downstream of the burner in the secondary chamber.

Liquid injection incinerators may be either horizontal or vertical units. A liquid waste has to be converted into a gas before combustion. The liquid is atomized by passing through the burner nozzles while entering the combustor. This is necessary to ensure complete evaporation and oxidation. If viscosity precludes atomization, mixing and heating or other means should be applied prior to atomization to reduce waste viscosity.

TABLE 4-6
COMMONLY USED INCINERATION TECHNOLOGIES IN WASTE TREATMENT

Type	Process principle	Application	Combustion temperature (°F)
Liquid injection	Wastes are atomized with high-pressure air or steam and burned in suspension	Liquids and slurries that can be pumped	1300–2500
Rotary kiln	Waste is burned in a rotating, refractory cylinder	Any combustible solid, liquid, or gas	1500–2500
Fluidized bed	Waste is injected into an agitated bed of heated inert particles; heat is efficiently transferred to the wastes during combustion	Organic liquids, gases, and granular or well-processed solids	1400–1800
Multiple hearth	Wastes descend through several grates to be burned in increasingly hotter combustion zones	Sludges and granulated solid wastes	1400–1800

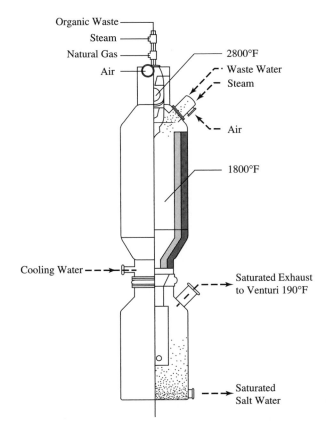

FIGURE 4-29
DOWN-FIRED
COMBUSTOR. (*Source:*
M. D. LaGrega, et al.,
*Hazardous Waste
Management.* Copyright ©
1994. Reprinted with
permission of The McGraw-
Hill Companies.)

The operating temperatures vary from 1300°F to 2500°F, with the most common temperature being about 1600°F. Residence times vary from less than 0.5 second to 2 seconds. The process usually requires 20 to 60 percent excess air to ensure complete combustion.

Liquid injection can be used to destroy virtually any pumpable waste or gas. Unlikely candidates for destruction include heavy metal wastes and other wastes high in inorganics. It does not need a continuous ash removal system, but requires air pollution control devices.

Liquid incinerators have no moving parts and require the least maintenance of all types of incinerators. The major limitations of liquid injection are its ability to incinerate only wastes that can be atomized in the burner nozzle and the burner nozzle's susceptibility to clogging. The process needs a supplemental fuel.

Liquid injection incinerators are highly sensitive to waste composition and flow changes. Therefore, storage and mixing tanks are necessary to ensure a reasonably steady and homogenous waste flow.

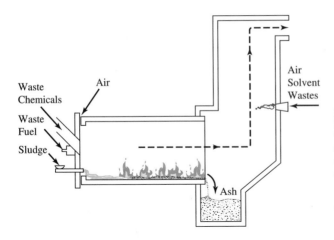

FIGURE 4-30
ROTARY KILN WITH
VERTICAL SECONDARY
COMBUSTION CHAMBER.
(*Source:* M. D. LaGrega, et
al., *Hazardous Waste
Management.* Copyright ©
1994. Reprinted with
permission of The McGraw-
Hill Companies.)

Rotary kiln incinerators are cylindrical, refractory-lined shells (Fig. 4-30). They
are fueled by natural gas or oil. Most of the heating of the waste is due to heat
transfer with the combustion-produce gases and the walls of the kiln. The basic
type of rotary kiln incinerator consists of the kiln and an afterburner.

Rotary kilns are capable of handling a wide variety of solid and liquid wastes.
Wastes are injected into the kiln and passed through the combustion zone as the
kiln rotates. The rotation creates turbulence and improves combustion. Rotary
kilns often employ afterburners to ensure complete combustion. Rotary kilns are
usually equipped with wet-scrubber air emission controls. The residence time and
temperature depend upon combustion characteristics of the waste. Combustion
temperatures range from 1500°F to 2500°F.

Rotary kilns are capable of burning waste in any physical form. They can incinerate solids and liquids independently or in combination and can accept waste feed
without any pretreatment preparation. Because of their ability to handle waste in
any physical form and their high incineration efficiency, rotary kilns are the preferred method for treating solid wastes. The limitations of rotary kilns include susceptibility to thermal shock, the need for good maintenance, and high capital cost
of installation.

Fluidized beds are also used for the disposal of solids and sludges (Fig. 4-31).
They are particularly well suited for incineration of high-moisture wastes, sludges,
and wastes containing large quantities of ash.

The advantages of the fluidized bed incinerator include simple design; lower,
more uniform temperatures; minimal NO_x formation; long life of the incinerator;
high efficiency; simplicity of operation; and relatively low capital and maintenance costs. It also has the ability to trap some gases in the bed, reducing the need
for a complex air emission control system. The disadvantages include difficulty in
removing residual materials from the bed, a relatively low throughput capacity,

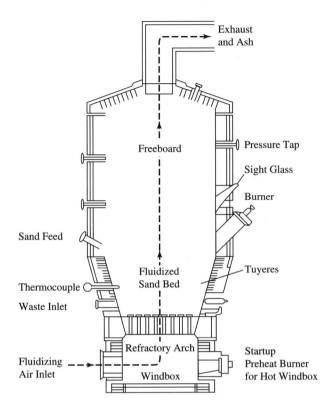

FIGURE 4-31
FLUIDIZED INCINERATOR.
(*Source:* M. D. LaGrega, et
al., *Hazardous Waste
Management.* Copyright ©
1994. Reprinted with
permission of The McGraw-
Hill Companies.)

the difficulty in handling residues and ash from the bed, and relatively high operating costs.

Multiple-hearth incinerators are used primarily for finely divided municipal trash and sewage sludge incineration (Fig. 4-32 on page 146).

AIR EMISSION CONTROLS

Air pollutants in waste incinerators include products of incomplete combustion and gases or particulate contaminates resulting from conversion of certain inorganic constituents. These air contaminants consist primarily of oxides of nitrogen, oxides of sulfur, and particulate matter. Trace levels of pollutants, such as chlorinated by-products, heavy metals, and acid gases, may also be of concern if they are not controlled.

Four types of air pollution equipment are employed for particulate control: electrostatic precipitators, venturi scrubbers, ionizing wet scrubbers, and baghouses. The selection procedure for these control devices depends on the inlet particulate

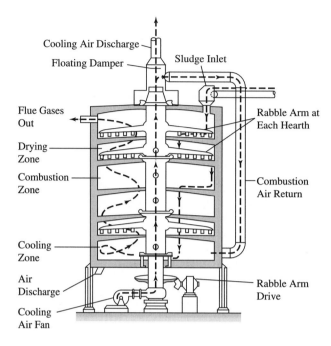

Cooling Air Discharge
Floating Damper
Sludge Inlet
Flue Gases Out
Rabble Arm at Each Hearth
Drying Zone
Combustion Zone
Combustion Air Return
Cooling Zone
Air Discharge
Rabble Arm Drive
Cooling Air Fan

FIGURE 4-32
MULTIPLE-HEARTH INCINERATOR. (*Source:* M. D. LaGrega, et al., *Hazardous Waste Management.* Copyright © 1994. Reprinted with permission of The McGraw-Hill Companies.)

loading, particle-size distribution, potential acid removal capability, gaseous control equipment, and regulatory requirements. [14]

BIBLIOGRAPHY

1 *Odor Thresholds for Chemicals with Established Occupational Health Standards,* American Industrial Hygiene Association, Akron, Ohio, 1989.
2 Buonicore, A. J., and W. T. Davis, *Air Pollution Engineering Manual: Air and Waste Management Association,* Van Nostrand Reinhold, New York, 1992.
3 Cralley, L. V., and L. J. Cralley, *In-Plant Practices for Job-Related Health Hazards Control,* Vol. 1: *Production Processes,* Wiley, New York, 1989.
4 Davis, M. L., and D. A. Cornwell, *Introduction to Environmental Engineering,* 2d ed., McGraw-Hill, New York, 1991.
5 Dravnieks, A., "Odor Perception and Odorous Air Pollution," *J TAPPI,* vol. 55, 1972.
6 Hunter, J. S., and D. M. Benforado, "Life Cycle Approach to Effective Waste Minimization," *JAPCA,* vol. 37, no. 10, 1987.
7 Moore, J. E., and R. S. O'Neill, "Odor as an Aid to Chemical Safety," *J Appl Toxicology,* 1983.
8 *Remedial Action at Waste Disposal Sites,* EPA/625, 6-85/006, U.S. Environmental Protection Agency, Cincinnati, Ohio, 1985.
9 Rich, L. A., "Solid Waste, An Overwhelming Problem Offers Business Opportunities," *Chemical Week,* vol. 143, no. 4, 1988.

10 Sullivan, R. J., *Preliminary Air Pollution Survey of Odorous Compounds,* U.S. Department of Health, Education and Welfare, Public Health Service, Raleigh, N.C., 1969.

11 Theodore, L., and J. Reynolds, *Introduction to Hazardous Waste Incineration,* Wiley, New York, 1987.

12 U.S. Environmental Protection Agency, *Compilation of Air Emission Factors,* AP-2, Research Triangle Park, N.C., 1988.

13 Water Pollution Control Federation, *Odor Control for Wastewater Facilities,* Washington, D.C., 1979.

14 Wentz, C. A., *Hazardous Waste Management,* 2d ed., McGraw-Hill, New York, 1995.

15 Whitehead, L., "Planning Considerations for Industrial Plants Emphasizing Occupational and Environmental Health and Safety Issues," *Appl Ind Hyg,* March 1987.

PROBLEMS

1 How do the diffusion rates of chlorine, ammonia, and dibutyl amine gases compare?

2 Determine the relative odor intensity for odorant A ($n = 0.3$) and odorant B ($n = 0.6$) when the concentration of each odorant decreases by a factor of 5.

3 Is the odor intensity of nitrobenzene or thiophenol more easily adjusted by air dilution?

4 Write a general stoichiometric air combustion equation for a typical hydrocarbon, C_xH_y.

5 Write a general air combustion equation for a typical hydrocarbon, C_xH_y, using 50 mol % excess air.

6 Bituminous coal with 2.0 wt % sulfur is burned with air by a utility company. The sulfur is 95 percent converted into SO_2 and 2.2 percent converted into SO_3. The remaining 2.8 percent of the sulfur is incorporated into the ash. Quantify the SO_2 and SO_3 generated per ton of coal burned.

7 A 10-wheel truck weighs 14 tons as it travels on an unpaved road at 45 mph. The road surface silt content is 2% ≤10 μm diameter particles. The location averages 25 precipitation days per year. What is the PM_{10} emission factor for this situation?

8 Bulk solids at a rate of 125 tons per day are being continuously dropped onto a belt conveyor. The solids have an average particle diameter of 25 μm with 32 wt % moisture content. The mean wind speed is 5 mph. What is the daily particulate emissions from the operations?

9 An organic waste has a BOD_3 of 35 mg/L and the rate constant is 0.13 per day. Estimate the ultimate BOD.

10 Why is it most desirable to eliminate the generation of wastes at the source?

11 Why will there never be unanimous approval and acceptance by local citizens for siting a new or expanded chemical processing facility in their locale?

12 Why is it important to involve the public of a local community early on in the siting decisions related to a new proposed processing plant?

13 Why is it generally more desirable to manage wastes on-site rather than to ship them off-site for management?

14 Describe a typical mass balance of a process related to the pollution prevention model.

15 Discuss relevant selections criteria that might be used in pollution prevention decision making.

16 How do economics and environmental regulations and laws often interrelate in hazardous waste management?

17 Discuss the steps you can take at a facility to minimize the generation of hazardous wastes.

18 What is the preferred hazardous waste treatment for acids and bases?

19 What are the advantages of biologically treating organic chemical wastes?

20 Discuss the advantages and disadvantages of using pure oxygen instead of air in biological treatment processes.

21 Discuss the role of pH in the separation of heavy metal hydroxides from aqueous waste streams.

22 Discuss the role of pH in the precipitation of heavy metal sulfides from aqueous waste streams. Contrast the effect of pH on the precipitation of hydroxides when compared to sulfides.

23 Why is it sometimes advantageous to polish the treated wastewater in a lagoon or basin prior to discharging the effluent?

24 Discuss the advantages and disadvantages of using sodium hydroxide versus calcium hydroxide to neutralize an acid waste stream.

25 Why is dewatering an important preliminary step in sending many sludges containing hazardous waste solids to the landfill?

26 Why are solidification and stabilization treatment processes often desirable for the treatment of heavy metal hazardous waste prior to final disposal?

27 Why is vitrification a preferred treatment process for nuclear solid waste prior to final disposal?

28 Why do rotary kilns have broad applicability in the incineration of hazardous wastes?

29 A new operation uses a hydrocarbon in the process system. Research shows that kerosene, No. 1 fuel oil, and gasoline are all promising. From a pollution prevention perspective, which hydrocarbons should be more strongly considered for the process?

30 Given the following average daily air emissions from an industrial process, what should be the priorities for pollution prevention?

Vinyl chloride	1 kg/d
Hexane	11 kg/d
Ethyl benzene	15 kg/d
Butane	25 kg/d

31 An industrial wastewater treatment plant is producing 160 kg of sludge per day. Given the following typical sludge analysis (wt %), is this sludge a hazardous waste? What are the opportunities for pollution prevention? Explain your answer.

Water	97.91
$Fe(OH)_3$	0.82
$Ni(OH)_2$	0.65
$Fe(OH)_2$	0.35
$Pb(OH)_2$	0.19
$Hg(OH)_2$	0.08

32 An industrial process uses cadmium and hexane raw materials. Are the wastes from this process hazardous? Why or why not?

33 A 25,000-m^2 lagoon has an average depth of 1.75 m; 650 m^3/d of 85 mg/L BOD_5 wastewater flows to the lagoon. There are three identical lagoons in series at the treatment facility. What is the combined retention time for the three lagoons and the BOD_5 loading?

34 Hazardous sludge from a treatment plant contains 1.75 wt % solids. It is drummed prior to disposal. A belt press could increase the sludge solids content to 23 wt %. Determine the annual sludge-volume reduction from the belt press if the sludge is generated at a rate of 11 m^3/d.

5

ELECTRICAL HAZARDS

The most versatile form of energy is electricity, which is widely used in our personal lives and in the workplace. We tend to take electricity for granted in the things we do and the many benefits we obtain from its wide availability. But electricity can also be hazardous to our health and well-being. The nervous system of the human body is sensitive to even low levels of electrical current. The improper use of electricity can result in bodily harm, fatalities, and property damage. Most potential electrical hazards can be controlled fairly easily and economically if safety procedures and control are introduced in the design stage.

ELECTRICAL TERMINOLOGY

The flow of electricity through a circuit is analogous to the flow of fluid through a pipe (see Fig. 5-1 and Table 5-1 on page 150). [7, 17] As the electricity flows through the conductor, it encounters resistance which causes heat to evolve. *Resistance* is measured in ohms (Ω). All materials have resistance. Good conductors, like copper and other metals and alloys, have low resistance. Poor conductors, like plastics, rubber, fabrics, and paper, have very high resistance and are used as insulators. All electrical circuits, machines, and appliances have conductors and insulators.

Electromotive force, or potential difference, is the force that makes electrons flow through a conductor. This electromotive force is called *voltage*. The difference in voltage between two points is measured in volts (V). Voltage between 24 and 600 V is potentially hazardous. [1]

Current, which is the quantity of electrons that passes through a conductor dur-

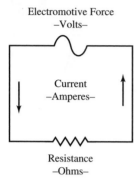

Electromotive Force
–Volts–

Current
–Amperes–

Resistance
–Ohms–

FIGURE 5-1
THE THREE PRINCIPAL CHARACTERISTICS OF AN
ELECTRICAL CIRCUIT.

ing a given time period, is measured in amperes (A). Electrical shock is normally measured in milliamperes.

$$E = IR$$

where E = voltage (V)
I = current (A)
R = resistance (Ω)

The resistance of an electrical conductor depends upon the dimensions and substance of the conductor. The resistance is directly proportional to the conductor length and inversely proportional to the cross-sectional area. The resistivity is the capacity of the conductor resistance at a given temperature and is based upon specified unit dimensions.

$$R = \rho \frac{L}{A}$$

where R = conductor resistance, Ω
ρ = resistivity, Ω.m
L = conductor length, m
A = conductor cross-sectional area, m^2

TABLE 5-1
THE ANALOGY BETWEEN FLUID FLOW AND ELECTRICAL
FLOW

Fluid flow	Electrical flow
Pressure	Voltage
Flow	Current
Friction	Resistance

As an illustration, copper conductors have a resistivity of 1.72×10^{-8} Ω.m or 1.72 mΩ.cm at 20°C. The resistivity of an aluminum conductor is 3.21 mΩ.cm at 20°C. The resistivity of insulators are very high by comparison, 5×10^{10} mΩ.cm for paper and about 1×10^{20} mΩ.cm for glass.

The resistance of a metallic conductor is usually directly proportional to increases in temperature. The change in the resistance of a conductor is usually based on the change from a 20°C reference value.

$$R = R_{20} \, (a + \alpha \Delta T)$$

where R = conductor resistance, Ω
R_{20} = conductor resistance at 20°C, Ω
α = temperature coefficient of resistance
ΔT = temperature change, °C

The temperature coefficient of resistance will vary with the conductor metal. For example, α for copper is 0.00393; for aluminum, it is 0.0038 and for carbon, it is –0.00025.

Resistors in an electrical circuit may be in series or in parallel. If the resistors are in series, the current has the same value throughout the circuit. In a parallel arrangement the current divides between the resistors, and the current in each branch is determined by its resistance. The parallel connection is normally used in electric lighting and power distribution so that a lamp or motor may be off or on and not disrupt the other operations of the circuit.

The sum of the series resistors of a circuit may be determined from the following:

$$R_{ser} = \sum_{i=1}^{n} R_i$$

The sum of the parallel resistors of a circuit may be determined with the following:

$$R_{par} = 1/ \sum_{i=1}^{n} (1/R_i)$$

The current in an electrical circuit is essentially the flow of electrons through it. The more electrons passing any point of a circuit in a given time, the greater the current. At a constant current,

$$q = It$$

where q = the charge or quantity of electricity passing a point of the circuit in coulombs (C)
I = current, A
t = time, s

An electron has a charge of 1.6×10^{-19} C, making 1 C equivalent to 6.24×10^{18} electrons.

For any given conductor with constant resistance, as the voltage increases, so does the current. In an analogous fashion, for any given voltage, the current increases as the resistance is lowered.

From a safety perspective, whenever the human body becomes an electrical conductor, any reduction in the resistance of the body to electrical flow can greatly increase the current flow to the body. It is the amount and duration of current flow to the body that determine the degree of hazard, not merely the magnitude of the voltage.

High voltage does not necessarily create great electrical danger. Many people are killed each year from 110-V and 220-V circuits, the most common electrical circuits in the United States.

There are other terms that are useful in the electrical field. The watt (W) is the unit of power and is what is required to do the work or perform the activity or operation. Power is the product of voltage times current.

$$P = EI$$

where P = power, W
$\quad\quad E$ = voltage, V
$\quad\quad I$ = current, A

Conductive parts or components are joined together by the process of bonding to form an electrical conductive path, which assures electrical continuity. Electrical equipment is grounded by connecting the equipment to earth or another conducting body by means of a ground.

ELECTRICAL INJURIES

The severity of electrical shock is based upon the current flow amount and duration, as shown in Table 5-2. Other factors include the parts of the body involved and the frequency of shock from alternative current exposure. In addition to electrical shock, heat generated can affect the body. [1]

The adverse effect of electrical shock will vary according to the individual, the setting, and the circumstances. These effects may vary based on whether the electrical source is direct current (DC) or alternating current (AC) (Table 5-3). Alternating current <100 mA is more hazardous than comparable direct current.

These values are approximate, and lower levels can be hazardous. An electrical shock may cause a person to lose control of all muscles. Prolonged exposure to electrical current is usually more serious than a brief momentary shock. Higher-frequency AC circuits usually generate heat rapidly, but have less shock severity. When an electric shock occurs, the results may be asphyxiation and partial paralysis of the lungs and diaphragm. When the heart has ventricular fibrillation with uncoordinated fluttering and spasms from an electrical shock >100 mA, a fatality usually results.

TABLE 5-2
ELECTRICAL CURRENT EFFECTS ON THE HUMAN BODY

Current (mA)	Effect on humans
3000	Heart stops Major burns to body tissue
700	Burns to body tissue
100	Ventricular fibrillation of the heart
70	Risk of death
50	Severe breathing
30	Severe shock
20	Difficult breathing
10	Muscle contractions
7	Painful electrical shock
3	Mild shock
1	Threshold of sensation

Source: National Safety Council.

There are a number of ways a person can receive an electrical shock (Table 5-4 on page 154). These shock sources all involve some part of the body coming into contact with an electrical source.

When a person receives an electrical shock, the body becomes a conducting path for the electric current. Electricity enters the body from a live source like a wire or metallic object. It is conducted through the body and leaves through another conducting surface, for example, a wet surface or other metallic object. The electrical path could be local, entering through one finger and leaving through another finger. It could also be general, entering through a hand and exiting through the feet. Some of the current may be conducted on the skin surface and some may be conducted internally. Electrical current will follow the path of least resistance or greatest conductivity. Moisture has a major effect on conduction of electricity. A

TABLE 5-3
ELECTRICAL SHOCK EFFECTS BASED ON THE TYPE OF CURRENT

Shock effects	Type of current	
	60 Hz AC (mA)	DC (mA)
Threshold of sensation	1	5
Threshold of muscle decontrol	6–9	70
Life threatening from heart and respiration failure	25	80
Threshold of heart fibrillation	100	100

Source: American Chemical Society.

TABLE 5-4
PRINCIPAL SOURCES OF ELECTRICAL SHOCK

A bare, live, energized conductor
A poorly insulated, live energized conductor
Failure of electrical equipment
Discharge of static electricity
Lightning strike

dry wooden floor is an insulator. The same floor, if wet, becomes an excellent conductor; the same is true for shoe soles. Hence the need to maintain dry conditions in and around electrical sources. The resistance of the human body and skin to electrical sources can be dramatic, as shown in Table 5-5. The degree of intimate contact with the incoming electrical source and the exit path also plays an important role in the electrical hazard. There have been many instances, for example, where dry gloves have prevented serious electrical shock when hands have accidently touched a live conductor such as a wire. [7]

In industrial accidents, people are usually electrocuted because they are careless near live conductors. Energized overhead electrical lines, enclosures with high-voltage equipment, shutdown electrical systems, and circuits with capacitors are hazardous and should be treated with care.

Normally electrical lines are insulated to protect people from coming in contact with bare conductors and for operating considerations. When the insulation becomes defective from damage or deterioration, a person could receive a shock. Any or all of the factors listed in Table 5-6 could create a hazardous electrical situation. Since most insulating compounds are plastic, rubber, fabric, or paper, the factors in

TABLE 5-5
ELECTRICAL RESISTANCE OF THE BODY

Part of the body	Resistance (Ω)
Dry skin	100,600–600,000
Dry finger of each hand on two electrodes	100,000
Moist finger of each hand on two electrodes	40,000
Wet (salt solution) finger of each hand on two electrodes	16,000
Tight grip of both dry hands on two electrodes	1,200
Wet skin	1,000
Tight grip of both wet (salt solution) hands on two electrodes	700
Internal body from hand to foot	400–600
Internal body from ear to ear	100

Source: National Safety Council; American Chemical Society.

TABLE 5-6
ENVIRONMENTAL CAUSES OF ELECTRICAL INSULATION DEGRADATION

Elevated temperature from current flow or ambient conditions accelerates degradation.
Moisture from humidity and operating conditions reduces resistance.
Oxygen or ozone causes oxidation.
Ultraviolet radiation from sunlight accelerates degradation.
Acids, bases, salt, and hydrocarbons cause chemical degradation.
Mechanical abrasion, crushing, bending, or cutting destroys insulation.
High voltage causes arcs, sparks, or the corona effect.
Rodents or other animals eat insulation.
Nonuniformity of the applied insulation can cause temperature differentials in the insulation.

Table 5-6 may affect a specific type of insulation in vastly different ways. It is advisable to avoid these environmental factors by whatever practical means in order to avoid an electrical hazard.

The electrical shock from current on the human body can cause injury or death because of many potential adverse health effects. The chest muscles may contract and interfere with normal breathing, causing asphyxiation. The nervous system may suffer temporary paralysis, causing breathing to cease. The normal rhythm of the heart may be disrupted, causing ventricular fibrillation. This occurs when the fibers of the heart muscles stop contracting and begin fibrillating, stopping blood circulation. At high current levels, muscle contractions of the heart cease, and the heat generated destroys tissues, nerves, and muscles, causing hemorrhaging. Burns can also result from lower currents with prolonged exposure. Usually industrial electrical accidents involve current flow from the hands to the feet, with the electrical path passing through both the heart and the lungs.

Electrical shock can stop the heart and lungs from functioning. Cardiopulmonary resuscitation (CPR) should be applied immediately to a victim of electric shock. The sooner CPR is applied, the better the probability of revival.

Eye and skin burns can occur from electrical flashes from the arcing of high voltages. The explosive violence of intense arcing may be caused by any of the following industrial situations: [1]

- Short-circuiting between bus bars or cables with high current loads
- Knife switch failure
- Opening knife switches under high current load
- Pulling fuses from energized circuits

When someone receives a shock from an electrical source, muscle contractions usually cause the worker to lose balance and fall. A fall from one level of height to a lower level under these circumstances could be even more hazardous than the electrical exposure by itself.

ELECTRICAL EQUIPMENT

Selection of electrical equipment should follow established codes and standards, which are based upon the type of intended service. A number of organizations have been involved in developing these codes and standards:

National Electrical Code (NEC)
National Electrical Safety Code (NESC)
National Fire Protection Association (NFPA)
American National Standards Institute (ANSI)
Code of Federal Regulations (29 CFR 1910, subpart S)
Underwriters Laboratories (UL)

There are also numerous state and local codes and standards.

HAZARDOUS ELECTRICAL ENVIRONMENTS

Electrical energy is an ignition source if sufficient oxygen is present in hazardous proportions and the temperature is equal to or greater than the flash point of the flammable mixture. Standard electrical apparatus, which is considered safe in ordinary environments, should not be installed in locations where flammable gases, vapors, dusts, and other easily ignitable materials are present. Electric sparks and arcs have caused many fires and explosions.

For any given location the flammable materials should be characterized prior to selection of electrical equipment, fittings, and wiring. Hazardous locations are classified based upon the properties of flammable gases, vapors, and liquids, or combustible dusts of fibers present in the environment. These hazardous locations are based upon the National Electrical Code (see Table 5-7). [18] This classification method is based upon the nature and degree of environmental hazards within a particular area. The requirements for electrical equipment and wiring for all voltages in locations where fire and explosion hazards may exist due to the presence of flammable gas, vapor, or liquid, combustible dust, or ignitable fibers or flyings are based upon the classifications in Table 5-7. [14] These electrical requirements are covered in the NEC of the NFPA 70, articles 500–504. [4, 12]

ELECTROSTATIC HAZARDS

Everyone has received an electrical shock from static electricity, by touching a doorknob after walking on a carpet. Static discharges are common when removing a garment. Usually these occur in the winter, when the air in heated buildings is dry. This phenomena is enhanced by low humidity, which permits the resistance of insulating surfaces to rise to a high level, allowing electrostatic charge accumulation to occur.

The accumulation of an electrostatic charge is caused by ionized impurities with an excess of positive or negative particles. Trace amounts of impurities, well below detectable limits, can cause electrostatic charging. For example, a high electrical

TABLE 5-7
NATIONAL ELECTRICAL CODE HAZARD CLASSIFICATION

Class and group	Division 1	Division 2
I: Flammable gases, vapors, and liquids Group A: acetylene Group B: hydrogen, butadiene, ethylene oxide Group C: ethylene, carbon monoxide, hydrogen sulfide Group D: gasoline, LPG, methane, acetone	May be present	Not normally present; may exist accidentally
II. Combustible dusts Group E: aluminum, magnesium Group F: carbon black, coal Group G: grain, food, plastics	May be present	Not normally present; may exist accidentally
III. Ignitable fibers of flyings Examples: textiles, woodworkings	Handled, manufactured, or used	In storage

Source: NFPA 70, Article 500 (1990).

charge, 1000 µC/m^3, requires an excess of only 1 ionized particle per 1.6×10^{12} molecules of hydrocarbon. [2]

The charging process takes place at an interface between dissimilar materials: metal and air, hydrocarbon and water, or metal and hydrocarbon. The charge separation occurs with molecules at the interface. So long as the dissimilar materials remain in contact, this charge separation is not significant, well below 1 V. When the charges are separated by moving the materials apart, the electrical potential difference rises. [10] A potential of 1 to 3 kV is not uncommon with sufficient separation.

Static electricity accumulation and discharge is a common ignition source in industry. Understanding the fundamentals of static charge is the best way to prevent its accumulation or at least recognize its potential buildup and subsequent control.

Whenever two dissimilar materials come in contact, electrons move from one surface to the other. As these materials are separated and more electrons remain on one surface than the other, one material takes on a positive charge and the other a negative charge. When one or both of the materials is a relatively poor conductor, the trapped electrons cause the charge buildup. [3, 21]

Mass transfer operations such as fluid flow of a liquid through a pipe or from an opening into a tank, the agitation and mixing of two immiscible liquids, the pneumatical conveying of solids, or filtration and screening are capable of causing static electricity and the buildup of large voltages. When these voltages discharge to a ground, the resulting spark can ignite flammable or explosive materials. Employees may suffer severe injury from reaction to the shock of a static spark.

The magnitude of the charge accumulation depends upon the nature of the material and the area and contact of the surfaces. Friction is not necessary to generate static electricity, but friction increases the electron release and the generation of ionized particles. [20]

The moisture level present in the ambient environment is inversely proportional to the generation of electric potentials. As the relative humidity decreases to ≤20 to 35 percent, static electricity can become significant. At humidity levels >65 percent, the air conductivity level increases to reduce the buildup of static electricity. [8]

For the flow of hydrocarbon liquids in pipelines, the pipe wall adsorbs some of the opposite-sign charges that are in the liquid adjacent to the interface (Fig. 5-2). When the flow begins, the charge in the liquid is carried away with the liquid flow and distributed throughout the fluid volume by turbulence. [6] The charge carried in the downstream liquid is called the streaming current.

$$I = 3.75 \times 10^{-6} \, (VD)^n$$

where I = streaming current, A
V = fluid velocity, m/s
D = inside pipe diameter, m
n = exponent, 1.9

Other studies have suggested slightly different constants and exponents. [2]

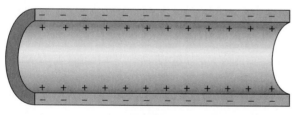

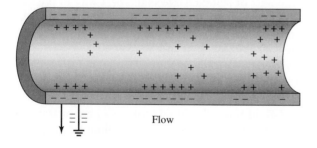

No Flow

Flow

FIGURE 5-2
PIPELINE ELECTRICAL CHARGE SEPARATION IN FLUID FLOW. (*Source: Occupational Safety Management and Engineering* by Hammer, Willie. Reprinted by permission of Prentice-Hall, Inc., Upper Saddle River, NJ.)

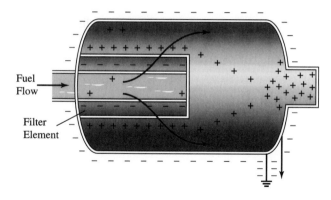

FIGURE 5-3
ELECTRICAL CHARGE
SEPARATION IN FLUID
FLOW THROUGH A
FILTER. (*Source:
Occupational Safety
Management and
Engineering* by Hammer,
Willie. Reprinted by
permission of Prentice-Hall,
Inc., Upper Saddle River,
NJ.)

Fuel
Flow

Filter
Element

Microfiltration of jet fuel to assure a clean fuel supply can generate much higher charge levels than a simple pipeline flow (Fig. 5-3). This results from all of the fluid being brought into intimate contact with the filter surface, where the charge separation occurs. For comparison, pipeline flow may produce 1 to 10 $\mu C/m^3$, while a microfilter may generate a charge as much as a factor of 100 higher. Additives are often used to reduce the charge. [19]

When there is spraying or splashing above the liquid surface as a liquid enters the tank, a charged mist may form (Fig. 5-4). A charged foam floating on the liquid surface also may be produced. [9]

Suspended particles or water droplets are normally charged. When they settle in a tank, it is possible to generate a significant sedimentation potential at the liquid surface.

When a charged fuel enters a metal tank, an equal but opposite charge is attracted to the inside of the tank shell (Fig. 5-5 on page 160). Simultaneously, an equal and same charge as the fuel is repelled to the outside tank surface. If the tank is properly grounded, this repelled charge is neutralized (Fig. 5-6 on page 160).

The accumulation of a static charge will become an ignition source only when

FIGURE 5-4
SPLASH FILLING OF A TANK TO GENERATE
STATIC ELECTRICITY.

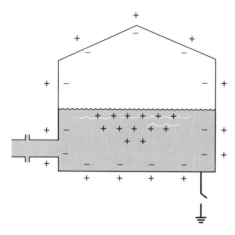

FIGURE 5-5
ELECTRICAL CHARGE INDUCTION IN
STORAGE TANK SHELL.

there is an electrical discharge capable of igniting a flammable mixture. There are two types of static electricity discharge, either spark or corona discharges.

A *spark discharge* is a nearly instantaneous passage of the static charge across an air gap between two electrodes. The spark channel extends completely across the gap and has a narrow diameter. Usually both electrodes are good conductors. If one of the electrodes is an insulator, the probability of a spark is greatly reduced because the rapid charge movement into the spark is impeded by the low conductivity of the insulator.

Corona discharges begin from a point source like a metal electrode and diffuse outward, either ending in space or spreading over a poor conductor. Corona discharges have a lower current and longer duration than spark discharges. Usually the energy released from a corona discharge is too diffuse and slow to become an ignition source.

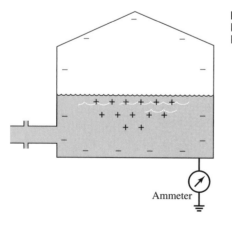

FIGURE 5-6
EFFECT OF GROUNDING AN INDUCED
POSITIVE CHARGE.

TABLE 5-8
MINIMUM IGNITION ENERGY FOR HYDROCARBON GASES

Hydrocarbon gas	Minimum ignition energy (mJ)
Hydrogen	0.017
Acetylene	0.017
Ethylene	0.08
Propane	0.25
Methane	0.29

The ignition of a flammable mixture by discharge of static electricity depends upon the release of sufficient energy. The minimum energy capable of igniting typical hydrocarbon gases is shown in Table 5-8. Hydrogen, acetylene, and ethylene require less ignition energy than saturated hydrocarbons. The electrical charge accumulation on an ungrounded conductor is a major safety risk because a conductor can discharge the stored electrical energy instantaneously as a spark.

The stored electrical energy on a conductor can be determined from the following:

$$J = \frac{CV^2}{2} = \frac{Q^2}{2C} = \frac{QV}{2}$$

where J = accumulated energy in joules (J)
C = capacitance in farads (F)
V = voltage, V
Q = electrical charge, C

The capacitance of many common objects in the workplace may range widely from 5 to 1000 pF, as shown in Table 5-9.

TABLE 5-9
TYPICAL CAPACITANCES FOR COMMON OBJECTS

Object	Capacitance (pF)
Hand tools	5
Buckets	20
Drum	100
Human	200
Automobile	500
Tank truck	1000

BONDING AND GROUNDING

The static electrical charge between two or more objects can be eliminated by bonding. The purpose of grounding is to eliminate a potential difference between an object and the ground of the earth. The principles of bonding and grounding two conductive bodies are shown in Figure 5-7.

When a charged body is insulated from ground, it has an electrical charge and voltage. An uncharged body insulated from ground has no electrical charge and voltage. When these insulated bodies are bonded together, the charge is shared and they have no potential difference. When a ground wire is added to one of the

FIGURE 5-7
PRINCIPLES OF BONDING AND GROUNDING TWO CONDUCTIVE BODIES. (*Source:* Used by permission of the National Safety Council, Itasca, Illinois.)

CHARGED AND UNCHARGED BODIES INSULATED FROM GROUND

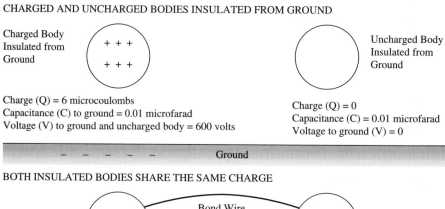

Charged Body Insulated from Ground

+ + +

+ + +

Uncharged Body Insulated from Ground

Charge (Q) = 6 microcoulombs
Capacitance (C) to ground = 0.01 microfarad
Voltage (V) to ground and uncharged body = 600 volts

Charge (Q) = 0
Capacitance (C) = 0.01 microfarad
Voltage to ground (V) = 0

Ground

BOTH INSULATED BODIES SHARE THE SAME CHARGE

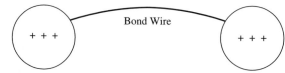

Bond Wire

+ + +

+ + +

Charge (Q) on both bodies = 6 microcoulombs
Capacitance (C) to ground for both bodies = 0.02 microfarad
Voltage (V) to ground = 300 volts

Both Bodies Bonded Together Will Share the Charge and Have No Potential Difference

Ground

BOTH BODIES ARE GROUNDED AND HAVE NO CHARGE

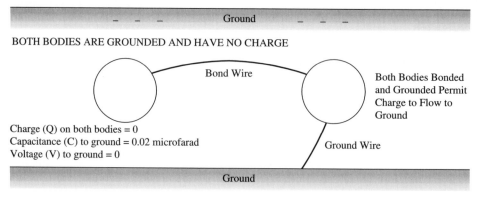

Bond Wire

Charge (Q) on both bodies = 0
Capacitance (C) to ground = 0.02 microfarad
Voltage (V) to ground = 0

Both Bodies Bonded and Grounded Permit Charge to Flow to Ground

Ground Wire

Ground

bonded bodies, the electrical charge flows to ground and neither body has an electrical charge or voltage. This bonding and grounding technique is effective only when both of the bonded objects are conductive materials. [1]

Bonding does not eliminate the static charge but will equalize the potential between the bonded objects so that a spark will not occur between them. Bonding will not eliminate a potential difference between the objects and the ground of the earth unless one of the objects has a conductive path to earth. Only an adequate conductive ground will continuously discharge a conductive body with an electrical charge.

Many electrical instruments and appliances are grounded. The case, or housing, is electrically connected to a special conductor to ground. Two of the three prongs on a three-prong plug carry the current, with the third connecting the case to ground. The electrical outlet into which the plug is inserted must have the corresponding three holes. A three-prong plug should not be bent or mutilated to fit a two-hole outlet. When using an adapter to ground the third prong independently from the two-hole outlet, be certain the ground wire of the adapter is securely grounded to the screw of the receptacle. Devices with the Underwriters Laboratories (UL) label can be depended upon to be safe for their intended use so long as they are not abused or worn to the point of becoming hazardous. The UL label should be on the appliance or instrument itself, and not merely on the electrical cord or plug. [7]

When flammable liquids are transferred from one container to another, a means of bonding is needed between the two conductive containers prior to pouring (Fig. 5-8 on page 164). Any static electricity generated from the filling process will be safely discharged. Providing a wire bond between the storage container and the container being filled will avoid a spark from the discharge of static electricity. Any other metallic connection path between the containers is acceptable, as long as it is secure and reliable. Some materials, like some plastics, cannot be as effectively bonded and should be used with caution. [1]

When a flammable-liquids tank car or truck is being loaded, the loading rack should be bonded and grounded as shown in Figure 5-9 on page 165. The downspout for loading liquids through open domes should be long enough to reach the tank bottom to avoid an electrostatic charge from fluid or splashing during the filling operation. The loading lines should be bonded to prevent a difference in static charge potential between the fill pipe and the tank car or truck. Fill pipes should be in constant contact with the rim of the tank opening. [5, 15, 16]

Above-ground storage tanks for flammable liquids that are on concrete or nonconductive supports should be grounded. The ground wires should be bare and not insulated so they may be visually inspected for damage. [11]

When liquids are transferred through or into piping, filters, or tanks, proper bonding and grounding usually drains off the static charge to ground as it is generated. Rapid flow rates can result in high electrical potentials on the surface of liquids, particularly for petroleum products, which are poor conductors, or hydrocarbons with moderate volatility, like ethanol.

This high static charge can be fairly well controlled by avoiding violent splash-

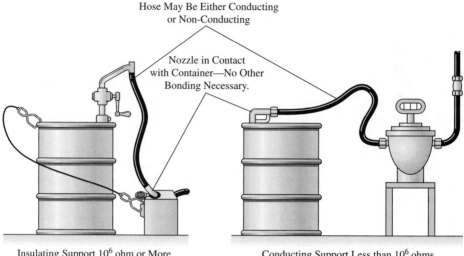

Hose May Be Either Conducting
or Non-Conducting

Nozzle in Contact
with Container—No Other
Bonding Necessary.

Insulating Support 10^6 ohm or More   Conducting Support Less than 10^6 ohms

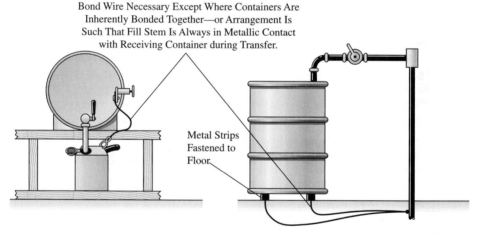

Bond Wire Necessary Except Where Containers Are
Inherently Bonded Together—or Arrangement Is
Such That Fill Stem Is Always in Metallic Contact
with Receiving Container during Transfer.

Metal Strips
Fastened to
Floor.

FIGURE 5-8
BONDING OF CONTAINERS DURING THE FILLING PROCESS BASED UPON NFPA
RECOMMENDATIONS.

ing during filling operations or by allowing sufficient relaxation time for the static
to at least partially dissipate. After filling a tank it is usually prudent to wait before
gauging the tank. The nature of the product, fill rates, and tank size affect the desirable lag time, up to 4 h for larger tanks.

Motor frames, control boxes, conduits, and switches should be bonded during
installation, as recommended by the equipment manufacturer. These recommendations should be consistent with the NEC in NFPA 70. [13]

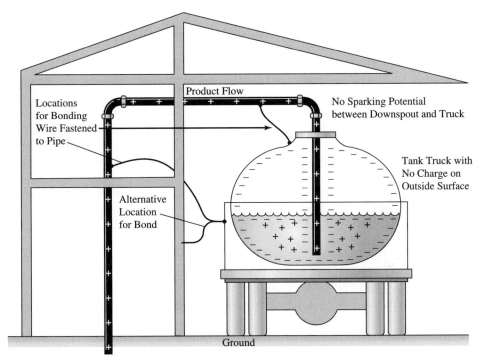

FIGURE 5-9
BONDING AND GROUNDING OF TANK TRUCK TO LOADING RACK. (*Source:* Used by permission of the National Safety Council, Itasca, Illinois.)

The bonding and grounding systems that have been installed should be regularly checked for electrical continuity. Corrosion, abrasion, and other deterioration or damage by careless personnel are usually the cause of bonding and grounding failure. Bare-braided flexible wire makes the best bonding and grounding components because they make for easy inspection and prevent concealed broken conductor wires.

Further discussion related to engineering and management controls will be found in Chapter 12.

FIRST AID FOR ELECTRICAL SHOCK

The shock of an electric current entering and leaving the body can knock a person down, cause unconsciousness, or stop breathing and heartbeat. In passing through tissues the current may cause widespread damage, even though the visible skin marks where the current entered and exited may be hardly noticeable.

When the victim of an electrical shock loses control of practically all muscles, that person needs instantaneous help. The person trying to help the victim must act properly to avoid being electrocuted also. Do not immediately touch the victim.

The victim first must be disconnected from the shock source. If possible, the best way to accomplish this is to switch off the current source. If the current switch is not convenient or practical, a live-wire source can be knocked off, dragged aside, or otherwise separated with a dry board, broom, wooden chair, or other well-insulated object. The victim must be rolled away from the current source, if the source cannot be removed. Until this is accomplished, the victim should be considered electrically live, and anyone trying to give first aid will receive a shock. Someone who has been struck by lightning is not electrically live and should be given first aid right away.

Once the victim has been freed from the electrical source, the unconscious victim needs immediate first aid. After you are certain the scene is safe for you to proceed, check the victim. Physically touch the victim and ask if he or she is all right.

If there is no response from the victim, call 911 or your local emergency phone number and request professional emergency help. Then go back to the victim; open the airway and look, then listen and feel for breathing.

If the victim is breathing, just make him or her comfortable until help arrives. If the adult victim is not breathing, give two breaths and check for a pulse.

If the victim has a pulse but is not breathing, start mouth to mouth resuscitation and give an adult one breath every 5 seconds until help arives.

If there is no pulse, begin chest compression. Continue at a rate of about 4 cycles per minute for adults. A CPR cycle consists of 15 chest compressions in 10 seconds and two breaths. Once you begin the CPR procedure you must continue it until professional help arrives, the victim begins breathing, you become exhausted, or the scene becomes unsafe for you.

There is an obligation on you to continue to provide CPR care, once you begin. You are not obligated to begin providing care, however. These guidelines must be followed in the United States.

If the victim is faint or pale or shows other signs of shock, lay the person down with the head slightly lower than the body trunk and elevate the legs. Treat any burns and wait for professional help to arrive. Be sure the victim is checked for internal injuries or bleeding even if the electric current exposure was slight.

BIBLIOGRAPHY

1 *Accident Prevention Manual for Business and Industry: Engineering and Technology,* National Safety Council, Itasca, Ill., 1992.
2 Bustin, W. M., and W. G. Dukek, *Electrostatic Hazards in the Petroleum Industry,* Research Studies Press, Letchworth, Hertfordshire, England, 1983.
3 Crowl, D. A., and J. F. Louvar, *Chemical Process Safety Fundamentals with Applications,* Prentice-Hall, Englewood Cliffs, N.J., 1990.
4 Deuschle, R., and F. Tiffany, *Barrier Intrinsic Safety, Fire Protection Manual,* vol. 2: *Hydrocarbon Processing,* Gulf Publishing, Houston, 1981.
5 Flammable and Combustible Liquids, NFPA 30.
6 Gavis, J., and H. E. Hoelscher, *Phenomenological Aspects of Electrification, Fire Protection Manual,* vol. 1: *Hydrocarbon Processing,* Gulf Publishing, Houston, 1985.

 7 Hajian, H. G., and R. L. Pecsok, *Working Safely in the Chemistry Laboratory,* American Chemical Society, Washington, D.C., 1994.
 8 Hammer, W., *Occupational and Environmental Safety Engineering and Management,* Van Nostrand Reinhold, New York, 1990.
 9 Herzog, R. E., et al., *Evaluating Electrostatic Hazard during the Loading of Tank Trucks, Fire Protection Manual,* vol. 1: *Hydrocarbon Processing,* Gulf Publishing, Houston, 1985.
10 Howard, J. C., *The Hazards of Static Electricity, Fire Protection Manual,* vol. 1: *Hydrocarbon Processing,* Gulf Publishing, Houston, 1985.
11 Kavianian, H. R., and C. A. Wentz, *Occupational and Environmental Safety Engineering and Management,* Van Nostrand Reinhold, New York, 1990.
12 Krigman, A., and R. Redding, *Test for Intrinsic Safety, Fire Protection Manual,* vol. 2: *Hydrocarbon Processing,* Gulf Publishing, Houston, 1981.
13 Mattson, R. E., *Electrical System Safety and Reliability, Fire Protection Manual,* vol. 2: *Hydrocarbon Processing,* Gulf Publishing, Houston, 1981.
14 Northrup, R. P., *Explosion-Proof Electrical Systems, Fire Protection Manual,* vol. 1: *Hydrocarbon Processing,* Gulf Publishing, Houston, 1985.
15 *Protection against Ignitions Arising out of Static, Lightning, and Stray Currents,* API RP2003, Washington, 1982.
16 *Static Electricity,* NFPA, Quincy, Mass., 1977.
17 Tenneco Oil Co., *Electricity in the Plant, Fire Protection Manual,* vol. 1: *Hydrocarbon Processing,* Gulf Publishing, Houston, 1985.
18 29 CFR 1910, OSHA, Washington, D.C., July 1, 1992.
19 Warren, J. H., et al., *Conductivity Additives Are Best, Fire Protection Manual,* vol. 2: *Hydrocarbon Processing,* Gulf Publishing, Houston, 1981.
20 Wright, L., and I. Ginsburgh, *What Experimentation Shows about Static Electricity, Fire Protection Manual,* vol. 1: *Hydrocarbon Processing,* Gulf Publishing, Houston, 1985.
21 Wright, R. V., *Fundamentals of Static Electricity, Fire Protection Manual,* vol. 1: *Hydrocarbon Processing,* Gulf Publishing, Houston, 1985.

PROBLEMS

 1 List the electrical applications in your home.
 2 How much DC current does an electric iron with 30 Ω of resistance take when connected to 110-V service? What is its power rating in watts?
 3 What is the body resistance to a 30-mA current being driven through a human body by a standard AC voltage supply of 240 V?
 4 What is the resistance of a 0.1-cm diameter conductor wire that is 20 m in length?
 5 An aluminum conductor wire is heated from 20°C to 150°C. What is the percentage increase in the conductor resistance?
 6 What is the resistance of a 110-V electrical circuit that draws a current of 4 A?
 7 What is the total electrical circuit resistance of three series resistors that are 5, 8, and 12 Ω, respectively?
 8 What is the total electrical circuit resistance of three parallel resistors that are 5, 8, and 12 Ω, respectively?
 9 A metal bucket is lowered on an insulated rope into a tank filled with fuel oil. The elec-

trostatic voltage at the fuel oil surface is 8000 V. What is the probability of an ignition spark from this activity?

10 What is the streaming current generated by liquid hydrocarbon flow at 8 m/s in a 0.5-m inside diameter pipe?

11 Why is liquid hydrocarbon flow through a microfilter more likely to produce a high electrical charge than the same flow through a pipeline?

12 A dry finger with a resistance of 100,000 Ω accidentally comes in contact with a 220-V electrical source. A wet finger with a resistance of 16,000 Ω accidentaly comes in contact with the same 220-V electrical source. Discuss the shock effects of these two situations.

13 What is the approximate range of electrical current shock that may cause death in humans?

14 Why is a single finger touching an electrical source less hazardous than both hands tightly gripping the same electrical source?

15 What are the principal sources of electrical shock?

16 What effect does moisture have on conduction of electrical current?

17 Discuss the adverse health effects caused by electrical shock.

18 Discuss how the flow of hydrocarbon fluids in a pipeline can generate a static electric charge.

19 Discuss how to bond and ground two steel drums that contain flammable liquids.

20 Discuss how to bond and ground two metal containers being used to transfer flammable liquids.

21 Why is relaxation time important after filling a hydrocarbon storage tank prior to gauging the tank?

22 Discuss the recommended first aid procedures for electrical shock.

FIRE AND EXPLOSION HAZARDS AND PROTECTION

Fire, which is a combustion process, is a complex chemical reaction between fuel, oxygen, and ignition sources. This exothermic reaction could involve a variety of fuels that may be in gas, vapor, liquid, or solid states of matter. The liquid and solid fuels normally vaporize before burning. These gases and vapors mix with oxygen or air to form the ignitable mixture. As these fuels are oxidized there is an emission of heat. If the combustion process results in a rapid pressure rise, an explosion may occur.

HEAT OF COMBUSTION AND HEAT TRANSFER

The typical heat of combustion for many fuels is shown in Table 6-1 on page 170. Generally, hydrocarbon gases like methane and ethane have higher heats of combustion than coal or hydrocarbon liquids. Hydrogen has a high heat of combustion. The ratio of hydrogen to carbon in the molecules is the critical factor in their heat content. The higher this ratio, the higher the heat content of the substance. Generally, the more hydrogen present in the fuel relative to the carbon available, the hotter the fuel will burn.

There are three heat transfer mechanisms. The mechanism of heat transfer for thermal conduction through homogeneous solids can be expressed by

$$Q = -kA\left(\frac{dt}{dx}\right)$$

where Q = rate of heat conduction along the x axis
A = cross section of the path normal to the x axis

TABLE 6-1
HEAT OF COMBUSTION FOR COMMON FUELS

Fuel	Heat of combustion (Btu/lb)
Coal	12,000–13,000
Hydrocarbon liquids	17,000–20,000
Hydrocarbon gases	20,000–23,000
Hydrogen	60,000

$-dt/dx$ = temperature gradient along that path
   k = thermal conductivity, which is a physical property of the substance

Heat loss by the conduction mechanism will increase proportionally with the refractory surface area and the incineration temperature. Heat loss by conduction can
be significantly decreased with improved insulation, for instance, with increased
thickness of refractory brick.

Heat transfer by convection is related to the properties of the convection
medium and the geometry of the system. The generation equation of the convection of heat is given by

$$Q = hA \, (T_2 - T_1)$$

where Q = heat transmission by convection
   h = film coefficient of heat transfer
   A = area of heat transfer surface
   T_1 = ambient temperature
   T_2 = temperature at the interface

Whereas conduction and convection heat transfer are affected primarily by temperature differences, the radiation mode of heat transfer increases with the level of
temperature. As a result, at low temperature levels conduction and convection represent a major contribution to heat transfer, while at higher temperatures radiation
becomes the controlling factor. When a particle is heated, radiant energy is emitted
at a rate that is primarily dependent upon the temperature of the particle and its
color and texture. Heat loss due to radiation in degrees Rankine (°R) can be expressed by the following equation:

$$Q = 0.174 \, A\varepsilon \left[\left(\frac{T_2}{100} \right)^4 - \left(\frac{T_1}{100} \right)^4 \right]$$

where Q = quantity of heat transferred
   ε = emissivity, which varies with color and texture of the particles

A = area of the heat transfer surface
T_1 = absolute temperature of the lower temperature element, °R
T_2 = absolute temperature of the higher temperature element, °R

Radiation is the most important mechanism for heat transfer in a fire because at these elevated temperatures the rate of heat transfer by radiation is proportional to the fourth power of the absolute temperature. Convection heat transfer, which takes place because of the motion of fluids, and conduction heat transfer, which involves heat being transferred from one molecule to another, are also important heat transfer mechanisms in considering fire safety.

FIRE TRIANGLE

The fuel must be present within desirable concentration limits for combustion. The oxygen supply must be above required minimum concentration levels. The ignition source must be above minimum temperature or energy levels. Together these parameters constitute the fire triangle, shown in Figure 6-1. The three elements of fuel, oxygen, and ignition are necessary to ignite ordinary burning and fires. They are interrelated by the fire triangle, which is used to create fires. The fire triangle concept is also used to prevent fires or control them once they have begun. [2, 17]

When a fuel substance is heated to its ignition temperature, it will ignite and continue to burn as long as fuel remains at the proper temperature and sufficient oxygen is present. Ignition temperatures are affected by the composition of the fuel-air mixture, the temperature and duration of the ignition source, and the ambient surroundings. The ignition temperature of a solid is also affected by the size of the solid, the rate of air flow, and the rate of heating.

FIRE TETRAHEDRON

The fire triangle contains fuel, oxygen, and an ignition source. These three components must be present for a fire to ignite. As long as sufficient quantities of fuel are present and the rest of the fire triangle is intact, the fire can be sustained.

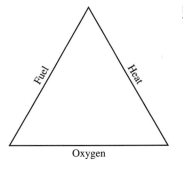

FIGURE 6-1
THE FIRE TRIANGLE.

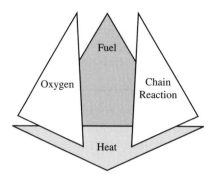

FIGURE 6-2
FIRE TETRAHEDRON OR FIRE PYRAMID.

The fire tetrahedron or fire pyramid, shown in Figure 6-2, adds a fourth component—chemical chain reaction—as a necessity in the prevention and control of fires. The combustion process releases heat through rapid oxidation of a fuel by oxygen in air. The chemical chain reaction involves free radicals, which are important intermediates in the initiation and propagation of combustion reactions. The removal of any one of the four faces of Figure 6-2 results in fire extinction. [9]

The formation of free radicals determines the flame speed. The formation and consumption of free radicals are necessary to sustain the flame reaction. If the free radicals are removed from the chain by chemical agents, the flame will cease to exist.

The autoignition temperature of a substance is the minimum temperature required to initiate self-sustained combustion in the absence of an ignition source.

The specific volume of a gas is a function of its temperature and pressure. The simplest such pressure-volume-temperature relationship is the ideal gas law.

$$PV = nRT$$

where V = volume of n moles of gas, l
$\quad\quad P$ = absolute pressure, atm
$\quad\quad T$ = absolute temperature, °K

and $\quad\quad\quad\quad\quad\quad\quad R$ = constant, $82.1\dfrac{\text{atm.l}}{\text{°K.gm mol wt}}$

The ideal gas law assumes that (1) the volume of the molecules is negligible compared with the total volume of the gas and (2) no forces act between the gas molecules.

Other PVT relationships that can be derived from the ideal gas law include the following:

$$\frac{P_1 V_1}{T_1} = \frac{P_2 V_2}{T_2}$$

At constant temperature, $P_1V_1 = P_2V_2$
At constant pressure, $V_1T_2 = V_2T_1$
At constant volume, $P_1T_2 = P_2T_1$

In an adiabatic process there is no heat transfer between the system and its surroundings.

The autoignition temperature may be exceeded by the adiabatic compression of a gas according to the following relationship:

$$\frac{T_2}{T_1} = \left(\frac{P_2}{P_1}\right)^{\frac{\gamma-1}{\gamma}} = \left(\frac{V_1}{V_2}\right)^{\gamma-1}$$

where T_1 = initial gas temperature, °R
T_2 = final gas temperature, °R
P_1 = initial gas pressure, psia
P_2 = final gas pressure, psia
V_1 = initial gas volume, in³
V_2 = final gas volume, in³
γ = gas ratio of C_p to C_v, γ_{air} = 1.4

Under adiabatic conditions the energy released by the pressure reduction of a pressurized gas is given by the following relationship:

$$W = -\frac{P_1V_1}{1-\gamma}\left[1 - \left(\frac{P_2}{P_1}\right)^{\frac{\gamma-1}{\gamma}}\right]$$

where W = the energy released, ft · lbs.

FUEL

The majority of industrial fires involve hydrocarbons, usually in the gaseous or liquid state. For alkane compounds with only carbon and hydrogen, the number of carbons in the hydrocarbon molecule determines its burning characteristics. [3] The number of carbon atoms in alkane compounds influence the state of matter, vapor pressure, vapor density, flash point, and boiling point. The same situation exists in the alkene, alkyne, and aromatic families of hydrocarbons. For example, gasoline with 4 to 12 carbons boils at 50 to 400°F and has a –45°F flash point. Kerosene with 12 to 15 carbons boils at 326 to 426°F and has a 126°F flash point. Greases, waxes, and asphalt with more than 20 carbons are either solids or semisolids.

The flammability of hydrocarbons is greatly affected by substitution of, particularly, halogens for hydrogen. For example, chlorine atoms in hydrocarbons greatly reduce the flammability of the substance by interfering with the combustion chemical chain reaction.

Example 6-1

What is the effect of the stepwise substitution of chlorine for hydrogen in methane?

Solution

Compound	Flammability
Methane (CH_4)	Highly Flammable
Chloromethane (CH_3Cl)	Flammable
Dichloromethane (CH_2Cl_2)	Less flammable
Chloroform ($CHCl_3$)	Not flammable
Carbon tetrachloride (CCl_4)	Fire extinguisher

In a fire chlorinated hydrocarbons may decompose to form toxic gases. Carbon tetrachloride will decompose in a fire to form phosgene and chlorine gases, which creates a hazard to people in the vicinity of the fire. As a result OSHA prohibits the use of carbon tetrachloride in fire extinguishers.

The most common criterion for measuring the ignitability of liquids is the flash point. The flash point is the minimum temperature that a liquid will give off sufficient vapor to ignite with the oxygen in air. Flammable and combustible liquids are characterized by flash points and boiling points, as shown in Table 6-2. These standards were developed by the NFPA No. 321 and are widely used in fire hazard identification. [7] While evaporation takes place below the flash point, insufficient quantities of vapor are present in air to form an ignitable mixture. The ignitable mixture within the flammable range is capable of the propagation of flame away from the ignition source.

An ignitable mixture can spread flame from an ignition source but will only sustain combustion when the temperature reaches the fire point. The fire point for sustainable combustion is usually only a few degrees above the flash point of the liquid.

Just because a flammable liquid is at a temperature below its flash point, a fire hazard still exists. The liquid could be a mixture with a small amount of low-flash-point liquid, which could make the entire mixture flammable. If the liquid is atomized into a mist or fog of liquid droplets, it can be ignited well below its flash point.

TABLE 6-2
NFPA FLAMMABLE AND COMBUSTIBLE LIQUID CLASSIFICATIONS

Class	Definition
Flammable IA	Flash point <73°F, boiling point <100°F
Flammable IB	Flash point <73°F, boiling point ≥100°F
Flammable IC	73°F ≤flash point <100°F
Combustible II	100°F ≤flash point <140°F
Combustible IIIA	140°F ≤flash point <200°F
Combustible IIIB	Flash point ≥200°F

FLAMMABLE LIMITS

When gases or vapors form flammable mixtures with air or oxygen, there is a minimum concentration of gas or vapor below which the propagation of flame does not occur on contact with a source of ignition.

There is analogous maximum concentration of gas or vapor in air, above which the propagation of flame does not occur. These boundary limits for mixtures of gas or vapor with air to ignite and propagate a flame are known as upper and lower flammable or explosive limits. A mixture below the lower flammable limit (LFL) is too lean to burn or explode. A mixture above the upper flammable limit (UFL) is too rich to burn or explode.

Generally the lower and upper flammable limits are based on normal ambient temperature and pressure (Table 6-3 on pages 176 and 177). There may be variation in flammable limits or range for temperatures and pressures above or below ambient. Usually when the temperature or pressure increases, the LFL is lowered and the UFL is raised. A decrease in the temperature or pressure usually narrows the range of the flammable limits.

Flammable gases have a lower flash point and, usually, a wider explosive range than flammable liquid vapors. Since the flash points of flammable gases are generally in the cryogenic temperature range, flash point is usually not a consideration in fire prevention related to flammable gases. Because flash point is not a consideration in defining a flammable gas, a gas is considered to be flammable if the lower flammability limit is ≤ 13 percent by volume at ambient conditions or if the flammability range is more than 12 percent at ambient conditions. [4] From a safety perspective the ventilation system and electrical equipment in a flammable-gas atmosphere must take into consideration that flammable gases are usually lighter than air, while most flammable vapors are heavier than air.

The LFL and UFL for a vapor mixture can be determined from the individual components of the mixture.

$$LFL_{mix} = \frac{1}{\displaystyle\sum_{i=1}^{n} \frac{y_i}{LFL_i}}$$

$$UFL_{mix} = \frac{1}{\displaystyle\sum_{i=1}^{n} \frac{y_i}{UFL_i}}$$

where y_i = volume or mole fraction of the individual vapor component in the mixture
LFL_i = LFL of the individual vapor component
UFL_i = UFL of the individual vapor component

The LFL of a flammable vapor can be estimated from its vapor pressure at its flash point at atmospheric pressure.

TABLE 6-3
TYPICAL FLAMMABLE LIMITS OF GASES AND VAPORS IN AIR

Gas or vapor	Lower limit (% by volume)	Upper limit (% by volume)	Closed-cup flash point (°F)
Acetaldehyde	4.0	57	−17
Acetone	2.5	12.8	0
Acetylene	2.5	80	—
Ammonia	15.5	26.6	—
Benzene	1.3	6.8	12
Butane	1.8	8.4	—
Butyl acetate	1.4	15.0	84
Carbon disulfide	1.2	50	−22
Carbon monoxide	12.5	74.2	—
Chlorobenzene	1.3	7.1	90
Crotonaldehyde	2.1	15.5	55
Cyclohexane	1.3	8.4	1
Cyclohexanone	1.1	—	111
Cyclopropane	2.4	10.5	—
Dichlorobenzene	2.2	9.2	151
1,2-Dichloroethylene	9.7	12.8	57
Ethane	3.1	15.5	—
Ether (diethyl)	1.8	36.5	−49
Ethyl acetate	2.2	11.5	28
Ethyl alcohol	3.3	19.0	54
Ethyl bromide	6.7	11.3	—
Ethyl chloride	4.0	14.8	−58
Ethyl ether	1.9	48	−49
Ethylene	2.7	28.6	—
Ethylene dichloride	6.2	15.9	56
Ethylene oxide	3.0	80	—
Furfural	2.1	—	140
Gasoline	1.4–1.5	7.4–7.6	−50
Heptane	1.0	6.0	25
Hexane	1.2	6.9	−15
Hydrogen cyanide	5.6	40.0	—
Hydrogen	4.0	74.2	—
Hydrogen sulfide	4.3	45.5	—
Isobutyl alcohol	1.7	—	82
Isopentane	1.3	—	—
Isopropyl acetate	1.8	7.8	43
Isopropyl alcohol	2.0	—	53
Kerosene	0.7	5	100
Linseed oil	—	—	432
Methane	5.0	15.0	—
Methyl acetate	3.1	15.5	14
Methyl alcohol	6.7	36.5	52
Methyl bromide	13.5	14.5	—
Methyl butyl ketone	1.2	8.0	—
Methyl chloride	8.2	18.7	—
Methyl cyclohexane	1.1	—	25
Methyl ether	3.4	18	—
Methyl ethyl ether	2.0	10.1	−35

TABLE 6-3
TYPICAL FLAMMABLE LIMITS OF GASES AND VAPORS IN AIR *(CONTINUED)*

Gas or vapor	Lower limit (% by volume)	Upper limit (% by volume)	Closed-cup flash point (°F)
Methyl ethyl ketone	1.4	11.4	16
Methyl propyl ketone	1.5	8.2	—
Mineral spirits No. 10	0.8	—	104
Naphthalene	0.9	—	176
Nitrobenzene	1.8	—	190
Nitromethane	7.3	—	95
Nonane	0.8	2.9	88
Octane	0.9	3.2	56
Pentane	1.4	7.8	—
Propane	2.1	10.1	—
Propyl acetate	1.8	8.0	58
Propyl alcohol	2.1	13.5	59
Propylene	2.0	11.1	—
Propylene dichloride	3.4	14.5	60
Propylene oxide	2.0	22.0	—
Pyridine	1.8	12.4	74
Toluene	1.3	7.0	40
Turpentine	0.8	—	95
Vinyl ether	1.7	27.0	—
Vinyl chloride	4.0	21.7	—
Xylene	1.0	6.0	63

$$LFL = \frac{\text{vapor pressure at flash point}}{760 \text{ mmHg}}$$

The Antoine equation can be useful in estimating the vapor pressure as a function of temperature

$$\log VP = A - \frac{B}{T + C}$$

where VP = vapor pressure, mmHg
T = liquid temperature, °C
A, B, C = empirical constants

FLAME PROPAGATION

Flame propagation is the reproduction of flames by a vapor-oxygen mixture within the flammable ratio of the mixture. The propagation can only take place when fuel is mixed with the correct volume of air or oxygen that is capable of causing combustion. For a flame to propagate through an opening in a vessel or container, the vapor on both sides of the opening must be within the flammable range.

The flame propagation rate in open air is dependent on the fuel-air ratio and the characteristics of the fuel. In open air, the average rate of flame travel through a vapor or gas-and-air mixture is 15 ft/s. [15] If the flames meet a fuel-air mixture above the UFL, the flames slow to pull more air into the mixture. When the flames meet a mixture below the LFL, the mixture is difficult to ignite. When a pressure wave compresses the mixture in front of the flames, the speed of flame propagation will increase dramatically. The pressure increases 6 to 10 times, with the speed of flame propagation increasing at an even greater rate. For example, when hydrogen mixed with oxygen ignites, the speed of flame propagation is high and a violent explosion occurs.

The propagation of flames into an enclosed vessel or area will occur only when the inside vapors are within the flammable range. If a flame propagation in open air leads to an opening in a tightly enclosed container with vapor outflow, the inside vapor will not ignite because it is too rich to burn and propagation will cease.

When there is a large hole in the container or vessel, the fire may continue to burn. A major explosion will not occur because the vapor-air mixture will not accumulate in sufficient volume.

The top vent on an enclosure should be small enough to not allow air to enter the tank, when the vent is burning. An explosion in the tank is highly unlikely, so long as the tank vent is burning because the tank vapor is too rich.

Carbon dioxide and steam, the inert gases from combustion along with nitrogen from air, will extinguish a fire. This principle can be used to suffocate a fire by shutting off the supply of fresh oxygen or air.

IGNITION SOURCES

There are many potential ignition sources at industrial operations:

- Electric sparks
- Smoking and matches
- Frictional heat
- Hot surfaces
- Overheated materials
- Open flames
- Spontaneous heating
- Welding and cutting
- Combustion particles

These ignition sources are capable of releasing sufficient energy for enough duration to cause the initial chemical reaction for combustion to occur. Temperatures of the fuel and the ignition source can greatly influence the ignition. [16]

Flame is present in many industrial operations. Flammable vapors and flames must be kept away from each other at operating facilities.

The principle of sparks as a source of ignition applies to all electrical sources. An electric spark involves the discharge of electric current across a gap between two charged objects when the objects come in close contact with each other. Static electricity and lightning are examples of electric sparks. [9, 10]

A pipe, vessel, or other surface can become hot enough to ignite a flammable vapor or gas. The size, shape, and surface area of the heated surface will influence the temperature, volume, and duration of exposure required to ignite flammable vapor or gas.

Heated surfaces can cause ignition and internal explosion in chemical storage tanks containing kerosene, diesel fuel, and fuel oil, even in small quantities. These higher-flash-point hydrocarbons are more likely to explode in heated tanks than lower-flash-point materials. When a fire occurs near a tank, there is usually little vapor present above the liquid in the tank. As a surrounding fire gradually heats the tank surface and contents, considerable time is required to vaporize the high-flash-point liquids to reach above the upper explosive limit. As these vapors reach the explosive range in the tank, there could be an internal explosion if the tank surface temperature is sufficiently high. Lower-flash-point liquids, when subjected to this same phenomena, produce a vapor phase too rich to explode in a much shorter time period. This greatly reduces the exposure duration for a possible internal explosion for liquids like propane, butane, or gasoline.

Hydrogen and acetylene are the two most easily ignited gases. What might be an ignition source for these gases might not be for butane or gasoline vapors.

At industrial operating facilities there are likely to be some fire hazard locations under normal conditions. The equipment that will be in this hazardous environment should be selected to avoid potential fires and explosions. Any equipment that normally produces sparks or arcs, like ordinary electric light switches, should not be placed in areas where flammable vapors or gases are present under normal conditions. If flammable vapors will only be present because of a leak, spill, or rupture, the elevation, vapor density, and quantity released should be considered for each location. Motors that create sparks or arcs should not be placed in a fire hazard area.

Liquified petroleum gases like propane and butane are heavier than air. They have a tendency to travel long distances and still be within their flammable range. Kerosene and other higher-flash-point hydrocarbons produce vapors that travel much shorter distances from the liquid source.

Figure 6-3 provides simple, readily recognizable, and easily understood mark-

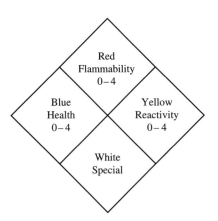

FIGURE 6-3
NFPA 704 SYSTEM FOR HAZARD IDENTIFICATION.

ings that will give at a glance a general idea of the inherent hazards of any material and the order of severity of these hazards as they relate to fire prevention, exposure, and control. It provides an appropriate alerting signal and on-the-spot information to safeguard the lives of both public and private fire-fighting personnel during fire emergencies. It also assists in planning for effective fire-fighting operations and may be used by plant design engineers and plant protection and safety personnel.

This system identifies the hazards of a material in terms of health, flammability, and reactivity. Order of severity in each of these categories ranges from 4, indicating a severe hazard, to 0, indicating no special hazard.

Health

The health hazard in fire fighting is a single exposure, which may vary from a few seconds up to an hour. The following is based upon protective equipment normally used by fire fighters.

4   Materials too dangerous for any exposure. The vapor could cause death. Normal full protective clothing and breathing apparatus will not provide adequate protection against inhalation or skin contact with these materials.

3   Materials extremely hazardous to health. Areas may be entered with extreme care. Full protective clothing, including self-contained breathing apparatus, coat, pants, gloves, boots, and bands around legs, arms, and waist should be provided. No skin surface should be exposed.

2   Materials hazardous to health. Areas may be entered freely with full-face mask and self-contained breathing apparatus that provides eye protection.

1   Materials only slightly hazardous to health. It may be desirable to wear self-contained breathing apparatus.

0   Materials which on exposure under fire conditions would offer no hazard beyond that of ordinary combustible material.

Flammability

The flammability hazard is based upon susceptibility to burning.

4   Very flammable gases or very volatile flammable liquids
3   Materials that can be ignited under almost all normal temperature conditions
2   Materials that must be moderately heated before ignition will occur
1   Materials that must be preheated before ignition can occur
0   Materials that will not burn

Reactivity

The reactivity hazard is based upon the susceptibility of materials to release energy either by themselves or in combination with water. Fire exposure has been considered along with conditions of shock and pressure.

4   Materials that are readily capable of detonation or of explosive decomposition or

explosive reaction at normal temperatures and pressures. Includes materials which are sensitive to mechanical or localized thermal shock. If a chemical with this hazard rating is in an advanced or massive fire, the area should be evacuated.

3   Materials that are capable of detonation or of explosive decomposition or of explosive reaction but which require a strong initiating source or which must be heated under confinement before initiation. Fire fighting should be done from an explosion-resistant location.

2   Materials that are normally unstable and readily undergo violent chemical change but do not detonate. Includes materials that can undergo chemical change with rapid release of energy at normal temperatures and pressures or violent chemical change at elevated temperatures and pressures. Also includes those materials that may react violently with water or form potentially explosive mixtures with water. In advanced or massive fires, fire fighting should be done from a safe distance or from a protected location.

1   Materials that are normally stable but may become unstable at elevated temperatures and pressures or react with water with some release of energy but not violently. Caution must be used in approaching the fire and applying water.

0   Materials that are normally stable even under fire exposure conditions and are not reactive with water. Normal fire-fighting procedures may be used.

Additional Markings

A fourth space in the identification symbol is reserved for additional information when such may be of value to the fire fighter. For example, any material which will react violently with water should carry the symbol $\overline{W}$ to indicate "avoid use of water." Radioactivity could be identified in this space as well as special information about extinguishing method. Materials that possess oxidizing properties can be identified by the letters ox.

EXPLOSIVE HAZARDS

An explosion is a rapid oxidation, resulting in pressure on a surrounding structure; it may cause structural damage and release considerable energy. Fire does not exert this pressure on surrounding structures.

The reaction rate of explosions is rapid, resulting in excessive pressure. This pressure is dependent on the composition of the vapor-air mixture, the initial pressure, and the initial temperature. During the initial explosion phase, the reaction rate is relatively slow. At normal ambient conditions the initial propagation rate is in the range of 10 to 309 ft/s. This means there is a delay before an explosion reaches the maximum pressure. This time delay is usually measured in milliseconds. Generally the larger the tank or container, the greater the time delay to reach the maximum pressure (Fig. 6-4 on page 182). [8] The size of the vessel does not affect the magnitude of the explosion pressure.

As shown in Figure 6-5 on page 182, the initial pressure greatly influences both the magnitude of the maximum pressure and the time delay for an explosion. [8] Generally, the higher the initial pressure, the higher the maximum explosion pressure and the less delay in reaching that pressure. Most chemical reactions in industrial operations are carried out at elevated pressure for productivity and other eco-

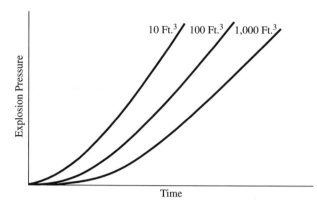

FIGURE 6-4
EFFECT OF ENCLOSURE VOLUME ON EXPLOSION PRESSURE AND TIME. The larger the tank or container, the greater the time delay to reach the maximum explosion pressure.

nomic reasons. These same process conditions require safety protection systems that are more stringent than those for normal atmospheric operating conditions.

The initial temperature of the vapor mixture greatly affects the flame speed. When an explosion occurs at a high initial temperature, the reaction rate is much higher than at a lower temperature (Fig. 6-6). [8] As the autoignition temperature of the initial temperature vapor is approached, safety systems become more sensitive to even minor changes in operating conditions because of this initial temperature phenomena.

FLAMMABLE VAPOR CLOUDS

The release of a flammable liquid may create a flammable vapor cloud, particularly if the ambient temperature is above the flash point of the liquid. Propane and butane liquified petroleum gases (LPG) are stored and processed at temperatures above their flash points; this would make any release of LPG capable of creating a flammable vapor cloud.

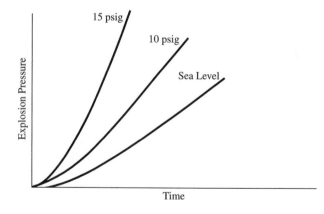

FIGURE 6-5
EFFECT OF INITIAL PRESSURE ON EXPLOSION PRESSURE AND TIME. The higher the initial pressure, the higher the maximum explosion pressure and less time delay in reaching that pressure. Many industrial chemical reactions are carried out at elevated pressures for economic and productivity reasons. This requires more stringent safety protection systems than at ambient pressure.

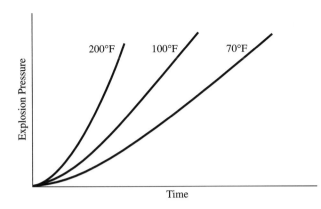

FIGURE 6-6
EFFECT OF INITIAL
TEMPERATURE ON
EXPLOSION PRESSURE
AND TIME.
When an explosion occurs
at a high initial temperature,
the reaction rate is much
higher than at a lower initial
temperature. When the au-
toignition temperature of the
vapor is approached, the
safety systems become
more sensitive to even minor
changes in operating conditions.

The flammable region of a vapor cloud is dependent on the vapor release rate, vapor properties, ambient weather conditions, and characteristics of the terrain. When LPG spills occur from pressurized storage, considerable adiabatic flashing will occur (Fig. 6-7). [10] Generally, lower boiling components exhibit greater flashing into the surroundings. The vaporization rate for propane is highly dependent upon the type of spill surface (Fig. 6-8 on page 184). [10] The propane, which is floating on top of the water, vaporizes at a high and constant rate because the water does not inhibit or attenuate the vaporation process. Soil and concrete absorb propane and then retain and inhibit its release into the atmosphere.

An LPG vapor cloud will be more dense than air. This cloud density will be re-

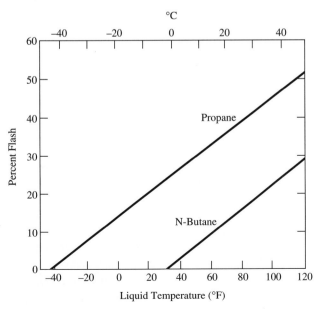

FIGURE 6-7
ADIABATIC FLASHING OF
LPG AS A FUNCTION OF
STORAGE
TEMPERATURE.
Generally, lower boiling
components exhibit greater
flashing into the surround-
ings.

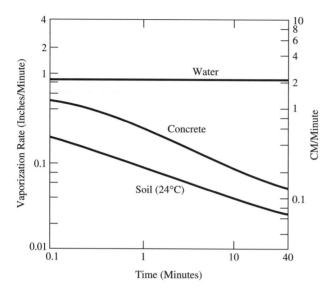

FIGURE 6-8
VAPORIZATION OF PROPANE SPILLS ON VARIOUS SURFACES AS A FUNCTION OF TIME. The vaporization rate of propane is highly dependent upon the type of spill surface. Propane floating on water has a high and constant rate of evaporation because the water does not inhibit or attenuate the vaporization process. Soil and concrete absorb propane and then retain and inhibit its release into the atmosphere.

duced as it is diluted with air, which will be entrained at the interfaces between the vapor cloud and the air. Figure 6-9 shows a model for predicting the vapor cloud characteristics. [10] This time-stepped profile is for a 300-gal/min pressurized propane spill with a 5-min duration on 75°F soil. There is a 50 percent flash of the propane into an average wind speed of 10 mph.

Figure 6-10 depicts the same size leak and ambient conditions as the previous Figure 6-9 except the vapor is caused by refrigerated propane. Both situations exhibit considerable vaporization and dispersion over time. The pressurized propane released produces a much larger vapor cloud and therefore creates more safety concerns.

If the vapor cloud contains a vapor-air mixture within the flammable limits and the cloud encounters an ignition source, the vapor cloud can ignite. The flame can then propagate back through the cloud to the vapor source if the flammable portion of the cloud is continuous. As the flame is propagated back, secondary fires may be ignited by materials in the flame path and anyone in the cloud could be severely burned.

Explosions of vapor clouds in the open air are infrequent. Their results can be devastating. A number of these explosions have occurred around natural gas and gas-liquids pipelines that were leaking. The ambient air at the location of the leak is usually quiet, causing the flammable vapor cloud to spread over a wide area before finding an ignition source. [1]

Another form of explosion is a boiling-liquid expanding-vapor explosion (BLEVE). It usually occurs when a pressurized storage tank containing a liquified gas is exposed to fire. The heat from the fire causes the gas to expand in the tank until the tank ruptures. The released liquid flashes and ignites instantaneously in a large fireball. The brief contact with the flame and radiation heat flux of the fireball can be catastrophic for people and structures in the vicinity.

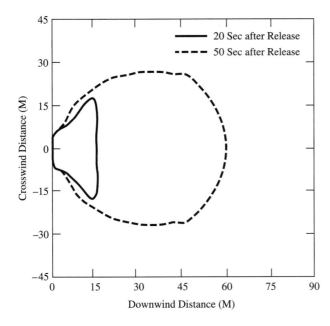

FIGURE 6-9
VAPOR CLOUD PROFILES
FOR A PRESSURIZED
PROPANE RELEASE AS A
FUNCTION OF TIME.
An LPG vapor cloud will be
more dense than air. The
cloud density will be re-
duced as it is diluted with
air, which will be entrained
at the interfaces between
the vapor cloud and the air.

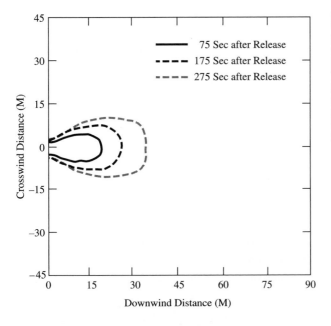

FIGURE 6-10
VAPOR CLOUD PROFILES
FOR A REFRIGERATED
PROPANE RELEASE AS A
FUNCTION OF TIME.
The refrigerated propane release covers a much smaller
area than the pressurized
propane release. Pressur-
ized propane produces a
much larger vapor cloud.

FIRE RADIATION HAZARDS

People can be injured and equipment can be damaged by flames from a fire or explosion. Injury and damage can also occur by thermal radiation well away from the actual flame propagation. The rate of thermal radiation an object receives from the fire is dependent upon the flame temperature and the geometric relationship between the exposed object and the flame. The radiant energy from the flame can be modeled accurately as a surface heat emitter with a radiant heat flux based upon the type of vapor fuel and the flame size.

A pool fire, which is based on an accumulation of flammable liquid on the ground, can be modeled to predict the heat flux profiles. The flame from the pool fire is modeled as a cylinder, whose diameter equals the pool diameter and whose length depends on the burning rate and the pool diameter. As the wind causes the flame to tilt from the vertical, the geometric relationship between the flame and the exposed object will change. This tilt angle will relate to the flame size, wind speed, and vapor density.

Using the figures in Table 6-4, models were developed for 50-ft-diameter and a 15-ft^2 propane pool fire with a 10 mph wind and relative humidity of 50 percent. These models are shown as isopleths in Figures 6-11 and 6-12, respectively. The larger propane pool fire, shown in Figure 6-11, created much more damage and destruction to the surroundings and injury to personnel than did the smaller propane pool fire. This phenomenon demonstrates the need for spill containment in areas with fire hazards.

DUST EXPLOSIONS

Dust consists of solid particles ≤10 μm in diameter that settle slowly in air. If the solid particles are combustible, a flammable dust-air mixture may be formed. Carbon particles like coal, organic particles like feed grains, and oxidizable metals like aluminum or magnesium are among the more hazardous dust in air mixtures. Dust explosions can involve an enormous energy release. The rate of pressure rise is fast and the maximum pressure is high.

Dust particles have a large surface-area-to-mass ratio. The surface absorption of oxygen by the solid particles or the evolution of a combustible gas upon heating are factors in dust explosions. The explosive limits for dust in air usually focus on

TABLE 6-4
RADIATION HEAT FLUXES AND THE POTENTIAL ADVERSE EFFECTS

Radiation heat flux [Btu/h•ft^2]	Potential adverse effects
1600	Skin will burn after 30-s exposure
4000	Wood could ignite
10,000	Metal structures could be destroyed

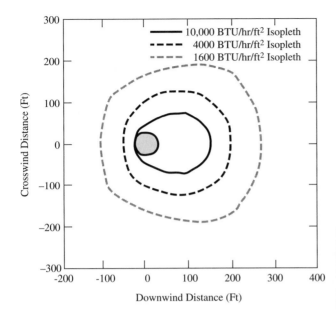

FIGURE 6-11
RADIATION ISOPLETHS
FROM A 50-FT DIAMETER
PROPANE POOL FIRE.
The 50-ft diameter propane
pool fire is affected by a 10-
mph wind speed.

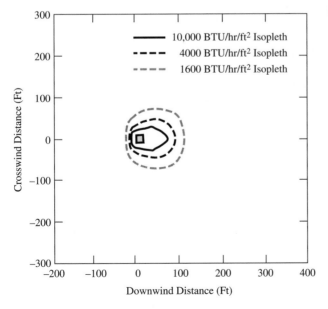

FIGURE 6-12
RADIATION ISOPLETHS
FROM A 15-FT² PROPANE
POOL FIRE.

the lower explosive limit, which has the appearance of a dense fog. Upper explosive limits for dust in air have not been as widely studied, but they do not have much practical significance.

The moisture content of air also plays a role because moisture reduces the tendency to ignite, promotes agglomeration of particles, and reduces the probability of ignition from static electricity.

SPONTANEOUS COMBUSTION

The chemical process of oxidation is an exothermic reaction; considerable heat is generated. The burning of flammable and combustible fuels is a typical example of this phenomena. When combustibles are below their ignition temperature, they will not ignite. When these same combustible materials are gradually oxidized and the heat generated allowed to accumulate, the temperature will increase until the materials ignite spontaneously. This situation is created whenever the heat generation rate exceeds the heat removal rate. Animal or vegetable oils are an excellent example of a material with a fairly high flash point and ignition temperature that could be gradually oxidized to create a fire hazard. Oil-soaked rags, insulation, or cotton materials are examples of potential sources of spontaneous combustion.

PRODUCTS OF COMBUSTION

Fire produces four types of products of combustion: flame, heat, fire gases, and smoke. Fire fatalities usually occur because of inhalation of fire gases and smoke; flame and heat cause thermal burns. The composition of the fire gases depends upon the chemical composition and structure of the fuel, the supply of oxygen, and the combustion temperature. Since most fuels contain carbon, carbon dioxide and carbon monoxide are usually formed in a fire. Oxygen supplies well in excess of stoichiometric requirements favor the formation of carbon dioxide, while carbon monoxide is usually the result of incomplete combustion of the fuel source. The formation of carbon dioxide causes an oxygen deficiency in air and a dramatic increase in the breathing rate of the affected personnel. Carbon monoxide is an affixient gas that reduces the oxygen carrying capacity of blood and interferes with oxygen exchange, thereby making it dangerous even at low concentrations. Water vapor in the form of steam is typically generated in fires that contain hydrocarbon fuels.

Depending upon the other elements that are present in the fuel, hazardous fire gases may include sulfur dioxide, nitrogen dioxide, hydrogen chloride, hydrogen sulfide, hydrogen cyanide, and ammonia.

Incomplete combustion products such as solid particulate matter and finely divided liquid drops appear as smoke. These smoke particles scatter light and adversely affect human vision during a fire because the average size of smoke particles is the same wavelength as that of visible light.

FUNDAMENTALS OF FIRE PREVENTION

If flammable gases or vapors are mixed with oxygen or air in flammable concentrations, eventually the mixture is likely to catch fire or explode. It is impossible to eliminate every conceivable source of ignition. The first principle of fire or explosion control is to prevent the formation of a fuel-air mixture that is within the appropriate flammable limits.

For example, industrial facilities that process hydrocarbons have an inherent potential for fire from materials being handled, the processes used, and the chemical reactions taking place. Usually, the flammable materials are so necessary and play such an integral role in the operations and products that they cannot be eliminated or substituted. As a result, the facility design and operating procedures should reduce the probability of a fire occurrence and minimize the consequences of the fire. [1]

Effective fire prevention must consider all possible potential fire causes and provide design and operating features to reduce or eliminate the causes. It is usually impractical to anticipate all potential fire situations. In addition, human error, random mechanical or electrical failures, and acts of nature cannot be totally controlled.

Large industrial fires often begin with an explosion. This initial explosion results in damage that releases flammable materials into the surroundings. The initial explosion may also damage protective design features and equipment, making them ineffective. In fire prevention at facilities that process large quantities of flammable materials, a great deal of emphasis should be placed on preventing explosions, thereby preventing a resultant fire.

An explosion produces an outward pressure wave, ranging up to several atmospheres. This pressure wave causes most of the initial damage. The pressure waves may be the result of a flammable vapor cloud, the sudden rupture of a vessel or an enclosure, the failure of equipment, valves of fittings, or human error. [18]

The damage effects are influenced by the peak overpressures, measured in pounds per square inch gauge (psig), and their associated shock waves, as shown in Table 6-5 on page 190.

Detonations usually involve unstable or highly reactive materials. The pressure wave from a detonation is much higher, ranging up to 50,000 atm or more. The elimination of the conditions for initiation of the chemical decomposition or reaction is the best way to prevent detonations. Their disastrous results can be minimized by carefully controlling and limiting the quantities of the unstable materials involved. Diluents and stabilizers can be used to improve the stability of these materials.

Flammable materials can normally be stored safely in environments that are free of oxygen or air. The fire or explosion hazard usually exists only when the flammables are released into the air or otherwise become mixed with air or oxygen. Fire prevention includes minimizing the potential for an accidental release or accumulation of flammable air mixtures in or around tanks, equipment, or confined spaces. Inert gas padding in equipment and tanks is often used to displace or exclude air. [2, 4]

TABLE 6-5
DAMAGE EFFECTS OF PEAK OVERPRESSURE OF AN EXPLOSION

Peak Overpressure (psig)	Damage effect
0.1	Small windows broken
0.4	Minor structural damage
1.0	Partial building collapse
3.0	Buildings pulled from foundations
4.0	Liquid storage tanks ruptured
5.0	Wooden telephone poles break
6.0	Houses destroyed
10.0	Buildings and machinery destroyed or heavily damaged

Chemical reactions, particularly exothermic reactions, may cause explosions unless they are carefully controlled. The reaction variables of temperature, pressure, residence time, and heat of reaction must be under control at all times or a dangerous situation could develop. These reaction kinetics parameters need to be understood to identify potential hazards and the severity of their occurrence. Normal and emergency venting should be provided for reaction systems. [9]

SYSTEM DESIGN AND LAYOUT

The proper layout of industrial facilities is an important factor in the prevention of catastrophic fires. The plant location, operating and storage units, and buildings should have sufficient physical separation to limit the property, facilities, and materials that could become involved in a fire. By planned isolation through facility layout the severity and frequency of single fires can be eliminated. Fire-resistant walls, doors, and other barriers, as well as dikes, diversion curbs, and adequate drainage are important fire prevention considerations. The normal prevailing wind, locations of emergency escape routes, and accessibility for the fire brigade must also be considered in the plant layout.

Once a fire is initiated, it may spread through a combination of factors. Heat transfer, direct burning, and the release of flammable vapor or liquid may contribute to a fire spreading. In buildings, fire mainly spreads by convection and radiation heat transfer and by direct burning. Hot gases spread throughout a building by natural convection, and elevated temperatures radiate considerable heat. Ignition of materials greatly increases the volume of hot gases and smoke particles, thereby providing more fuel for combustion. The flue effects from openings into the atmosphere draw additional air into the flames. Flying sparks carried by the convection currents may ignite the other fuel sources in the building.

In more open areas like process facilities a fire spreads by radiation heat transfer and the release of flammable vapor or liquids. Releases may occur because of vessel or piping rupture. A flashback often results because dense hydrocarbon vapors may travel considerable distances before finding an ignition source.

Any large spill or leak of flammable materials has the potential to fuel a major fire. The greatest fire hazard at a facility is usually where the largest amount of flammable liquid and gas is stored. The largest volume in any single tank or vessel represents the greatest potential fire unless a domino effect is present. When flammable liquids are contained under normal temperatures and pressures, the main fire hazard is the spread of flames. When flammable gas or liquid is stored or processed at elevated temperatures and pressures, there is the potential for a fuel rich vapor cloud to ignite, resulting in a fireball.

When large quantities of flammable vapor escape rapidly into the surroundings, a serious fire can result. This may occur from a ruptured tank, containing liquified flammable gas at ambient conditions or flammable liquids at elevated temperatures. The dispersion of a released vapor cloud is influenced by many factors:

- Wind velocity and direction
- Discharge velocity and direction
- Vapor density
- Ambient temperature
- Presence of nearby structures

If the vapor cloud disperses rapidly, it may be harmless and fail to ignite. If it does not disperse, it will search for and eventually find an ignition source. Once the cloud is ignited, an explosion may occur. If the vapor cloud is too rich to explode, the cloud may ignite around its outer fringe and take off as a fireball. The fire hazard of such a mobile fireball includes intense heat transfer by radiation that can ignite or burn anything in its vicinity. Also, convection heat transfer in the wake of the fireball could further ignite anything in its wake.

A boiling-liquid expanding-vapor explosion (BLEVE) happens when a liquified gas escapes from a pressurized vessel that has been exposed to fire. The relief of the internal pressure of the vessel causes a blast wave that results in severe damage to the blast surroundings. The resulting fireball causes associated thermal radiation and carries projectiles of large fragments. If the expanding vapor is flammable and mixed with air, the vapor ignition could cause a sudden expansion with an explosion.

A flammable or explosive gas, vapor, or dust may ignite in a confined space like a tank or an enclosed building. The rapid expansion of the products of combustion will be restrained by the walls of the tank or building. The rapid pressure rise may exceed the rupture pressure of the walls, which could burst with a loud noise. The overpressure wave from the explosion is a sudden and narrow pressure spike with time. Such a confined vapor cloud explosion can cause tremendous damage to property and injure or kill people in the vicinity.

The proper selection of materials of construction can reduce the damage from an industrial fire by limiting the ability of the fire to spread. Wood, most plastics, and other combustibles can contribute fuel to a fire. The products of combustion of some of these combustibles can be toxic and hazardous to fire fighters and endanger personnel. If combustibles are avoided as materials of construction, less fuel is available to sustain the fire.

Fire-resistant material is capable of withstanding the adverse effects of elevated temperatures and flames for a finite period of time. These fire-resistant materials have been tested by a controlled temperature versus duration of fire. The material of structural components is then rated in time increments of 0.5, 1, or 2 h for withstanding the test fire. Fire-resistant materials of construction provide additional time to control and extinguish the fire. Fire-resistant construction materials are commonly used in many critical construction applications:

Structural members like beams and columns that directly support piping, equipment, and vessels containing flammable or combustible liquids or gases

Control rooms or central control buildings that could be exposed to fires in process units

Enclosed exit stairways or corridors that are needed for the safe escape of personnel

Walls and barriers that separate hazardous storage and process areas

When cooling from fire water is available from fixed automatic sprinklers, sprayers, or deluge systems, generally 1- to 2-h fire ratings are sufficient. When fire water is available only from manual hoses or other manual systems, generally 3- to 4-h fire ratings may be necessary.

The most common, useful, and viable fire protection medium is water. It is inexpensive and abundant and has a high cooling capacity. A reliable water supply and piping distribution system must be capable of delivering the water to the fire protection equipment and systems.

This water delivery system must maintain sufficient water pressure and flow rate throughout a fire emergency situation. There are several types of systems that meet these requirements, including city water mains, elevated storage tanks, and dedicated pumping capacity. Fire water pumps, which are commonly used, follow the industry NFPA 20, Standard for Centrifugal Fire Pumps. [11] Diesel engines are preferred for these centrifugal fire pumps because electric power may be lost during a fire. It is also preferable for fire pumps to start automatically if there is a pressure loss in the fire water distribution system. These fire pumps have rated capacities from 250 to 5000 gal/min at rated heads of 90 to 580 ft. Industry standards require the fire pumps to be capable of delivery of 150 percent of rated capacity at not less than 65 percent of rated head. [16]

The distribution system for fire water usually consists of a network of large-diameter pipe usually at least 6 in in diameter. Large hose water streams from water mains typically require water pressures of 100 to 150 psig.

The main fire water system at an industrial plant should be able to provide continuously 3000 to 4000 gal/min. The main system could be supplemented with hoses from storage tanks or rivers. A major fire could require 4 to 8 hours of water supply. If the fire has not been extinguished within 8 hours, the plant would probably be completely destroyed. Fire hydrants should be about 150 ft apart in process areas and about 300 ft apart in storage tank farms. A grid system must provide sufficient block valves to isolate the hydrants for needed repairs.

Water sprinkler systems require 50 to 100 psig. Fire trucks with pumping systems have lower water pressure requirements, 20 to 50 psig. These pumping units are capable of achieving discharge pressures of 125 to 150 psig. The maximum velocity of a stream of fire water is 10 ft/s.

Sprinkler systems are the most common fixed fire protection system at industrial facilities. Sprinklers provide highly efficient water delivery to the fire and greatly reduce the fire damage. Fires from combustible solids and liquids can usually be extinguished by sprinkler systems. Fires involving flammable gases and liquids are normally not extinguished by sprinklers. However, the water from the sprinkler system would cool the fire, thereby reducing its intensity and ability to spread and protect equipment and surrounding structures from thermal damage. Sprinkler systems are commonly used in office buildings, hotels, warehouses, and other construction with combustibles.

Deluge systems feed fire protection water through a network of smaller-diameter piping from a control valve to the water main supply. These open-spray nozzles are usually actuated by automatic fire detection and control devices. Deluge systems are placed in situations where intense fires could develop rapidly. These include chemical process equipment, units, buildings, storage vessels, and pipe racks. Deluge water is usually applied at rates in the range of 0.1 to 0.5 gal/min per square foot of exposed surface area. The NFPA 15, Standard for Water Spray Fixed Systems for Fire Protection, is commonly used in the design of deluge systems.

Sprinkler systems have multiple nozzles that are individually sealed with a heat-sensitive device. Each sprinkler head will individually open when exposed to the heat of a fire. Sprinklers may be either wet- or dry-pipe systems.

Wet-pipe systems contain pressurized water and commence water flow on the fire the instant the sprinkler head is opened. They should not be used where the ambient temperature is below freezing. Dry-pipe systems are filled with pressurized dry air, thereby preventing water from entering the pipes because of a differential check valve. When the dry-pipe nozzle is opened by fire heat, the air pressure drops and the check valve allows fire protection water to enter the system. The time lag caused by the dry air should be taken into consideration in designing a dry-pipe system.

The design of sprinkler systems usually follows NFPA 13, Standard for the Installation of Sprinkler Systems. Each sprinkler head or nozzle should protect a 90 to 160 ft^2 floor area.

FIRE DETECTION AND CONTROL

In order for the fire protection system to operate effectively, the fire must be detected early and planned actions be taken as soon as possible. The primary purpose of fire detection, alarm, and control systems is to minimize injury and property damage. The fire should be detected by the quickest, most reliable means. An alarm should alert personnel to escape and take action to control the fire. The fixed fire protection systems should be activated. Any equipment or processes that could contribute fuel or other hazards should be shut down and isolated.

There are numerous fire detection devices that relate to a specific characteristic of fire. These fire detectors are classified by the particular fire phenomena.

- *Smoke detectors* react to concentration of particles in air.
- *Heat detectors* react to elevated temperature or rapid rates of temperature rise.
- *Flame detectors* react to the light spectrums of fires.
- *Fire gas detectors* react to the products of combustion.

The type of fire expected influences the type of detector. Other considerations include the surroundings where the detector is placed and the relative cost of the detector. The proper locations for the detector must take into consideration the distance limitations between devices, the physical arrangement of the area, the point of fire origin, and the expected air movement pattern. The Standard on Automatic Fire Detectors, NFPA 72E, provides useful information and guidance on fire detectors and their placement.

The selection of alarm devices is usually based on the expected fire location, the location of trained personnel, and the nature of the surrounding environment. Audible alarms, such as horns and bells, produce a distinctive signal unless the normal noise level is excessive. In those high-noise areas, lights are more commonly used. The alarm must be provided in the area of the fire occurrence and at a location that is constantly attended by trained fire protection personnel. [12, 13]

Both the fire detection and alarm devices are tied together by a control system, which monitors the fire detection devices and activates the alarm. The control system must provide a constant and reliable power source to the detection and alarm devices. It is common for the control system to activate fixed fire protection systems and to shut down critical process operations at the facility.

The fire detection, alarm, and control system should be selected based upon the manual and automatic responses desired. They can vary in complexity from simple local alarms to highly automated detection and alarms that summon the local fire department. The trend has been to more complex and sophisticated computerized systems after the alarm has been activated.

If the fire department is notified about the location of the fire, the presence of hazardous materials, the particular fire conditions, and recommended actions to be taken prior to its arrival on the scene, fire protection personnel are afforded an additional measure of safety.

STORAGE TANKS FOR FLAMMABLE LIQUIDS

Many techniques have improved the safe storage of flammable liquids in tanks. Vapor conservation or reduction is a common fire safety technique that offers an attractive economic return on the incremental investment. The valuable flammable liquids are conserved and the potential for fugitive air emissions is reduced.

When an inert gas like nitrogen replaces air in the tank, vapor space and adequate insulation is provided if the vapor phase is maintained at a fairly constant

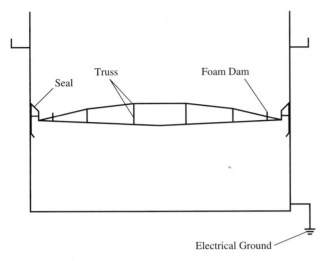

FIGURE 6-13
PAN-TYPE FLOATING
ROOF TANK. (*Source:*
Reprinted with permission
from NFPA 30, *Flammable
and Combustible Liquids
Code,* Copyright © 1996,
National Fire Protection
Association, Quincy, MA
02269. This reprinted
material is not the complete
and official position of the
National Fire Protection
Association, on the
referenced subject which is
represented only by the
standard in its entirety.)

temperature. The effect of ambient temperature variations is greatly reduced, thus increasing vapor conservation. [5]

The floating roof tanks shown in Figure 6-13 eliminate the large vapor space that is present in the standard cone roof tank shown in Figure 6-14. This provides effective vapor conservation. The floating roof tank has the best fire safety rating of all storage tank designs. The only drawbacks are potential leaks around the seal and higher maintenance costs. [4]

The floating diaphragm tank, shown in Figure 6-15 on page 196, also has a high degree of fire safety. It greatly reduces vapor losses. If vapors should leak past the diaphragm, a potentially dangerous vapor-air mixture would result.

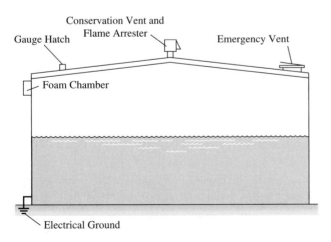

FIGURE 6-14
STANDARD CONE ROOF
TANK. (*Source:* Reprinted
with permission from NFPA
30, *Flammable and
Combustible Liquids Code,*
Copyright © 1996, National
Fire Protection Association,
Quincy, MA 02269. This
reprinted material is not the
complete and official
position of the National Fire
Protection Association, on
the referenced subject which
is represented only by the
standard in its entirety.)

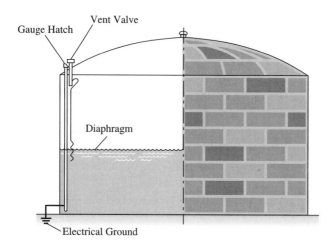

Gauge Hatch

Vent Valve

Diaphragm

Electrical Ground

FIGURE 6-15
FLOATING DIAPHRAGM
TANK. (*Source:* Reprinted
with permission from NFPA
30, *Flammable and
Combustible Liquids Code,*
Copyright © 1996, National
Fire Protection Association,
Quincy, MA 02269. This
reprinted material is not the
complete and official
position of the National Fire
Protection Association, on
the referenced subject which
is represented only by the
standard in its entirety.)

FIRE EXTINGUISHMENT

Water has been and always will be the preferred medium for most fire extinguishment, particularly for combustibles like wood and paper. The application of water to fires has undergone considerable evolution. Bucket brigades, hoses and pumps, improved nozzles for hoses, placement of fixed water mains, and portable water supplies have improved the use of water for fire fighting over the years.

Smaller water particles will increase the heat absorption efficiency, thereby extending the ability of a given volume of water to extinguish a fire. The water addition reduces the temperature of the fire to a level that inhibits the combustion of the fuel source. Also the water spray and steam replaces air as the necessary oxygen source in the vicinity of the fuel.

When a burning fuel is cooled below the ignition temperature of its vapor, the fire will be extinguished. The water rate to accomplish this must be at least sufficient to cool the vapor below this temperature. This phenomenon makes water the logical choice for combustible liquids with a flash point exceeding 200°F. Fuel sources with a specific gravity greater than water can be readily extinguished with water because the oxygen in air is prevented from reaching the fuel. Flammables that are soluble in water or are hydrophilic may also be extinguished by water, which serves to dilute the fuel concentration level.

The use of a water fog spray is the most common fire-extinguishing agent. It is efficient, requires minimal personnel to operate, provides quick coverage, is low-cost, and is not toxic. There are a few potential drawbacks to water extinguishment. Water may damage products or materials and could cause a hazard from electrical conductivity. [6]

The primary water supply source must be capable of supplying all of the necessary fire-fighting water for at least 4 hours. The main water line should be cast iron or steel pipe in the 175- to 225-psig cold-pressure design range. [3] The water should be supplied by centrifugal pumps because they provide a steady constant flow of water at a uniform pressure. Centrifugal pumps also have the advantage of not increasing line pressure if the downstream valves are closed or the line is otherwise restricted. This situation may occur during a fire emergency and could result in line rupture or equipment damage.

The water-pumping requirements for a facility should be based on a single fire concept, that is, only one major fire at a given time. Fire water is often available based upon the process operating requirement. The water spray rate for most ordinary combustible solids and flammable liquids is 0.2 to 0.5 gal/min per square foot of protected surface area.

Carbon dioxide extinguishes fire by diluting the vapor-air mixture below its flammable limit. This starves the fire of its fuel and oxygen sources. [20] Carbon dioxide is an easily stored, noncorrosive gas that retains its effectiveness in storage for extended time periods. It is clean and inert, will not conduct electricity, and provides some cooling to the fire. It also has some disadvantages: carbon dioxide is ineffective for Class A fires or large fires in general; Class B fires extinguished by CO_2 are subject to flashback; and the low discharge temperatures may cause freezing.

Carbon dioxide systems require little if any maintenance or special precaution after installation. So long as the CO_2 cylinders do not leak or otherwise lose pressure the system integrity has been maintained. A central CO_2 storage supply can be used to protect multiple areas at a facility.

The application of CO_2 for fire extinguishment may be accomplished by several procedures: local application, total flooding, and prolonged discharge. Local application of CO_2 involves the discharge of CO_2 directly on the hazard. Total flooding is accomplished by completely filling the entire area near the hazard with CO_2, usually in an enclosed space. The prolonged-discharge technique usually has a high initial CO_2 discharge, followed by a somewhat lower rate for an extended time period.

Usually CO_2 systems are activated by automatic sensing devices that are triggered by an emergency. The CO_2 is only effective as long as it is contained in the area of the hazard. Windows, doors, and ventilation systems must be closed whenever CO_2 is in use to assure its effectiveness.

Carbon dioxide is effective in extinguishing Class B combustibles and Class C energized electrical equipment. High concentrations of CO_2 should be avoided by fire safety personnel because loss of consciousness and possible death could result if the concentration approaches 35 volume percent in air.

Electrical fires are a good application for CO_2 extinguishment. The carbon dioxide will not harm the sensitive electrical equipment. The CO_2 will completely vaporize and be dispersed into the surroundings. Unlike water, the vaporized CO_2 leaves no residue on the hazard. Also the use of water on an electrical fire would cause a shock hazard that could be lethal to nearby personnel.

Dry chemicals in a powder form can extinguish fires quickly by local application or total dispersion surrounding the hazard. The dry chemical is usually discharged by compressed carbon dioxide or nitrogen. These units must be recharged after each use with the addition of dry chemical and replacement of the carbon dioxide or nitrogen cylinder. This is a low-maintenance technique, requiring only a periodic pressure check of the nitrogen supply for leakage.

Sodium bicarbonate ($NaHCO_3$), potassium bicarbonate, and potassium oxalate ($K_2C_2O_4H_2O$) are types of dry chemicals used for fire extinguishment. The powder particle size and application rate are important parameters in the effectiveness of these dry chemicals. Generally, small particles are more effective than larger particles, and sufficient dry chemicals must be used to extinguish the fire.

Dry chemicals are most effective in electrical hazard fires, where the residual powder will not clog or damage the electrical equipment. The dry chemical decomposes upon contact with the heat of the fire to liberate carbon dioxide, which replaces the surrounding air. Dry chemicals are most effective in enclosed areas.

Foam can be an effective extinguishing agent for flammable liquids that have a lower specific gravity than water. The buoyancy of the foam gives it the ability to form a barrier to separate the flammable liquid surface from the air.

Most foam applications are based upon mechanical foam that contains air-filled bubbles. These closed-cell bubbles are interconnected to form an airtight blanket on the flammable liquid surface. This foam blanket keeps flammable vapors from permeating the fire and insulates the heat of radiation from the flammable liquid.

Mechanical foam is formed by passing a solution of foam compound in water through an agitation device in the presence of air. The compound or foam stabilizer produces expansion of the water by a factor of 10 to 16. The compound makes up about 3 percent of the water solution prior to the air addition. The more gentle the foam application, the quicker the extinguishment of the fire. The amount of foam required may vary somewhat, but 2 to 6 in of foam blanket is usually sufficient to extinguish most liquid hydrocarbon fires.

The use of chemicals to produce carbon dioxide–containing foams has declined except for some storage tank applications. A chemical reaction forms bubbles that contain carbon dioxide.

EMERGENCY FIRE PROTECTION SYSTEMS

A fire situation is created at a facility by the combination of fuel, oxygen, ignition source, and a chain reaction. The removal of one or more of these components of the fire tetrahedron provides desirable fire protection. The fuel and oxygen concentration must also be within the flammable range. There are a number of systems that provide fire protection including fire fighting and snuffing systems. [19]

Fire plugs are usually placed strategically throughout a facility. The main water lines should be looped and be at least 4 in in diameter. Block valves should be placed so that fire water is available at all times during repairs to the water lines.

The water pump should be capable of discharging at least 500 gal/min at 125 psig. There should be at least a 4-h supply of water in storage to support the system. At least 150 gal of liquid foam should also be available in storage.

The fire hose is a basic piece of fire-fighting equipment at most industrial facilities; it must be well maintained to function properly. If the fire hose fails in use, the consequences can be catastrophic. The fire control efforts would be disrupted and the flailing hose could cause personal injury. Polyester hose is stronger and more abrasion-resistant and will not mildew compared to cotton hose.

The fire hoses on all fire trucks in the vicinity must have threads that are compatible with the fire plugs in the plant. This should be verified with the local fire department because adapters can be obtained to assure compatibility. If fire hoses are located at the fire plug, they should always be hooked up to the fire plug. Once the plug is opened with this permanent hookup and the pump started, fire fighters can readily proceed to the fire with the hose.

The injection of liquid foam at the fire hydrant is more desirable than at the hose nozzle because it provides more freedom of movement for fire fighters. The personnel handling the foam can be removed from the fire itself. It also eliminates the need for these personnel to carry foam to the end of the hose. The special foam nozzles should always be available at the hose storage area.

Most fires begin as small fires that can be easily controlled at the initial stage. Portable handheld fire extinguishers should be readily available at operating facilities to provide an early defense against a major fire. Carbon dioxide, foam, and dry-chemical extinguishers are most commonly used. [17]

Carbon dioxide extinguishers contain pressurized CO_2 in a heavy-duty gas container. This type of fire extinguisher is most effective during the initial seconds of operation, when the gas pressure is at the higher range of at least several hundred psig. If the CO_2 is prematurely released or improperly aimed, the effectiveness is greatly reduced. It is most effective at 5 to 8 ft from the fire. The CO_2 displaces oxygen to a level that will not support combustion. It is particularly effective on electrical and flammable liquid fires. It should be used with caution in confined areas, where fire fighters may lose consciousness from lack of oxygen.

Foam extinguishers contain two chemical solutions that are held in separate compartments. When these inert chemicals are mixed together, a pressurized frothy foam is created and flows from the tank hose. The foam with CO_2 bubbles in a liquid film will insulate the fire and cut off the air.

The foam should be applied so that it creates a uniform and contiguous blanket. If the liquid fire is in a tank or open container, the foam should be applied by banking off the walls to allow it to spread slowly over the surface. Another option is to allow the foam to gently fall on the surface of the burning liquid. If the fire is the result of a spill, the foam application should begin at one side of the burning area and progressively blanket more and more of the fire area until it is all covered. The foam stream should have sufficient pressure to reach 30 to 40 ft from the extinguisher.

The typical foam extinguisher has less than a 1-min capacity, making the foam application critical to obtaining uniform coverage. The foam is corrosive to metals and machinery, making water cleanup necessary after the fire extinguishment. Water-based foams should be avoided in freezing ambient conditions. Specialty foams are available for below-freezing temperatures. Foam should never be used on electrical fires because it is an excellent electrical conductor.

Dry-chemical fire extinguishers are filled with a free-flowing dry powder like sodium bicarbonate that liberates CO_2 at elevated temperatures. The heavier-than-air CO_2 stays low to the fuel and reduces the air available for combustion, thereby extinguishing the fire. This mechanism is similar to foam extinguishment. The dry-powder application is critical to effective fire extinguishment. The powder should be applied to the base of the flame with a fan-shaped movement from an upwind position. The use of powder will be limited to about 15 s from a typical 20-lb extinguisher. Dry-chemical extinguishers must be recharged with new dry chemical and a new gas cartridge after each use.

A snuffing system may use water or steam to extinguish fire. A water fog is a typical example; it removes sufficient heat to reduce the temperature below the ignition temperature. The water is discharged from fog nozzles to generate finely divided water droplets that are directed into the hottest portion of the fire. The water becomes steam and this heat removal causes fire extinguishment.

When steam is applied to the fire, the air concentration is reduced by steam displacement. As the steam reduces the air concentration, the air-fuel mixture flammability is reduced below the flammable range. At this point extinguishment of the fire occurs.

BIBLIOGRAPHY

1 Brauer, R. L., *Safety and Health for Engineers,* Van Nostrand Reinhold, New York, 1990.
2 Bugbee, P., *Principles of Fire Protection,* NFPA, Quincy, Mass., 1984.
3 Chandnani, M. K., *Design Fundamentals, Fire Protection Manual,* vol. 1: *Hydrocarbon Processing,* Gulf Publishing, Houston, 1985.
4 Cote, A. E., *Fire Protection Handbook,* NFPA, Quincy, Mass, 1990.
5 Duggan, J. J., *Five Safer Methods of Chemical Storage, Fire Protection Manual,* vol. 1: *Hydrocarbon Processing,* Gulf Publishing, Houston, 1985.
6 English, W. P., *Basic Tools of the Trade, Fire Protection Manual,* vol. 1: *Hydrocarbon Processing,* Gulf Publishing, Houston, 1985.
7 *Fire Protection Guide on Hazardous Materials,* 8th ed., NFPA, Quincy, Mass., 1984.
8 Grabowski, G. J., *Theoretical and Practical Aspects of Explosion Protection, Fire Protection Manual,* vol. 1: *Hydrocarbon Processing,* Gulf Publishing, Houston, 1985.
9 Kavianian, H. R., and C. A. Wentz Jr., *Occupational and Environmental Safety Engineering and Management,* Van Nostrand Reinhold, New York, 1990.
10 Martinsen, W. E., et al., *Spill Protection for Liquified Gas, Fire Protection Manual,* vol 1: *Hydrocarbon Processing,* Gulf Publishing, Houston, 1985.
11 *National Fire Codes,* National Fire Protection Association, Quincy, Mass., 1990.
12 *Accident Prevention Manual for Business and Industry: Administration and Programs,* National Safety Council, Itasca, Ill., 1992.
13 *Accident Prevention Manual for Business and Industry: Engineering and Technology,* National Safety Council, Itasca, Ill., 1992.
14 Risinger, J. L., *Burning Characteristics of Hydrocarbon Products, Fire Protection Manual,* vol. 1: *Hydrocarbon Processing,* Gulf Publishing, Houston, 1985.
15 Risinger, J. L., *Flame Propagation of Hydrocarbon Products, Fire Protection Manual,* vol. 1: *Hydrocarbon Processing,* Gulf Publishing, Houston, 1985.
16 Risinger, J. L., and C. H. Vervalin, *Sources of Ignition, Fire Protection Manual,* vol. 1: *Hydrocarbon Processing,* Gulf Publishing, Houston, 1985.
17 Schultz, N., *Fire and Flammability Handbook,* Van Nostrand Reinhold, New York, 1985.

18 Soden, J. E., *Basics of Fire Safety Design, Fire Protection Manual,* vol. 1: *Hydrocarbon Processing,* Gulf Publishing, Houston, 1985.

19 Tenneco Oil Co., *Emergency Systems in the Plant, Fire Protection Manual,* vol. 1: *Hydrocarbon Processing,* Gulf Publishing, Houston, 1985.

20 Vervalin, C. H., *Cationic and Ampholytic Surfactants, Carbon Dioxide and Inhibition of Combustion, Fire Protection Manual,* vol. 1: *Hydrocarbon Processing,* Gulf Publishing, Houston, 1985.

PROBLEMS

1 A gas pipeline exploded with earthquake force in a rural southeastern Texas town, killing 1 person, injuring 21, and igniting a blaze that burned out of control for several hours. Accident investigators said they could not immediately determine the cause of the explosion but suspected that leaking liquified petroleum gas had collected in a low-lying ravine and may have been ignited by sparks from a vehicle or a home stove pilot light. If you were the design engineer of this pipeline, how would you reduce or eliminate the risk of such an explosion occurring?

2 A gun powder manufacturing facility explodes instantaneously, killing all personnel in the vicinity. How would you try to determine the cause of this fatal accident?

3 A carbon steel, flammable-liquid storage drum has a surface temperature of 85°F and a wall thickness of 0.5 in. The 55-gal drum contains a flammable liquid with a flash point of 75°F. What is the maximum allowable heat transfer rate by conduction to keep the temperature of the liquid 10°F below its flash point?

4 A flammable liquid has a heat of combustion of 10,000 Btu/lb and is consumed in a fire at a rate of 3 lb/min. The rate of transfer of heat is 8000 Btu/min by radiation and convection. The heat vaporization of the flammable liquid is 6000 Btu/lb. What is the minimum amount of heat that must be removed in order to bring this fire under control?

5 A flammable liquid has a vapor pressure of 200 mm Hg at its flash point. Calculate the LFL for the vapor generated from the liquid.

6 A cylinder has a pressure of 3000 psig. The dimensions of the storage room are $15 \times 20 \times 30$ ft. The hydrogen gas cylinder has a volume of 110 ft.[3] The control valve on a cylinder of hydrogen in a flammable storage area is broken, releasing hydrogen into the storage room. After the valve broke on the cylinder, were the contents of the room flammable?

7 In a room that contains 4 lb mol of oxygen in air, 15 lb of combustible carbon is burned. Assume that 60 percent of the carbon forms CO_2 and the remaining 40 percent forms CO. Assuming a room atmosphere with less than 19.5 percent oxygen will endanger life, determine whether the air in the room is at a dangerous level.

8 Why is the heat of combustion of hydrocarbon liquids higher than the heat of combustion of coal?

9 Why do gasoline and toluene have similar upper and lower flammability limits?

10 Hydrogen is being processed in a room. Where should the ventilation system for this room be located?

11 Butane is being processed in a room. Where should the ventilation system for this process be placed?

12 Describe the fire tetrahedron.

13 In addition to the fire triangle components, what additional component must be present to have a major industrial fire?

14 Describe NFPA 704, System for Hazard Identification.

15 The leading cause of home fires is (a) smoking materials, (b) arson, (c) heating equipment, or (d) electrical equipment.

16 The leading cause of home fire fatalities is (a) lightning, (b) smoking materials, (c) cooking equipment, or (d) electrical equipment.

17 The majority of fire deaths occur at (a) school, (b) home, (c) work, or in (d) vehicles.

18 The greatest fire-hazardous roofing material is (a) untreated wood shakes, (b) asphalt, (c) slate, or (d) tile.

19 The largest number of home fires and associated fatalities occurs in the (a) spring, (b) summer, (c) fall, or (d) winter.

20 The following time segment accounts for the largest number of home fire deaths: (a) midnight to 4 A.M., (b) 4 to 10 A.M., (c) 10 A.M. to 6 P.M., (d) 6 P.M. to midnight.

21 The largest number of home fires starts in (a) a closet, (b) a utility room, (c) the kitchen, or (d) the attic.

22 Fire deaths result mostly from burns. True/False

23 The smoke from a fire will wake you up. True/False

24 When your clothing catches on fire, you should (a) run to the bathtub or shower, (b) sit still and yell for help, (c) stop, drop, and roll, or (d) put baking soda on the fire.

25 If a small cooking fire starts you should not (a) escape and then call the fire department, (b) pour water on it, (c) slide a lid over the pan, or (d) turn off the heat.

26 Estimate the LFL and UFL of a 1 vol % hexane, 3 vol % methane and 2 vol % ethylene mixture in air. The LFL and UFL for each of these chemicals are shown below.

	LFL (vol %)	UFL (vol %)
Hexane	1.2	6.9
Methane	5.0	15.0
Ethylene	2.7	28.6

27 A safe level ≤ 75 percent of the LFL for a flammable vapor in air is generally used for toluene storage. The LFL of toluene is 1.3 vol % in air. What is the maximum toluene storage temperature that will maintain a safe LFL in air?

28 What would be the effect of adiabatic compression of hexane and air at 100°F from 14.7 to 500 psia. The autoignition temperature of hexane is 500°F and $\gamma = 1.4$ for air.

29 How much energy is released when a 2-ft^3 nitrogen gas cylinder at 1900 psia fails? Assume isentropic conditions for the expansion; $\gamma = 1.4$.

30 Estimate the LFL for methane based upon the rule of thumb of one-half of the stoichiometric air requirement. Compare the estimate with the experimental LFL of 5 mol %.

31 Estimate the LFL for hydrogen based upon the rule of thumb of one-half of the stoichiometric air requirement. Compare the estimate with the experimental LFL of 4 mol %.

32 Using a normal butane LFL of 1.9 mol %, estimate the minimum permissible oxygen concentration for complete combustion.

33 A process furnace heater is operated at 2300°F and is insulated with 4-in-thick fire brick on the inside. The fire brick has an average thermal conductivity of 0.25 Btu/h·ft^2·°F and the 0.35-in steel furnace wall has a thermal conductivity of 24 Btu/(h·ft^2·°F). The exterior steel surface is 210°F and is a safety hazard to plant personnel, requiring insulation with a 0.3 Btu/(h·ft^2·°F) thermal conductivity to be added to reduce the exterior surface temperature to 115°F. The maximum steel operation temperature cannot exceed 450°F. What should be the thickness of the exterior insulation, assuming a negligible temperature drop across the steel wall? Is this a valid assumption? What is the furnace heat loss reduction from the insulation?

34 The safety of a flammable-liquid storage tank requires an inert nitrogen atmosphere with less than 2 vol % oxygen. How could this inert blanket be assured?

35 How could leaking flammable gas that is lighter than air create an explosion hazard in a facility?

36 Compare the flammability of carbon disulfide to ethyl alcohol.

	Carbon Disulfide	Ethyl Alcohol
Flash point, °C	−30	13
Ignition temperature, °C	100	423
Flammable range, %	1.3–44	4.3–19
Vapor density	2.2	1.6

37 Identify several fundamental methods for extinguishing a fire.

38 Why is a water spray or fog more desirable than a water jet in fighting flammable liquid fires?

39 Why is foam effective in extinguishing liquid hydrocarbon fires?

40 Why is foam not recommended for running liquid fires?

41 Why is carbon dioxide desirable for fire extinguishment when a major goal is minimum damage to the property or materials at risk?

42 What design safety considerations could be used to provide fire protection for pressurized vessels?

43 The design pressure for a flammable liquids storage tank must be capable of a 50 percent reduction or being reduced to 8 bar in 15 min, whichever is lower. If an LPG tank is designed to withstand 10 bar, which of these criteria apply?

44 What is the most suitable fire extinguishment substance for a class A fire?

45 What type of substances should be used to extinguish a class B fire?

46 What is the most suitable substance to extinguish a class C fire, which is based upon electrical equipment?

47 Compare the equivalent amount of TNT (1100 kcal/kg heat of combustion) required for peak overpressure damage with 1 ton of a hydrocarbon (12,000 kcal/kg heat of combustion). Assume a hydrocarbon efficiency factor of 3 percent.

48 An unconfined vapor cloud explosion can damage a building based on the distance from the center of the explosion and the mass of flammable vapor according to the following relationship:

$$\text{Damage} = (\text{overpressure distance}) (\text{mass})^{1/3}$$

A concrete block control building is predicted to fail when subjected to a 4.0 overpressure at a radius of 250 m from a UCVCE of 25 tons of vapor. If the control building is subjected to a UCVCE of 11 tons of vapor, determine the critical building distance to failure. What could be done to improve the building safety?

49 What types of industries have the greatest risk of dust explosions?

50 Why is acetylene a potential source of explosive energy?

51 Why are organic peroxides potential explosion hazards?

52 When an industrial fire occurs, the possibility of a domino effect arises. What are examples of such domino effects?

53 The top of a flammable-liquid storage tank is on fire and must be continuously wetted with water to avoid overheating nearby facilities. The desired water flow rate on the

tank cross-sectional surface is 0.2 gal/min (ft²). If the cylindrical storage tank is 150 ft in diameter and 45 ft high, what is the water requirement?

54 There are three cylindrical flammable-liquid storage tanks in a diked tank farm. One of the tanks is on fire and threatens to ignite the other two tanks. All of the tanks are 200 ft in diameter and 50 ft high. The water requirement on the burning tank is 0.25 gal/min (ft²) cross-sectional area. The water needed on the other two half-full tanks is 0.2 gal/min (ft²) surface area. How much water is required?

55 Determine how many gallons of liquid nitrogen are needed to purge the methane in a 6,000,000-gallon LNG storage tank.

56 The vent area needed to prevent major damage to a low-pressure building from an internal explosion is as follows:

$$A_v = \frac{CA_s}{\sqrt{P}}$$

where A_v = vent area required, ft²
  A_s = internal surface area, ft²
  C = constant for material dust
  P = maximum internal pressure allowable, psi

A building that can withstand a maximum internal overpressure of 0.3 psi has an internal surface area of 25,000 ft². Cornstarch is being processed in the building, creating a potential dust hazard. The cornstarch C value is 0.12 psi$^{0.5}$. Determine the venting area needed for this building.

57 A gas compressor building with 50,000 ft² of surface area has been designed to withstand a maximum overpressure of 0.35 psi. Using the vent area equation in the previous problem, what is the required vent area to prevent major damage to the building? The equation constant for methane is 0.37 kPa$^{0.5}$.

58 Estimate the TNT equivalent energy release rate when a 400-ft³ compressed-air tank at 190 psig fails. Assume the air is an ideal gas ($\gamma = 1.4$) and behaves according to the following relationship:

$$W = -\frac{PV}{\gamma - 1}\left[1 - \left(\frac{P_a}{P}\right)^{\gamma - 1/\gamma}\right]$$

where P = vessel pressure, psia
  V = vessel volume, in²
  P_a = atmospheric pressure, psia
Also assume TNT energy release is 2000 Btu/lb TNT.

59 Will an explosion occur when an air-ethylene gas mixture at atmospheric pressure and 100°C is compressed to 1100 psia? The autoignition temperature for ethylene is 490°C, γ is 1.2, and the mixture is within the flammable range. The following relationship will apply for the gas compression.

$$\frac{T_2}{T_1} = \left(\frac{P_2}{P_1}\right)^{\gamma - 1/\gamma}$$

60 A flammable-liquid storage cabinet at 80°F contains the following potential fire hazards:

Liquid	Vapor pressure at 80°F (mm Hg)	Flammable range (mol %)	
		LFL	UFL
Acetone	237	2.6	12.8
Methanol	132	6.7	36.0
Carbon disulfide	380	1.3	50.0
Isopropyl alcohol	47	2.0	12.0
n-Pentane	540	1.5	7.8
n-Octane	15	1.0	6.5
p-Xylene	10	1.1	7.0

If a spill occurs and equilibrium is reached, which of these chemicals would form a flammable mixture in air? What type of fire extinguisher would be appropriate for each of these chemicals?

7

RADIATION HAZARDS

With the development of nuclear power and x-ray equipment, the human race has received many benefits. Along with these benefits there are hazards that can cause considerable harm if uncontrolled. There are many naturally occurring radiation hazards in our environment that are also highly hazardous, for example, radon, radium, uranium, and the sun. There is considerable cancer lifetime risk to people who have been exposed to radiation. [5] The human body is ill-equipped to protect itself against radiation and cannot even detect it with its natural sensing mechanisms.

The control of radiation in the workplace is based mainly upon prevention techniques. Many types of radiation can be found in the workplace and in our environment. The two main categories are ionizing radiation and nonionizing radiation (Fig. 7-1). These two categories are generally used to show the effect of radiation wavelength and energy on living organisms. These potential hazards are regulated by the federal government. [9, 12, 13]

IONIZING RADIATION

All matter is composed of atoms. Atoms have a nucleus of protons and neutrons; electrons spin in orbit around the nucleus. When ionization occurs, there is an energy transfer that changes the normal electrical balance in an atom. Atoms that have an unstable nucleus become radioactive from the emission of particles from the nuclei. Some elements are naturally radioactive and others, like isotopes, become radioactive after bombardment with neutrons or other particles.

The most common types of ionizing radiation are particulate radiation of alpha

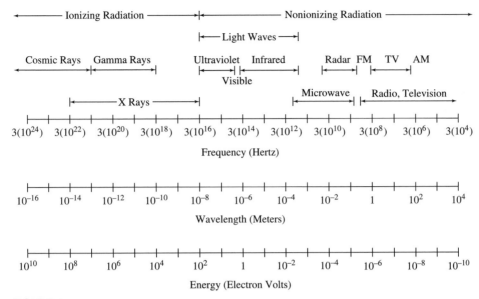

FIGURE 7-1
ELECTROMAGNETIC SPECTRUM AND THE CATEGORIES OF ELECTROMAGNETIC
RADIATION.

particles, beta particles and neutrons, electromagnetic gamma radiation, and x-rays. All of these ionizing radiation sources can be hazardous to human health unless adequate safeguards are provided. [3, 4]

Alpha particles are emitted from the nuclei of radioactive atoms. They have high speeds and high energy and are the least penetrating. Their maximum range in air is about 10 cm and can be stopped by a sheet of paper or the outer layer of skin. If they enter the body, alpha emitters are hazardous to human health. They tend to concentrate in bones and body organs like the lung, liver, or kidney. As they disintegrate, they can damage tissue. The normal routes of body entry by alpha emitters are inhalation, digestion, and through wounds. [8]

Beta particles are also emitted from the nuclei of radioactive atoms, but they are much smaller in mass than alpha particles and have considerably more penetrating power. Beta emitters are capable of penetrating wood to 4 cm and the human body to 1 cm. Like alpha emitters, they are internal radiation hazards and have the same routes of body entry. They can be stopped by an ordinary wall or a 1.3-cm-thick sheet of aluminum.

Neutrons are particles released upon the disintegration of radioactive isotopes. Neutron particles are highly penetrating and require heavy shielding. They are capable of penetrating the human body to several centimeters. Within the body the neutrons release excess energy which can cause tissue damage. Secondary releases of energy may also occur from alpha, beta, and gamma emitters released from the neutrons. [1]

Gamma radiation originates in the nucleus of an atom. It can penetrate deeply into tissue, and may produce burns similar to deep sunburn, alter genes to cause mutations, and reduce the white blood cell count and encourage infections.

X-rays cause cancer by inducing mutations. [7] X-rays are a form of electromagnetic radiation that is produced when high-speed electrons strike the target material inside the x-ray tube. The energy of the x-ray is inversely proportional to the wavelength (Fig. 7-1). The quality of the x-ray, which is the power to penetrate through matter, depends upon the wavelength and the material being irradiated. Hard x-rays with short wavelengths will penetrate thick steel plate. Soft x-rays with long wavelengths are less penetrating. The most common safeguard for x-ray equipment is protective shielding, as shown in Table 7-1. Shielding is used to reduce the x-ray penetration in terms of half-value layers, which is the material thickness to reduce the incident radiation by 50 percent. [8]

Radioactive materials retain their radiation properties for extended periods of time. This reduction in radiation occurs naturally and results from the decay of radioactivity. The half-life is the time rate of decay for a radioactive material to lose one-half of its beginning intensity or strength. Some materials lose their radioactivity in days or even minutes, others in years. The half-life for a radioactive isotope is specific for that isotope (see Table 7-2). A linear relationship is normally assumed in the average rate of decay of radioactive isotopes, making the half-life constant for each material.

High-energy radiation will penetrate and produce effects deep into the body, while low-energy radiation has its maximum effect on the surface of the body. Alpha and beta radiation will be completely absorbed by tissue. Gamma rays will be partially absorbed. Radiation is usually measured by the ionization produced by the passage of radiation through a medium.

Several basic units are used to quantify radiation (see Table 7-3). Ionizing radiation can produce biological damage to cells or tissue. The type of radiation, the radiation energy, and the substance being irradiated are factors in the total amount of radiation absorbed in a tissue.

Radiation exposure time and intensity determine the probability of tissue injury from electromagnetic radiation. The degree of injury is directly proportional to

TABLE 7-1
HALF-VALUE LAYERS FOR CERTAIN SHIELDING MATERIALS

Shielding materials	Type of radiation	
	Cobalt 60 (cm)	Cesium 137 (cm)
Lead	1.2	0.6
Copper	2.1	1.6
Iron	2.2	1.7
Zinc	2.7	2.1
Concrete	6.6	5.3

TABLE 7-2
HALF-LIFE OF SELECTED RADIOACTIVE ISOTOPES

Element	Symbol	Mass number	Half-Life (yr)	Radiation emission*
Carbon	C	14	5730	P
Cesium	Cs	137	30	P, R
Cobalt	Co	60	5.3	P, R
Iodine	I	131	0.02	P
Iron	Fe	55	2.6	R
Nickel	Ni	63	92	P
Polonium	Po	210	8000	P
Radium	Ra	226	1602	P, R
Radon	Rn	222	3.82 d	P, R
Selenium	Se	75	0.3	P
Sodium	Na	22	2.6	P
Strontium	Sr	90	28	P
Sulfur	S	35	0.2	P
Uranium	U	238	4.5×10^9	P

*P = radioactive particles; R = radioactive rays.

these two variables. Radiation intensity is dependent upon the energy of the source, the distance from the source, and the type of shielding present, if any. The more powerful the source, the greater the intensity. The intensity from a point source is inversely proportional to the square of its distance. Shielding requirements depend upon the type of radiation present.

Alpha particles can be shielded with paper sheet or aluminum foil; beta particles with lead, aluminum, or glass sheeting; gamma rays with sheeting or walls based upon the materials in Table 7-1.

TABLE 7-3
COMMON UNITS TO MEASURE RADIATION

Rad	The unit of absorbed dose of ionizing radiation that is equal to the absorption of 100 ergs/g of irradiated material or tissue.
Roentgen	The unit of exposure dose of ionizing radiation from x-radiation or gamma radiation. It is equal to the amount of radiation whose absorption by a standard cubic centimeter of air will result in one electrostatic unit of charge.
Rem	The dosage of ionizing radiation that will cause the same biological effect as a dose of 1 roentgen of x-, or gamma radiation, 1 rad of x-, gamma, or beta radiation, 0.1 rad of neutrons or high-energy protons, and 0.05 rad of particles heavier than protons with sufficient energy to reach the lens of the eye.
Curie	The rate at which radioactive material emits particles: 1 curie corresponds to 3.7×10^{10} disintegrations per second.

BIOLOGICAL EFFECTS OF IONIZING RADIATION EXPOSURE

Humans and other animals are exposed to naturally occurring ionizing radiation, resulting in natural background radiation to individuals of about 80 mrem/yr in a normal environment. There is a typical dose-response relationship to this background exposure, as well as other radiation exposures from the workplace or lifestyles. [10]

Synthetic (manmade) radiation is somewhat higher than background radiation. The greatest sources are diagnostic tests—x-rays, CAT scans, and magnetic resonance imaging (MRI)—and tracers using injected, inhaled, or orally administered radioactive isotopes.

Radiation treatment may also result in adverse effects such as hair loss, skin and eye reactions, and skin cancer. Ionizing radiation is also used to examine welds and internal equipment, packages, and baggage. The nuclear power industry contributes an estimated 1 percent of the total synthetic radiation in the United States. [10]

Whenever ionizing radiation passes into or through a substance, energy is transferred. The radiation is absorbed, metabolized, and distributed throughout the body. The *dose* is the amount of radiation absorbed by the substance. The *effects* depend on the type of radiation and exposure time. Ionizing radiation destroys the reproductive capacity of cells or causes them to mutate.

There is usually a latent period between initial radiation exposure and observable adverse body effect. The latent period may be long or short, depending on the size of the dose. The response of the various body tissues and organs to identical irradiation may vary considerably. The hazards of radiation exposure have been well known for many years, but quantification has been more difficult because meaningful data are usually gained from laboratory research and studies about population exposure. As more information about radiation exposure has become available, allowable radiation levels have been reduced over the years and acceptable risks have become more conservative.

PROTECTION FROM IONIZING RADIATION

Exposure to ionizing radiation produces a higher than normal incidence of cancer, bone damage, and cataracts and a reduced life expectancy as well as damage to the human reproductive system. In 1975, the National Council on Radiation Protection published occupational exposure limits for radiation (Table 7-4). OSHA has established more restrictive and conservative radiation exposure limits in the workplace (see Table 7-5). [12]

The most important somatic effect of ionizing radiation is cancer. It is difficult to attribute cancer solely to radiation, however. The estimation of excess radiation risk at low dosage levels usually involves extrapolation from observations at higher dosages.

Cancer may occur in nearly all tissues of the human body. The natural incidence of cancer from radiation depends on age, sex, and the type and site of neoplasm origin. Solid tumors are more common than leukemia, but they have a longer la-

TABLE 7-4
MAXIMUM PERMISSIBLE OCCUPATIONAL RADIATION EXPOSURE LIMITS (1975)

Type of exposure	Upper exposure limit
Whole body occupational exposure	
Prospective annual limit	5 rem/yr
Retrospective annual limit	10–15 rem/yr
Long-term accumulation	$(N - 18) \times 5$ rem*
Skin	15 rem/yr
Hands	75 rem/yr (25 rems per quarter)
Forearms	30 rem/yr (10 rems per quarter)
Other organs, tissues and organ systems	15 rem/yr (5 rems per quarter)
Pregnant women	0.5 rem in gestation period
Population dose limit	0.17 rem/yr
Emergency dose limits: life saving	
Individual	100 rem
Hands and forearms	200 rem
Total	300 rem
Emergency dose limits: less urgent	
Individual	25 rem
Hands and forearms	100 rem
Total	100 rem
Family of radioactive patients	
Individual (< age 45)	0.5 rem/yr
Individual (> age 45)	5 rem/yr

*N is age in years.
Source: National Council on Radiation Protection, 1975.

TABLE 7-5
FEDERAL GOVERNMENT IONIZING RADIATION LIMITS FOR OCCUPATIONAL
EXPOSURE

Type of exposure	Recommended exposure limit (Rems per calendar quarter)
Whole body, head and trunk, active blood-forming organs, lens of eyes, or gonads	1.25
Hands and forearms, ankles and feet	18.75
Skin of whole body	7.5

Exceptions: An employee may permit whole-body doses of 3 rems in any calendar quarter. Whole-body doses plus the accumulated occupational dose to the whole body should not exceed 5 $(N - 18)$ rems, where N is the age of the individual on the last birthday.
Source: OSHA, 29 CFR 1910.96.

tency period. Cataracts, skin damage, hematologic deficiencies, and impairment of fertility or the fetus are examples of other somatic radiation effects. [10]

Most of the conclusions regarding genetic or hereditary radiation risk to humans are based upon animal studies. The genetic effects of radiation are usually gene mutations and chromosome aberrations, coupled with their effect on future generations. Some results of genetic change are tragic and conspicuous, while others are trivial and invisible. These genetic results usually occur in the next generation, but may be delayed until several generations into the future. The general population is exposed to much more significant levels of ionizing radiation, when compared to occupational exposure levels. Table 7-6 identifies natural radiation and patient x-ray exposures.

A radiation safety program should establish safe working procedures, identify and quantify radiation, decontaminate and dispose of hazardous materials and wastes, and maintain proper records. Although ionizing radiation cannot be sensed by sight, touch, hearing, taste or smell, it can be detected and measured with instrumentation.

Radiation measurements are usually performed on the radiation source, surface

TABLE 7-6
SOURCES OF IONIZING RADIATION GONADAL EXPOSURE
TO THE GENERAL POPULATION

Radiation source	Equivalent dose rate (mrem/yr)
Medical and dental x-rays	
Patients	20
Occupational	0.4
Radioactive isotopes	
Patients	3
Occupational	0.15
Weapons testing	4.5
Consumer products	4.5
Nuclear power industry	
Environmental	1.0
Occupational	0.15
National laboratories	0.2
Military applications	0.04
Industrial applications	0.01
Air travel	0.5
Natural radiation	
Cosmic	28
Body radionuclides	28
Terrestrial	26
Total	116.45

Source: National Research Council, 1980.

contamination, and airborne radioactivity. These radiation surveys may use a variety of detectors, depending upon the type of measurement.

Ionizing radiation can be detected and monitored with film badges, dosimeters, thermoluminescence detectors, ionization chambers, and Geiger-Mueller counters. These radiation meters should be calibrated with standard laboratory procedures by qualified personnel to assure their reliability. [8]

In protecting humans from ionizing radiation, there are three main considerations: time, distance, and shielding. These three variables are interrelated in any well-defined radiation safety program.

The longer the time of exposure to ionizing radiation, the higher the probability of damage to tissue and organs. There is a direct proportional relationship between exposure time and dose. If the dose rate at a work location and the maximum allowable dose for the given time period are known, the maximum acceptable exposure time can be obtained. With proper planning, only necessary employee exposures should be permitted. The radiation levels and the necessary time to achieve the required work task in a restricted area may result in higher-than-allowable exposures to an employee. This situation can usually be safely managed by spreading the total exposure over several employees to permit individual exposures below the allowable limit.

As an approximation, the inverse square law can be used to calculate the change in external radiation exposure as the distance from a point source in open space is varied. When a person moves away from the radiation source, the radiation intensity falls off by the square of the distance. A safe distance from radiation sources must be maintained by employees, based upon a workplace survey. Hazardous radiation areas in the workplace should be restricted to employee entry and signs should be posted (Figure 7-2). The standard colors for the sign are magenta or purple on a yellow background.

The use of shielding to protect employees from radiation is common; the greater

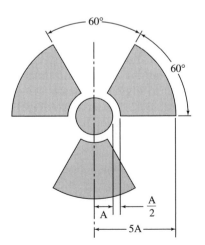

FIGURE 7-2
STANDARD SYMBOL FOR RADIATION WARNING SIGNS.

the shielding mass, the greater the protection afforded the employees. There are many forms of shielding: cladding on radioactive material, heavy-walled containers, concrete cells, and deep layers of water. High-density materials like lead attenuate x-rays or gamma radiation more effectively for a given thickness than less dense materials like concrete. The penetration of x-rays through shielding materials is normally expressed in value layers, as shown in Tables 7-1 and 7-7.

The establishment of a controlled area is basic in radiation protection. Any work area where radiation dose rates are excessive should be guarded by barriers and warning signs (Figure 7-2), have security guards to restrict entry, and otherwise be isolated from the rest of the workplace.

Work performed in the controlled area should be planned and managed to minimize radiation exposure to employees. Radiation can easily contaminate air, water, and solid materials, potentially exposing employees and the public. Contaminated clothing and shoes can be hazardous, particularly if employees unknowingly leave the workplace while wearing them.

The degree of toxicity of radioactive sources depends upon the amount and kind of radiation material. Radioactive materials are classified according to their relative toxicity (Table 7-8). Class 1 isotopes have very high toxicity, class 2 have high toxicity, classes 3 and 4 have moderate and slight toxicity, respectively.

The use of radionuclides to measure materials is practiced in the industrial and medical fields. This practice may create hazardous exposures in the workplace. X-ray machines use remote actuation and shielding for employee protection. Workers should wear film badges and use dosimeters when operating sealed radiation sources such as gauges. Sealed sources can be subject to breakage and spillage, creating severe hazards; they should be tested at regular standardized intervals.

Laboratory and medical use of radioisotopes create hazards of various degrees, depending on the types of compounds, their activity and volatility, the procedures involved, and the dosages to organs and tissues. [8]

TABLE 7-7
ONE-TENTH AND HALF-VALUE THICKNESSES FOR COMMON
RADIATION SHIELDING MATERIALS

Shielding material value thickness (cm)	Type of radiation		
	Cobalt 60	Radium 226	Cesium 137
Lead			
0.1 value	4.11	4.70	2.13
0.5 value	1.24	1.42	0.64
Iron			
0.1 value	7.36	7.70	5.72
0.5 value	2.21	2.31	1.73
Concrete			
0.1 value	22.9	24.4	18.0
0.5 value	6.90	7.37	5.33

Source: Nuclear Regulatory Commission.

TABLE 7-8
RELATIVE TOXICITY OF SELECTED RADIOACTIVE ISOTOPES

Class 1: very high toxicity	Sr-90, Pb-210, Ra-226, U-233
Class 2: high toxicity	Ca-45, Fe-59, Sr-89, I-131
Class 3: moderate toxicity	Na-22, S-35, Mn-52, Co-58, Zn-65
Class 4: slight toxicity	H-3, C-14, F-18, Cr-51

Source: International Atomic Energy Agency.

Radiation should be treated with respect because of both external and internal hazards. External hazards are usually measurable through continuous monitoring devices. They can be controlled by limiting the exposure time and rate. Providing additional shielding or locating personnel away from the source are the main ways of limiting external radiation hazards.

Internal radiation hazards usually occur from inhalation of radioactive materials. Detection of hazardous air levels with associated operating conditions can be used to alert employees to minimize exposure. Whenever a potential air inhalation hazard exists, personnel should evacuate to a safe location. Radioactive spills should be isolated and cleaned up quickly.

Internal radiation may also occur through skin cuts or abrasions; this is serious because the blood disperses radiation quickly throughout the entire body. The radionuclide content of body fluids and wastes is important in determining the intake of radioactivity. The unwanted effects of radiation exposure often occur many years after exposure, making it difficult to identify the causal effect. Record keeping should be practiced to maintain related information for future reference.

The treatment, storage, and disposal of radioactive wastes have become a serious problem in many countries. With the end of the cold war both the United States and Russia have experienced an escalation in both public and government concern on this issue. [6] The management of radioactive waste in the United States must be in compliance with 10 CFR, part 20, to limit public and occupational exposure. [2] The U.S. Department of Energy has generated huge quantities of radioactive wastes in its nuclear weapons complex. [14]

NONIONIZING RADIATION

Nonionizing radiation has longer wavelengths and lower frequencies (Fig. 7-3). Nonionizing radiation includes radio, television, radar, microwave, infrared, visible light, and ultraviolet areas of the electromagnetic spectrum.

ULTRAVIOLET RADIATION

The ultraviolet radiation spectrum is between x-rays and visible light on the electromagnetic spectrum. Photochemical reactions in the skin can be initiated by UV radiation, producing vitamin D_3, which helps to prevent ricketts. UV radiation ex-

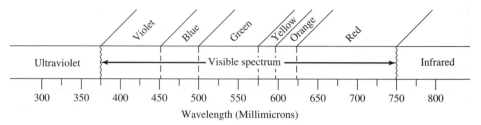

FIGURE 7-3
EXPANDED ELECTROMAGNETIC SPECTRUM IN THE VISIBLE LIGHT REGION.

posure has decreased the incidence of certain infectious diseases. Large dosages of UV radiation may destroy eyes and skin. Suntan or pigmentation of the skin also results from UV exposure. [8]

The ultraviolet spectrum has several UV regions (Table 7-9). The UV-A, or black light, ranges between 400 and 320 nm on the electromagnetic spectrum. The most important application of UV-A radiation is the generation of visible fluorescent light. A long glass tube with mercury vapor is coated with a material called phosphor, for example, using magnesium tungstate coating for a blue-white light. An electrical discharge through the vapor results in UV radiation that energizes the phosphor, which becomes a source of visible light. Most of this UV radiation is absorbed in the glass bulb and the phosphor coating.

UV-A radiation is also responsible for suntan, the pigmentation of skin. The sun's rays damage skin cells from exposure to UV-A and UV-B radiation, leading to early wrinkles, skin cancer, and other skin problems. Suntan is the body's attempt to protect itself from the harmful rays of the sun. Risk factors associated with skin cancer are shown in Table 7-10.

The UV-B region, the erythemal region, is the most biologically active region and potentially the most harmful. When UV-B radiation is absorbed by the cornea of the eye, keratitis can result.

The UV-C region, between 280 and 100 nm, has bacterial or germicidal effects on the body. This region is usually present in germicidal lamps and welding operations. Welding arcs, which contain the entire spectrum of UV radiation, have the potential to adversely affect exposed employees. Eye protection and safe distances

TABLE 7-9
REGIONS OF ULTRAVIOLET RADIATION

Ultraviolet designation	Electromagnetic spectrum range (nm)	Observable effect
UV-A	400–320	Suntan
UV-B	320–280	Eye damage, sunburn
UV-C	280–100	Bactericidal

TABLE 7-10
COMMON RISK FACTORS FOR SKIN CANCER

Fair skin
Blond or red hair
Moles, freckles, or birthmarks
Frequent sun exposure
Frequent sunburn
Light-colored eyes

are the most widely used safety controls in welding operations. The severity of irritation or burn to the eye from the flash of electric arc welding depends on the exposure time, the energy level, and the UV wavelength. The luminance and radiance of any arc weld is directly proportional to the square of the amperage.

VISIBLE ENERGY

Ordinary window glass is nearly opaque to UV radiation from sunlight and transparent to visible light. The visible portion of the electromagnetic spectrum, shown in Figure 7-3, is between 380 and 750 nm. Visual comfort is the most important criterion in setting safe exposure levels. The violets and blues have shorter wavelengths than the oranges and reds. Illumination in the workplace should be well balanced for safe working conditions. The goal of industrial lighting is to provide a safe working environment and comfortable and efficient visual effects and to reduce losses from poor visual performance.

INFRARED RADIATION

The infrared radiation region extends from the red visible light region to microwaves, which is 750 nm to 3 mm. IR radiation occurs from differences in temperature between the higher-temperature energy source and the receiver. Most IR applications involve heating for baking, drying, heat treating, and dehydration of industrial products. Heating rates are controlled by the radiant energy intensity, absorption of the exposed surfaces, and heat loss to the surroundings.

IR radiation heats the skin of exposed people, depending upon the wavelength, the quantity of energy transmitted, and the length of exposure. The main safety hazards of IR radiation are damage to skin tissues and the eyes. To avoid adverse eye effects, exposure to IR wavelengths greater than 770 nm should not exceed 10 mW/cm^2 averaged over any 0.1-h period. [11, 13]

MICROWAVES

Microwave ovens and dryers are widely used to heat, cook, and defrost food. They are convenient, clean, and quick and react instantaneously to controls. The human

TABLE 7-11
RECOMMENDATIONS TO MINIMIZE EMPLOYEE EXPOSURES DURING LASER
OPERATIONS

---

1 Train personnel on laser hazards and the importance of limiting exposure.

2 Post conspicuous warning signs against looking directly into the primary beam and specular reflections.

3 Allow only authorized personnel to operate lasers; never leave lasers unattended during operation.

4 Secure operating lasers in enclosed, brightly lighted areas that are regularly monitored for ozone and stray radiation.

5 Provide personnel with baseline and periodic eye examinations.

6 Ensure that lasers in operation are not directed at personnel and that laser targets have dull, nonreflective surfaces.

---

body exhibits thermal effects when the microwave energy is transferred to the body. The higher the frequency, the lower the health hazard. At frequencies less than 3000 MHz, if the intensity and exposure time are sufficient, body temperatures can rise and localized skin damage may occur. From a microwave oven safety viewpoint, the power tube and wave guide should be shielded at all times. Microwave leakage usually occurs whenever there are cracks or breakage in the power tube or oven. The removal of interlocks on the oven doors can create a serious hazard.

LASERS

Lasers produce high-intensity light radiation with a narrow wavelength band. They have numerous industrial, military, and communication applications. Lasers are serious hazards to the human eye when uncontrolled because they are a point source of tremendous brightness near the source and this light is a narrow beam. The laser beam is capable of inflicting serious injury to an unprotected eye from thermal and biological effects.

Laser hazard control should minimize the opportunity for ocular exposure by either the direct beam or specular reflections. Recommended precautions for the protection of operating personnel are shown in Table 7-11.

BIBLIOGRAPHY

1 Anton, T. J., *Occupational Safety and Health Management,* 2d ed., McGraw-Hill, New York, 1989.

2 Berlin, R. E., and C. C. Stanton, *Radioactive Waste Management,* Wiley, New York, 1989.

3 Hammer, W., *Occupational Safety Management and Engineering,* 4th ed., Prentice-Hall, Englewood Cliffs, N.J., 1989.

4 Kavianian, H. R., and C. A. Wentz, *Occupational and Environmental Safety Engineering and Management,* Van Nostrand Reinhold, New York, 1990.

5 Klaassen, C. D., et al., *Casarett and Doull's Toxicology,* 5th ed., McGraw-Hill, New York, 1996.

6 Krieger, G. R., *Accident Prevention Manual for Business and Industry: Environmental Management,* National Safety Council, Itasca, Ill., 1995.

7 Ottoboni, M. A., *The Dose Makes the Poison,* 2d ed., Van Nostrand Reinhold, New York, 1991.

8 Plog, B. A., et al., *Fundamentals of Industrial Hygiene,* 3d ed. National Safety Council, Itasca, Ill., 1988.

9 Slote, L., *Handbook of Occupational Safety and Health,* Wiley, New York, 1987.

10 Tang, Y. S., and J. H. Saling, *Radioactive Waste Management,* Hemisphere, New York, 1990.

11 *Threshold Limit Values and Biological Exposure Indices,* American Conference of Governmental Industrial Hygienists, Cincinnati, Ohio, 1995.

12 29 CFR 1910.96, OSHA, Washington, D.C., 1992.

13 29 CFR 1910.97, OSHA, Washington, D.C., 1992.

14 U.S. Environmental Protection Agency, *Approaches for the Remediation of Federal Facility Sites Contaminated with Explosive or Radioactive Wastes,* EPA/625/R-93, 013, Washington, D.C., 1993.

PROBLEMS

1 When a radium atom emits an alpha particle, radioactive radon gas is formed. How long will it take for the radon to lose at least 99 percent of its radioactivity?

2 Radioactive waste with carbon 14 is to be stored in an underground repository. How long must this repository maintain its geological integrity to assure the safe decay of at least 90 percent of the activity of this waste?

3 The radiation level from a gamma-ray source is 50 mrem/h. What is the radiation from this source in 3 h?

4 If the radiation exposure time in problem 3 is increased by a factor of 4, what dose level will result?

5 An ionizing radiation exposure rate of 3 mrem/h for 24 h is increased to 17 mrem/h. At the higher rate how long should the exposure be to achieve the same dosage as at the lower rate?

6 A restricted area has ionizing radiation at a rate of 136 mrem/d. It is necessary for a 5-h maintenance task to be performed in the restricted area by two, always-present maintenance personnel. The allowable radiation exposure limit for an 8-h work day is 13 mrem/d per person. How would you safely schedule the maintenance work?

7 A point source of gamma radiation has an intensity of 3 mrem/h at 2 ft. What would be the radiation level at a distance of 11 ft?

8 If the distance from an ionizing radiation source is decreased from 25 m to 3 m, how is the exposure affected?

9 What are the OSHA-established radiation exposure limits in the workplace?

10 What are the main sources of ionizing radiation risk for most people?

11 What are common risk factors for skin cancer in humans?

12 What procedures can help reduce workplace hazards from lasers?

8

NOISE HAZARDS

As industry has become more mechanized, noise levels have usually become more intense and higher than those normally experienced outside the workplace. Workers' compensation laws now cover the loss of hearing because of noise exposure. These compensation laws have encouraged employers to control industrial noise.

Occupational hearing loss is a partial or complete hearing impairment of one or both ears as a result of workplace exposures. Noise-induced hearing loss and, to a lesser extent, acoustic trauma from sudden blasts or explosions are the main causes of occupational hearing impairment. Noise-induced hearing loss is the result of chronic, cumulative exposure to hazardous noise over many years. It usually affects both ears equally and permanently. Acoustic trauma is severe damage to the inner ear from a single incident of sudden, intense acoustic energy. [4]

ANATOMY OF THE EAR

The human ear is a highly complex organ that enables humans to hear sound, which is based on variations in air pressure above and below the surrounding atmospheric pressure. The air pressure fluctuations, which are the sound waves, vary in intensity, frequency, harmonic motion, and direction. The sound waves create sensation experienced when the air pressure fluctuations reach the ear. The normal human ear can detect sound waves at 20 to 20,000 cycles per second, or hertz. The sound waves are converted within the ear into electrical impulses, which are passed to the brain for interpretation and reaction. The ear is divided into three parts, the external, middle, and inner ear (Fig. 8-1 and 8-2). [4]

The *external ear* consists of the outer portion attached to the head, called the

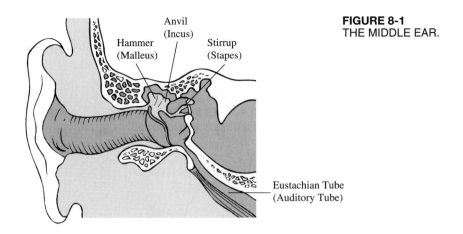

FIGURE 8-1
THE MIDDLE EAR.

pinna, and the external auditory or ear canal. The pinna acts as a funnel to capture and collect the sound waves so that they can be transported via the ear canal to the eardrum, or tympanic membrane, which separates the external ear from the middle ear. The eardrum has a mucous membrane and a layer of fibrous tissue (see Fig. 8-3 on page 222).

Particulate matter is filtered in the outer third of the ear canal by small hairs and cerumen, a wax secreted by the ceruminal glands. The cerumen also removes bacteria to keep the ear canal free from infection.

The *middle ear* is the cavity between the eardrum and the bony wall of the inner ear. The ossicles are located in the middle-ear cavity and link the eardrum to an opening in the wall of the inner ear called the oval window. Three ossicles, or bones, called the hammer, anvil, and stapes, form the sound-conducting mechanism in the middle ear. The middle ear is lined with mucous membrane that is similar to the lining of the mouth. When the handle of the hammer is set into motion by the movement of the eardrum, there is a mechanical transfer of the action through the ossicular chain to the oval window. The oval and round windows are

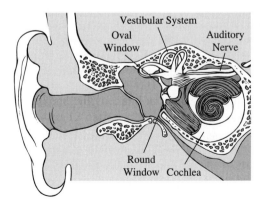

FIGURE 8-2
THE INNER EAR.

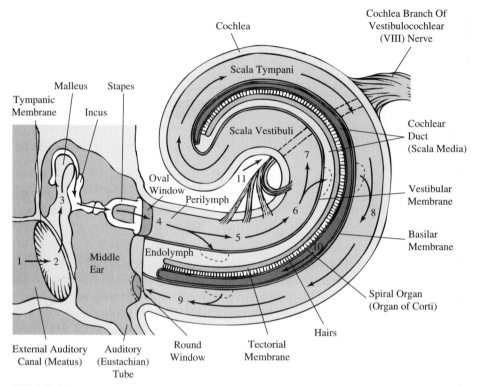

FIGURE 8-3
THE PHYSIOLOGY OF HEARING. (*Source: Principles of Human Anatomy,* 4th. ed. by G. J. Tortora. Copyright © 1986 by Biological Sciences Textbooks, Inc. Reprinted by permission of Addison-Wesley Educational Publishers.)

located on the inner wall of the middle ear. The oval and round windows are the movable barriers between the middle and inner ears. As the footplate of the oval window moves inward, the membrane on the round window moves outward. When the condition reverses, the round-window membrane moves inward and the footplate of the oval window is pulled out.

The eustachian tube equalizes the middle-ear pressure with the external ambient pressure. When the ambient pressure is increasing, the eustachian tubes tend to close. When the ambient pressure is decreasing, the eustachian tubes tend to open. Swallowing and yawning cause pressure differentials that also cause the eustachian tubes to open.

The *inner ear* contains bony and membranous labyrinths where receptors for hearing and sensing are located. There are three divisions to the inner ear: the vestibule, the cochlea, and the semicircular canals. The bony labyrinth, which contains a fluid called perilymph, surrounds the membranous labyrinth, which contains a fluid called endolymph. The bony labyrinth protects the delicate membranous labyrinth and houses the cochlea and the semicircular canals.

The cochlea is shaped like a snail shell with two and a half turns and three channels that are partitioned from each other. [7] The channel above the bony partition is called the scala vestibuli; the channel below the bony partition is called the scala tympani; the third channel is the membranous labyrinth called the cochlear duct. The cochlear duct contains the membranous labyrinth, which is separated from the scala vestibuli by the vestibular membrane. The membranous labyrinth is separated from the scala tympani by the basilar membrane. The spiral organ, which is the actual hearing organ, rests on the basilar membrane. The spiral organ has a number of hair cells that are receptors of auditory sensations. This complex structure of hair cells extends into the endolymph of the cochlear duct. The mechanical sound vibrations are transformed into neural activity by an electrochemical impulse, which travels to the brain via the cochlear or acoustical nerve. The delicate tectorial membrane is over and around the hair cells of the spiral organ. High-intensity noise can easily damage the hair cell patterns of the spiral organ. Chronic elevated noise exposure can permanently distort and degenerate the function of the hair cells. [7, 8] The damage and distortion is somewhat reversible with reasonable rest periods between exposures.

MECHANISM OF HEARING

Sound waves are the result of the alternate compression and decompression of air molecules. When sound waves reach the ear, they are directed by the pinna into the ear canal. As the sound waves strike the eardrum, their alternating compression and decompression causes the tympanic membrane to vibrate. The membrane vibrates more rapidly in response to higher frequencies than to lower frequencies.

The central area of the eardrum is connected to the middle ear with the hammer, or malleus, which also vibrates in response to the eardrum. These vibrations are passed along to the anvil, or incus, and in turn to the stapes, or stirrup. As the stapes vibrates, the oval window is pushed in and out to transmit the vibrations to the perilymph of the bony labyrinth in the inner ear. This mechanical motion of the eardrum is thereby effectively transmitted through the middle ear and into the fluid of the inner ear.

As the oval window is pushed inward, the perilymph of the scala vestibuli propagates pressure waves through the scala vestibuli and into the perilymph of the scala tympani. These pressure waves push the vestibular membrane inward and increase the pressure of the endolymph inside the cochlear duct. The basilar membrane responds to the pressure by being pushed outward into the scala tympani. The high-frequency resonance of the basilar membrane occurs near its base. The low-frequency resonance of the basilar membrane occurs near its apex. The pressure change in the scala tympani causes the perilymph to exert pressure on the round window, causing it to push back into the middle ear. As the sound pressure waves diminish, the stapes moves backward to reverse the procedure, with the basilar membrane bulging into the cochlear duct.

As the basilar membrane vibrates, the hair cells of the spiral organ are moved against the tectorial membrane. The movement of the hair cells generates the electrochemical potential that ultimately results in nerve impulses. The differences in sound frequencies cause intensity variations in the regional vibrations of the basilar membrane. Higher-intensity sound waves will cause greater vibration of the basilar membrane and increase the sensation of loudness. Quiet sounds cause less vibration of the basilar membrane and lower vibration of the membrane. The nerve impulses are passed on to the cochlear nerve for signal transmission to the brain. [7]

HEARING DISORDERS

Sensorineural hearing loss is commonly associated with disorders of the inner ear. Heredity, aging, disease, noise, and ingested toxic substances can influence sensorineural hearing loss. It usually proceeds as a progressive loss of hearing. Trauma, an injury to living tissue caused by an extrinsic agent like a blast from a sudden explosion, is an example of noise-induced hearing loss. So is the more common chronic hearing loss as a consequence of repeated noise injury over a long time period. Industrial noise and lifestyle noise are the major causes of chronic hearing loss, which is usually irreversible.

Presbycusis, a hearing loss of the elderly, gradually changes the delicate labyrinth structures as people age. It may begin to become apparent at middle age. Presbycusis cannot be cured but may be treated with a hearing aid.

In addition, hearing impairments that are not induced by noise include the following:

- Physical blockage of the auditory canals by excessive wax or foreign objects
- Diseases that infect the inner ear, for example, smallpox
- Punctured eardrums or displacement of the ossicles by traumatic damage
- Damage from drugs such as streptomycin, furosemide (lasix), or quinine

The external ear has a thin, sensitive skin that is subject to sunburn or frostbite. It needs to be protected from weather extremes. The ear canal contains wax to protect it from water during bathing and swimming and at other times from drying and scaling. Foreign objects that become accidentally lodged in the ear should be removed by a physician. Most ear infections are enhanced by elevated skin temperature and humidity. The ear canal wax is usually self-cleansing, but abnormal wax buildup may occur, causing wax impaction. The cleaning of ear wax with cotton-tipped swabs may pack the wax in the ear canal. As a result, wax impaction is best cleaned by an otologist because of the risk of either injury or infection from the cleaning process.

A heavy blow to the ear may compress the air in the ear canal sufficient to rupture the eardrum. Sudden pressure changes from explosions or underwater diving can also cause perforations. Such perforations usually heal spontaneously.

If the eustachian tube fails to open and ventilate the middle ear properly, fluid could be pulled into the middle ear through the oval window, or the eardrum could be pulled inward, causing hearing loss. Infectious diseases of the middle ear may result in fixation of the ossicles, causing them to vibrate ineffectively or not at all. Disease may also produce a gap in the chain of the ossicle that interrupts hearing.

PROPERTIES OF SOUND

Sound is the pressure variations that a human ear can detect. The medium for sound transmission may be air, water, or solid media. Sound frequency is measured by the number of sound pressure variations per second, in cycles per second or hertz.

The difference between sound and noise is in the hearing of the receiver. When sound carries no useful information, it is noise. In other words, the sound is unwanted by the listener because it is unpleasant noise. The way humans react differentiates between sound and noise.

Sound and vibration forms of sound energy have similar physical characteristics. The difference between them is that sound produces a hearing sensation. Vibration results in the sensation of touch.

Sound is generated and propagated as back-and-forth vibrations that compress the air and cause a slight increase in pressure, followed by a partial vacuum of the air. The alternate compression and rarefaction of the air cause fluctuations in the atmospheric pressure that propagate outward from the sound source. As the pressure fluctuations strike the eardrum, it vibrates based on the atmospheric pressure changes. This vibration is translated into a neural sensation in the inner ear. The sound is then interpreted by the brain.

The longitudinal vibration of the conducting medium is called a sound wave. Sound waves may vary in amplitude (intensity) and frequency (Fig. 8-4). Intensity measures the loudness of the sound. The sound frequency determines the pitch. Sound wave A is a lower pitch than sound wave B because the sound waves of B have a shorter frequency. The intensity of sound wave C is greater than sound wave A because the distance Y is greater than the distance X (see Fig. 8-4). [4]

Sound frequency is the number of times per second that a sound source point is

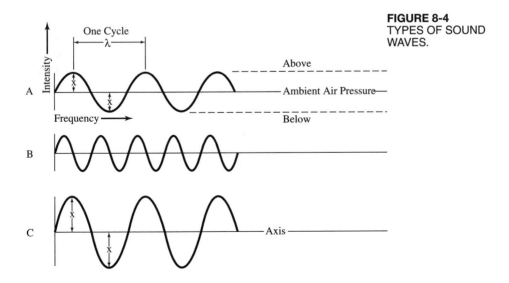

FIGURE 8-4
TYPES OF SOUND WAVES.

displaced from equilibrium, goes through a complete cycle, and returns to its equilibrium position. This means frequency is the number of times per second a vibrating body requires for one complete cycle of motion. The reciprocal of the frequency is called the sound wave period, which is the time required for a complete cycle to be completed. The audible range for normal humans with good hearing is about 20 to 20,000 Hz.

The distance that a sound wave travels in one cycle is the wavelength. The speed of sound is the velocity. The speed of sound is the product of the wavelength and the frequency.

$$v = \lambda f$$

where v = speed of sound, m/s
  λ = wavelength, m
  f = frequency, Hz

The frequency of sound is one factor in the annoyance level to humans. High-frequency noise is usually more annoying than lower-frequency noise. Narrow-frequency bands of noise are less desirable than broad-frequency bands with multiple frequencies. Sound wavelength and frequency are inversely proportional to each other for any given speed of sound. In air at 20°C the speed of sound is 344 m/s or 1130 ft/s. This results in a considerable delay between what you see at a distance and what you hear at the same distance. Common examples of this phenomena include lightning and thunder, firearm discharge and report, shout and echo, or explosion and blast. The speed of sound varies with the surrounding media. As the medium becomes more dense and less compressible, the speed of sound increases.

Air	344 m/s
Water	1433 m/s
Wood	3962 m/s
Steel	5029 m/s

The sensitivity of the human ear is truly remarkable. Minute changes in sound pressure can easily be detected. Sound consists of a series of positive pressure disturbances or compressions, alternating with negative pressure disturbances or rarefactions. Because of these alternating disturbances, the mean value of the sound pressure is zero, not a useful measurement from a practical viewpoint.

The root-mean-square (rms) of these alternating disturbances is useful because it is the square root of the arithmetic mean of their squares. The rms converts both the negative and positive sound pressures to positive values for the magnitude of the sound wave. The pressure units used to measure sound are newtons per square meter (N/m^2) or atmospheres (atm). At a sound frequency of 1000 Hz a barely audible sound has an rms pressure of about 2×10^{-10} atm, while a loud jet engine sound has an rms pressure of about 0.02 atm. This enormous range between barely audible and painful sound pressure is difficult to envision. The sound pressure of the threshold of pain is about 1×10^8 times greater than the sound pressure of the threshold of hearing. These large differences in sound pressures are neither convenient nor practical to use in the workplace, making a relative logarithmic scale more desirable.

The decibel (dB) is the unit that is used to measure these relative sound differences. It is proportional to the logarithm of the ratio of two amounts of acoustical power. Arbitrarily, zero decibels is the sound pressure at the level of the threshold of hearing for healthy young humans. As the sound level increases, the decibel level increases, as shown in Table 8-1. This table covers the entire audible range.

The sensation in loudness is proportional to the logarithm of the stimulus, which is the sound intensity or energy. As the stimulus is increased, so is the sensation.

$$\alpha = C \log \frac{I}{I_0}$$

where α = sensation
$\quad C$ = constant
$\quad I$ = final intensity
$\quad I_o$ = initial intensity

TABLE 8-1
COMMON EXAMPLES OF SOUND PRESSURE LEVELS

	Sound pressure		
	(dB)	(N/m²)	(atm)
Rocket motor	195	100,000	1.0
Jet engine	160	2,000	0.2
Threshold of pain	140	200	2×10^{-3}
Heavy metal band	130	60	6×10^{-4}
Riveter, chipper	120	20	2×10^{-4}
Punch press	110	6	6×10^{-5}
Passing truck	100	2	2×10^{-5}
Lawn mower	90	0.6	6×10^{-6}
Garbage disposal	80	0.2	2×10^{-6}
Vacuum cleaner	75	0.1	1×10^{-6}
Conversation	60	2×10^{-2}	2×10^{-7}
Private office	50	6×10^{-3}	6×10^{-8}
Quiet room	40	2×10^{-3}	2×10^{-8}
Whisper	20	2×10^{-4}	2×10^{-9}
Threshold of good hearing	10	6×10^{-5}	6×10^{-10}
Threshold of exceptional hearing	0	2×10^{-5}	2×10^{-10}

The sensation value is a bel when $C = 1$, and a decibel when $C = 10$. The decibel is the standard unit of measure for sound intensity.

$$\alpha = 10 \log \frac{I}{I_0}$$

There are 1×10^{-5} N/m² in 1 atmosphere of pressure. One N/m² equals one pascal (Pa) of pressure, which equals 10 µbar. The hearing threshold then has a reference point of 2×10^{-5} Pa or 20 µbar. The mean square pressure level may be described by the following relationships:

$$L_{SP} = 10 \log \frac{P^2}{P_o^2} = 20 \log \frac{P}{P_o}$$

where L_{SP} = sound pressure level, dB
  log = log to base 10
  P = rms sound pressure, N/m²
  P_o = reference sound pressure, N/m²

The reference sound pressure is the threshold of hearing at a reference tone of 1000 Hz. It is normally 2×10^{-5} N/m² or 20 µPa. A scale of 0 to 140 dB is the broad practical scale for normal human hearing.

The audible frequency range for the average healthy human is about 20 to 20,000 Hz. This means that human response to sound depends upon the frequency. The human ear is most sensitive to hearing between 500 and 5000 Hz. The basis for this phenomena is shown in Figure 8-5, which is from a study of the response

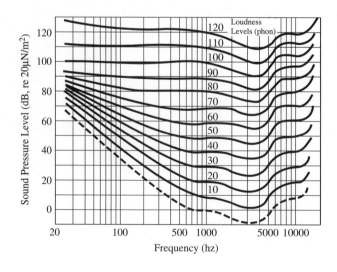

FIGURE 8-5
EQUAL LOUDNESS CONTOURS OF THE HUMAN EAR.

of the human ear to sound. [4] The contours were developed by having the loudness level of a test tone at a given frequency adjusted by young men until it sounded as loud to them as the reference tone at 1000 Hz. The more pronounced variations in this study occurred at frequencies <500 Hz and >5000 Hz. It is particularly noticeable at lower frequencies. For sounds at the lower frequencies to appear as loud as those at higher frequencies, they have to be heard at a greater sound pressure level. The lowest contour in Figure 8-5 represents the loudness level for the threshold of hearing and varies ±10 dB among individuals. The term *phon* for the loudness level in Figure 8-5 is the unit on the loudness scale beginning at zero for the faintest audible sound and corresponding to the decibel scale with the number of phons for a given sound being equal to the decibels of a pure 1000-Hz tone judged by the listener to be equal in loudness to the given sound.

Sound pressure weighting to vary frequency like the human ear is accomplished with three different standardized weighting networks, A, B, and C (see Fig. 8-6). The A network approximates the equal loudness curves at low sound pressure levels, the B network at medium sound pressure levels, and the C network at high sound pressure levels. [4] Sound level meters incorporate these networks to attenuate various frequency ranges. The A scale network provides high attenuation for low frequencies, moderate attenuation for medium frequencies, and minimal attenuation for high frequencies. The B scale network filters out minimal middle-range frequency sound. The C network provides very little attenuation of low-frequency sound. As a result of these weighting networks, the A scale–weighted sound level measurement has become the most commonly used scale in the assessment of overall noise hazard levels for the human ear. The A scale network is the most similar in frequency to the human ear, as shown in Figure 8-5. This A-weighted sound level, which is expressed as dBA, has been adopted by the ACGIH and OSHA for assessing workplace noise.

The frequency content of the noise can be approximated by measuring and comparing the level of noise on all three scales, dBA, dBB, and dBC weighting networks. The difference between the scale readings will show differences between

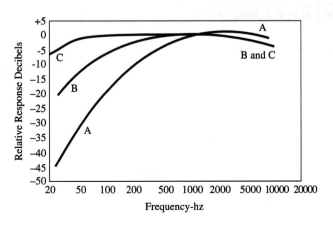

FIGURE 8-6
FREQUENCY RESPONSE
ATTENUATION FOR THE A,
B, AND C WEIGHTING
NETWORKS.

low-, medium-, and high-frequency sounds. If the measurement variations for the same work area are compared with the different weighting networks, the frequency filtration will make the dominate frequency spectrum readily apparent, particularly at low frequencies. If all of the weighting scales show comparable readings, the area noise is probably fairly well balanced between medium and high frequencies.

OCCUPATIONAL NOISE EXPOSURE

Hearing is adversely affected by exposure to sound over the human lifetime. Normal, healthy children possess the best hearing capability. Gradually, over many years our hearing capabilities decline because of exposure to sound and noise. To protect our hearing, we should limit both occupational and lifestyle sound and noise exposure to an acceptable level during our lifetime. Risk assessment can be used to develop criteria for acceptable small noise changes in the hearing levels of exposed workers during their working careers. Practical, reliable, and equitable risk exposure data and information are needed to define these acceptable exposures.

When the human ear is subjected to relatively high levels of noise over a period of time, hearing may be impaired. Many factors will influence this hearing loss, including the factors shown in Table 8-2. The complex relationship of these numerous hearing loss factors requires many years of study to draw meaningful conclusions.

Exposure to high noise levels may cause a temporary reduction in hearing ability, called a temporary threshold shift. When the person is removed from the noisy environment, the temporary threshold shift is gradually diminished. If the noise exposure is repeated over time, the hearing loss could become a permanent threshold shift. The risk of hearing damage is a function of the total noise exposure and the exposure duration at each noise level. The OSHA regulations have an 85-dBA action level and a 90-dBA permissible exposure limit. These levels were believed to be consistent with avoiding a permanent threshold shift.

TABLE 8-2
FACTORS THAT INFLUENCE OCCUPATIONAL HEARING LOSS

Age of the employee
Preemployment hearing impairment
Diseases of the ear
Sound pressure level of the noise
Sound frequency of the noise
Length of daily exposure
Duration of employment
Ambient conditions of the workplace
Distance from the noise source
Employee lifestyle outside the workplace

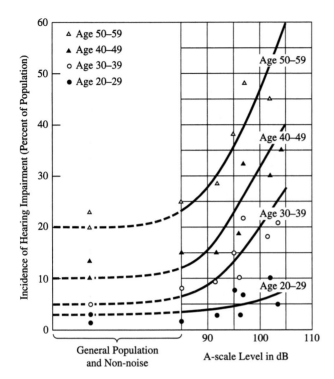

FIGURE 8-7
THE EFFECT OF AGE ON HEARING.

The age of the employee is an enormous factor in hearing loss, as shown in Figure 8-7. This family of curves relates the incidence of significant hearing loss to age and occupational noise exposure over working careers. About 20 percent of people in the general population who are not exposed to occupational noise will experience hearing loss in their lifetime. The steady-state exposure to industrial noise over a career will increase the incidence of hearing loss. The magnitude of occupational hearing loss is directly proportional to the decibel exposure level in the workplace. This study identified a threshold for hearing loss in the workplace in the range of exposure of 85 to 90 dBA. The OSHA noise regulations, further discussed in Chapter 10, were based in part on this information.

Unprotected exposure to noise levels >115 dBA should be avoided because this is definitely a hazardous level. Sound levels <70 dBA are safe to unprotected ears. Since most industrial exposure falls into the 70- to 115- dBA range, additional information about the type and duration of the noise exposure is usually needed.

Frequency is important because above 500 Hz there is a greater potential for hearing loss than at the lower frequency ranges. Noise with a sharp frequency peak usually is a greater hazard to hearing than the same intensity levels spread over a broad frequency range.

Total exposure time at a given noise level is a major factor in noise-induced hearing losses. Steady, long-term, continuous noise exposure is more damaging to hearing than intermittent exposures at the same or even somewhat higher sound pressure levels. Periods of rest between such noise exposure allow the hearing mechanism to recover and recuperate.

The harmful effects of noise exposure and sound pressure level are related, but this is not a linear relationship. The greater the loudness, the less exposure time required to produce an equivalent loss of hearing. For example, for a given sound pressure level, a doubling of the exposure time will not produce twice the hearing loss. For a given exposure time, doubling the sound pressure level will not result in twice the hearing loss. The steady-state noise from a turbine compressor probably has a different effect on hearing than the impact noise of a punch press.

MEASUREMENT OF WORKPLACE SOUND

The most widely used instrument to measure sound pressure variations is the sound level meter. This instrument has a microphone, an amplifier with a calibrated attenuator, appropriate A, B, and C weighting networks, and an indicating meter. It is a sensitive electronic voltmeter, which measures the electric signal from the microphone and amplifies the electrical signal to deflect the needle on an indicating meter. The frequency-response characteristics are controlled by the appropriate weighting networks. Sound level meters have a broad dB range, such as 50 to 120 or 40 to 140 dB, relative to 20 $\mu N/m^2$. The meter for field use should be reliable, rugged, lightweight, and reasonably stable in operation.

Industrial noise is made up of differing sound intensities at various frequencies. Engineering control of noise is best achieved by identification of pure tone components to find where the noise energy lies within the frequency spectrum. Octave-band analyzers are used to determine the location of the sound pressure levels within the frequency spectrum. The center frequencies increase by a factor of 2; for example, 31.5, 63, 125, 250, 500, 1000, 2000, 4000, 8000, 16,000 Hz. Usually a sound level meter is used to first find the excessive noise, followed by the use of the octave-band analyzer. The analyzer readings are obtained by filtering out the frequencies outside of the octave to be measured. The octave-band sound pressure levels are converted to the equivalent A-weighted sound level (Figure 8-8). [9] This equivalent A-weighted sound level may differ somewhat from the actual A-weighted sound level.

Workers may move about to various locations during the work day, making fixed-location noise measurements inadequate. Dosimeters are used when the personal exposure of an employee must be made at several employee work locations during the work day. The dosimeter is worn by the employee to provide an integrated time-weighted-average (TWA) exposure of the employee during the workday. This sound-measuring instrument has a microphone in the hearing zone of the employee.

SOUND MEASUREMENT SURVEYS

Source noise measurements are often made in the presence of background or ambient noise levels. Any survey should describe the procedures followed, including

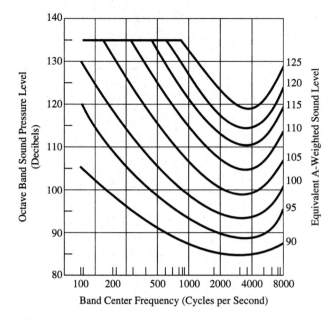

FIGURE 8-8
EQUIVALENT SOUND
LEVEL CONTOURS.

the measurement techniques and depositions, operating conditions, exposure time, and instrument calibration. [5, 6]

It is usually difficult to accurately evaluate the noise exposure to an employee because of exposure fluctuations throughout the work day. It is not practical or reliable to define noise exposure to an employee based on only a single reading on a sound level meter. The TWA is a much more meaningful measurement for noise exposure. TWA exposure for noise is calculated in an analogous manner to TWA exposure for toxic chemicals.

Whenever the employer believes the possibility exists in the workplace for exposure to hazardous noise, a hearing conservation program should be voluntarily considered and probably implemented by the employer. There are OSHA regulations in Chapter 10 that discuss the need for a mandatory hearing conservation program in the workplace based on ≥85 dBA over an 8-h TWA.

Any hearing conservation program should begin with a preliminary facility-wide noise level survey. Appropriate sound level meters and personnel should be used to locate areas or operations where workers may be exposed to hazardous noise levels. The preliminary noise survey should focus on those work areas where it is difficult to communicate in a normal tone of voice. Another indicator of the need for such a survey would be complaints by workers of speech or hearing problems.

The quality of the information developed during the noise survey should be reproducible by other personnel, who may wish to verify noise exposure situations at the facility. The preliminary survey should provide sufficient data and information to identify noise exposure problems and their extent in the workplace. A more detailed quantitative noise survey should follow into the identified trouble areas to help define the noise problem in depth and determine employee exposure time in detail.

A detailed quantitative noise survey should follow at the targeted locations from the preliminary survey. The objectives of a detailed noise survey, shown in Table 8-3, should identify noise problems and assist in the development of sound, practical solutions to those problems. An effective hearing conservaton program should be based upon solving the problems from the survey. Each noisy work area is the result of equipment or processes that are generating noise. The employees in these work areas should be identified, along with the duration of noise exposure to them.

Noise exposure measurements should be made at about ear level. The preferred microphone location should be in accordance with ANSI S1.13-1971, *Methods for the Measurement of Sound Pressure Levels.* The microphone should be at least 1 m from the noise source. The noise survey should be done with a sound level meter that meets the standards of ANSI S1.4-1971 (R1976), *Specification for Sound Level Meters, Type 1 or 2.* The A-scale slow response setting should be used on the sound level meter.

The detailed noise survey consists of area measurements, workstation measurements, and related exposure durations. The first step is to take measurements that are performed for the normal maximum and minimum noise levels at the center of each work area.

Next, the workstation measurements should be performed wherever the measurements at the center of the work area were at least 80 dBA. Maximum and minimum noise levels should be recorded. Consistent noise levels of 90 dBA are unsatisfactory. Noise levels that never exceed 85 dBA are satisfactory. These workstations with readings between 85 and 90 dBA are potential problem areas, requiring further analysis.

The third step in the detailed noise survey is analysis at those workstations that vary above and below 85 dBA.

If an employee has noise exposure in different work areas, or if the noise level varies at a single workstation, it is necessary to measure those variations in noise level and exposure duration. The exposure duration times may be obtained by con-

TABLE 8-3
OBJECTIVES OF A DETAILED NOISE SURVEY

---

1 Obtain detailed, specific information on the noise levels present at each identified workstation of employees.

2 Determine regulatory compliance.

3 Define which employees and workstations require audiometric testing and determine additional areas where audiometric testing is desirable.

4 Identify recommendations for management and engineering controls.

5 Determine required areas for hearing protection.

---

sulting the employee and the supervisor or through visual minitoring. An employee log of time and work area is usually the most accurate approach to obtaining reliable exposure duration information. Noise dosimeters may also be desirable for quantitative accuracy.

When two individual sound sources, for example, two nearby machines, are present in the same work area, there is an additive effect for the combined noise level of the two sources that is greater than either individual source. The combined effect of two noise levels may be determined with Table 8-4. The difference in the two noise levels determines the decibel addition to be added to the higher of the two noise sources.

Sometimes there will be more than two different noise sources present in a work area. When this occurs, the combined effect of these three or more sounds should also be calculated with the aid of Table 8-4. First, place all of the noise sources in order of ascending decibels. Then, combine the two lowest decibel sources, and then combine that total with the result of the next higher decibel source. Continue up the decibel levels until all of the sources have been combined for the final combination level.

The OSHA-permissible noise exposures, shown in Table 8-5, are based on 29 CFR 1910.95. The values not included in Table 8-5 can be calculated with the following relationship:

$$T = \frac{8}{2^{(L-90)/5}}$$

TABLE 8-4
DECIBEL ADDITION FOR MULTIPLE SOURCES OF
SOUND PRESSURE LEVELS

Difference in levels (dB)	Addition to higher level (dB)
0	3.0
1	2.6
2	2.1
3	1.8
4	1.4
5	1.2
6	1.0
7	0.8
8	0.6
9	0.5
10	0.4
11	0.3
12	0.2

TABLE 8-5
OSHA-PERMISSIBLE NOISE EXPOSURES FOR AN 8-H
WORK DAY

Exposure duration allowable (hours per work day)	Permissible sound level (dBA, slow response)
8	90
6	92
4	95
3	97
2	100
1.5	102
1	105
0.5	110
≤0.25	115

Source: 29 CFR 1910.95.

where: T = exposure duration allowable, hours per work day
L = permissible sound level, dBA

When there are two or more periods of noise exposure at different levels, their combined effect should be determined by the following relationship.

$$D = 100 \sum_{i-1}^{n} \frac{C_i}{T_i}$$

where: D = accumulated noise exposure, percent per work day
C = exposure time at specific noise level, dBA
T = exposure time permitted in Table 8-4 at that level, h
n = number of exposure periods

If the sum of the exposure periods equals or exceeds 100 percent, the permissible exposure limit has been exceeded. Noise levels below 90 dBA are not considered in this assessment procedure for an 8-h work day.

When a dosimeter is used to measure sound pressure levels over an 8-h work day, the readout will be in terms of a dose. The dosimeter reading can be converted to a TWA with the following equation:

$$\text{TWA} = 16.61 \log\left(\frac{D}{100}\right) + 90$$

where: TWA = 8-h TWA sound level, dBA per 8-h work day
       D = accumulated noise exposure, percent per 8-h work day

In addition to this calculated evaluation, the exposure to impulsive or impact noise should never exceed a 140-dBA peak sound pressure. Impact noise is a sharp burst of sound and differs from the more common steady-state noise. The peak levels for impact noise are more difficult to measure. Noise that is <0.5 s in duration and not repeated more than once per second is considered to be impulse or impact noise. Explosions or hammers are examples of impact noise.

OTHER VIBRATION EFFECTS

Vibration can have many adverse effects, with noise being the most common. Other adverse effects of vibration result mainly from low-frequency sounds that may contribute to serious disorders in humans. Employees can tolerate vibrations for a short period of time, but prolonged exposure can result in pain and aggravation of disorders.

Vibration can cause metal fatigue, which can cause failure of mechanical parts and equipment under stress. Steering mechanisms, motor parts, fluid lines, pressure vessels, storage tanks, and liquid containers are all subject to vibration failure or rupture.

Vibrating tools, particularly handheld vibrating tools, can increase the severity of rheumatoid arthritis, bursitis, injury to soft tissues of the hands, and restrict flow of blood in the exposed part of the body. For example, a vascular disease like Raynaud's disease is characterized by abnormal vasoconstriction usually of the hands upon exposure to cold or emotional stress. Raynaud's disease, while not being caused by vibrating hand tools, can certainly be aggravated by holding vibrating tools for extended periods of time.

BIBLIOGRAPHY

1 *Threshold Limit Values and Biological Exposure Indices,* American Conference of Governmental Industrial Hygienists, Cincinnati, Ohio, 1994.
2 Hammer, W., *Occupational Safety Management and Engineering,* 4th ed., Prentice-Hall, Englewood Cliffs, N.J., 1989.
3 Kavianian, H. R., and C. A. Wentz, *Occupational and Environmental Safety Engineering and Management,* Van Nostrand Reinhold, New York, 1990.
4 Plog, B. A., et al., *Fundamentals of Industrial Hygiene,* 3d ed., National Safety Council, Itasca, Ill., 1988.
5 Royster, J. D., and L. H. Royster, *Hearing Conservation Programs,* Lewis, Chelsea, Mich., 1990.
6 Slote, L., *Handbook of Occupational Safety and Health,* Wiley, New York, 1987.
7 Tortora, G. J., *Principles of Human Anatomy,* Harper & Row, New York, 1986.
8 Sigler, B. A., and L. T. Schuring, *Ear, Nose and Throat Disorders,* Mosby, St. Louis, 1993.
9 29 CFR 1910.95, OSHA, Washington, D.C., July 1, 1992.

PROBLEMS

1 What is the gain in sound intensity caused by a 1-dB increase in loudness?
2 What is the change in sound intensity caused by the following changes in loudness?
  a +3 dB
  b +10 dB
  c +20 dB
  d −5 dB
  e −10 dB
3 A sound pressure is 10 times the threshold of hearing. How many decibels are generated?
4 What is the decibel level for the following sound pressures, using 2×10^{-5} N/m² as a reference?
  a 12 N/m²
  b 1.6 N/m²
  c 248 N/m²
5 What is the approximate occupational hearing loss for a group age 50 to 59 that has had steady-state noise career exposure to 90 dBA in the workplace?
6 What is the approximate occupational hearing loss for the following age groups and steady-state noise career exposures in the workplace?
  a age 20 to 29, 95 dBA
  b age 30 to 39, 100 dBA
  c age 40 to 49, 90 dBA
  d age 50 to 59, 100 dBA
7 What is the equivalent A-weighted sound level for an octave-band sound pressure level of 100 dB? The band center frequency is 500 Hz.
8 What is the equivalent A-weighted sound level for the following octave-band analyzer readings?
  a 110 dB at 2000 Hz
  b 105 dB at 4000 Hz
  c 95 dB at 200 Hz
9 What is the work day noise exposure for an employee who is exposed to the following noise levels during the work day? How does it compare to the OSHA standard?
  80 dBA for 1 h
  85 dBA for 1.5 h
  90 dBA for 2 h
  97 dBA for 3.5 h
10 What is the noise exposure compared to the OSHA standard for the following employee exposure levels?
  84 dBA for 6.5 h
  92 dBA for 0.5 h
  100 dBA for 0.5 h
  105 dBA for 0.5 h
11 What is the noise 8-h TWA based upon a dosimeter reading of 105 percent per 8-h day?
12 What is the allowable noise exposure duration for an 8-h work day if the permissible exposure limit is 98 dBA?
13 What is the permissible noise exposure limit for an 8-h work day, if the allowable exposure duration is 1.2 h?
14 What is the combined effect of two sound sources, 95 dBA and 90 dBA?

15 Three different pieces of equipment operate in the immediate vicinity of each other. What is the combined noise effect, given the following noise readings for each machine?

Noise source	Sound pressure level (dBA)
Machine A	87
Machine B	92
Machine C	96

9

HAZARD COMMUNICATION TO EMPLOYEES

The OSHA Hazard Communication Standard, 29 CFR 1910.1200, is meant to protect employers and employees by having producers, users, and importers evaluate the hazards of their chemicals. The information about these chemical hazards must be transmitted to the employer and employees. This information includes material safety data sheets, container labeling, and employee training. [8, 11]

GOAL OF HAZARD COMMUNICATIONS

The intent of the Hazard Communication Standard is to provide employers and employees with chemical hazard information and appropriate protective measures. Some of the typical items that may be involved in hazard communication are shown in Table 9-1. Many of these items are among the most common OSHA violations that are cited under the OSHA General Industry Standards, as shown in Table 9-2. This makes 29 CFR 1910.1200 the most often violated OSHA standard for industry and worthy of considerable attention from a safety perspective. Employees must be provided with sufficient information about potential health hazards from exposure to workplace chemicals to allow them to make more knowledgeable decisions about the personal risks of their work and to impress upon them the need for safe work practices. Worker psychology and human error can play an important role in occupational health and safety risk. [10] The employer must provide employees with comprehensive training that includes those areas shown in Table 9-3 on page 242. [17]

The Hazard Communication Standard applies to any hazardous chemical as established by either of the following:

TABLE 9-1
TYPICAL ITEMS IN HAZARD EVALUATION AND COMMUNICATION

1 Developing and maintaining a written hazard communication program
2 List of all hazardous chemicals present in the workplace
3 Labeling of containers that contain chemicals in the workplace
4 Labeling of containers that contain chemicals for shipment
5 Preparation and distribution of material safety data sheets (MSDSs) to employees and downstream employers
6 Development and implementation of employee training programs for chemical hazards and protective measures

- OSHA 29 CFR 1910, pt Z, "Toxic and Hazardous Substances"
- American Conference of Governmental Industrial Hygienists (ACGIH), *Threshold Limit Values for Chemical Substances and Physical Agents in the Work Environment*

Cancer is of special concern to employers and employees. A chemical can be established as a carcinogen or potential carcinogen under the Hazard Communication Standard by any of the following sources.

TABLE 9-2
MOST FREQUENTLY CITED OSHA 29 CFR 1910 GENERAL INDUSTRY STANDARDS FOR 1994

1 Hazard communication: Failure to develop an adequate written plan 1910.1200(e)[1]
2 Hazard communication: Failure to provide employee information and training 1910.1200(h)
3 Lockout/tagout: Failure to establish an energy control program (procedures and employee training) to prevent injuries from unexpected energization of equipment 1910.147(c) [1]
4 Machine guarding: Failure to guard employees from point of operation hazards 1910.212(a) [1]
5 Abrasive wheel machinery: Failure to guard abrasive wheels 1910.215(b) [9]
6 Hazard communication: Failure to label containers of hazardous chemicals with the identity of chemicals 1910.1200(f) [5] (i)
7 Hazard communication: Failure to obtain a material safety data sheet for hazardous chemical 1910.1200(g) [1]
8 Medical services/first aid: Failure to provide eye and body wash equipment where employees are exposed to corrosives 1910.151(c)
9 Hazard communication: Failure to label containers of hazardous chemicals with appropriate warnings 1910.1200(f) [5] (ii)
10 Mechanical power transmission: Failure to guard pulleys 1910.219(d) [1]

Source: National Safety Council, 1994.

TABLE 9-3
OSHA-REQUIRED AREAS TO BE INCLUDED IN EMPLOYEE TRAINING UNDER HAZARD
COMMUNICATION

- Overview of the requirements of 29 CFR 1910.1200
- Methods, procedures, and techniques to detect and characterize the presence of hazardous chemicals in the workplace.
- Physical and health hazard assessment of those hazardous chemicals present
- Measures employees can take to protect themselves from the hazards, such as appropriate work practices, emergency procedures, and personal protective equipment
- Location and availability of the hazardous material inventory, MSDSs, and other written hazard communication material
- Instruction in how to use MSDSs
- Explanation of the hazardous chemical labeling system

- National Toxicology Program, Annual Report on Carcinogens
- International Agency for Research on Cancer monographs
- OSHA 29 CFR 1910, pt Z, "Toxic and Hazardous Substances"

Industrial hygiene plays an important role in hazard communication, from material safety data sheet (MSDS) interpretation to risk assessment, through employee training.

Many workplace chemicals are not covered by 29 CFR 1910.1200 because they are regulated by other federal legislation. These include pesticides, foods, food additives, color additives, drugs, cosmetics, distilled spirits, wine or beer, and hazardous wastes.

Mixtures of chemicals are hazardous if the results of the mixture are hazardous. Without such mixture testing, the mixture has the same health hazards as those components with concentrations ≥1.0 percent by weight or volume. The mixture is carcinogenic if a carcinogen is present and has a ≥0.1 percent concentration.

In addition, if the hazardous or carcinogenic component of a mixture could be released and exceeds the OSHA PEL or the ACGIH TLV, then the mixture has the same hazard as the component.

The manufacturers and importers of chemicals must convey the hazard information to employers with labels on containers and material safety data sheets. The employer must in turn provide this hazard information on labels and MSDSs to the employees.

CONTAINER LABELS FOR HAZARDOUS CHEMICALS

Chemical manufacturers, importers, vendors, and distributors must label, tag, or mark containers of hazardous chemicals with appropriate hazard warnings for employee protection. The hazard warning can be a message, selected words, pictures, or symbols, as long as it conveys the hazards of the chemical in the container. Containers of hazardous chemicals being shipped should be labeled, tagged, or marked with the identity, appropriate hazard warnings, and the name and address of the manufacturer or other responsible party. [1]

In the workplace, each container must be labeled, tagged, or marked with the identity of the hazardous chemicals and show hazard warnings appropriate for employee protection. Written hazard warnings in the United States should be in English and be prominently displayed. If the employer believes it desirable, additional languages may also be used.

ANSI has a voluntary standard, ANSI Z129.1-1994, for the precautionary labeling of hazardous industrial chemicals. This standard was developed by a technical committee of the Chemical Manufacturers Association (CMA) and then received ANSI accreditation. The intent of Z129.1-1994 is to provide guidance for precautionary information on container labels and to complement an overall hazard communication program. The ANSI label information, shown in Table 9-4, is not intended to include all information on the properties and hazards of substances. It should have sufficient information for the safe handling of the substance. More detailed information should be available from MSDSs and other health and safety communications. [20]

The product or hazardous component identification item in Table 9-4 should be adequate to allow selection of proper action in case of exposure. The hazard signal word should indicate an immediate hazard in diminishing order of severity: DANGER, WARNING, and CAUTION. For highly toxic chemicals, a skull-and-crossbones symbol and the word POISON could be placed above the instructions for added emphasis.

The hazard statement should provide a notice of hazards that are present during

TABLE 9-4
ANSI-RECOMMENDED SUBJECT MATTER FOR INCLUSION ON PRECAUTIONARY
LABELS FOR HAZARDOUS INDUSTRIAL CHEMICALS

Product or hazardous component identification
Hazard signal word
Statement of hazard
Precautionary measures
First aid instructions
Antidotes
Notes to physicians
Fire instructions
Spill or leak instructions
Handling and storage instructions
Reference MSDS
Name and address of manufacturer, importer, or distributor
Telephone number for additional information

the use of the substance: for example, "Extremely flammable or harmful if absorbed through the skin." Precautionary measures should supplement the statement of hazard by providing protective measures: for example, "Avoid breathing dust or keep away from heat, sparks, and flame."

First aid instructions should be given whenever exposure or contact warrant immediate first aid treatment. Antidotes should be included on the label if they are known and easily administered. Notes to physicians should list the known antidotes and recommended medical practices.

Fire instructions for confining and extinguishing fires should be provided to people who handle containers during shipping and storage. Spill or leak instructions should give methodologies for managing spills or leaks to minimize exposure, injuries, and environmental contamination. Handling and storage instructions should include additional information for special handling and storage procedures. An MSDS reference should be made, if one is available.

The name and address of the manufacturer, importer, or distributor should include the street, city, state, and zip code. Imported substances should also provide the country and mail code. A telephone number for additional product information should include the applicable hours of operation.

Small containers or packages of substances may have label space limitations. The effective use of the applicable precautionary information may require prioritizing the items in Table 9-4. Label durability and legibility should also be considered for label effectiveness.

Another labeling system that is simple and easy to understand is the NFPA 704 system. It is particularly useful to provide serious hazards and an order of severity of those hazards.

There are several alternatives for container labeling in the workplace. Employers can post signs or placards that convey the hazard information for a number of stationary containers with similar contents and hazards. Various types of standard operating procedures, process sheets, batch tickets, and similar written materials can be substituted for container labels if the same information is conveyed to employees in the work area.

Portable containers used for the immediate transfer of hazardous chemicals from labeled containers are exempt from labeling.

MATERIAL SAFETY DATA SHEETS

Chemical manufacturers and importers should obtain or develop an MSDS for each hazardous chemical they produce or import. Employers, users, and distributors of hazardous chemicals should be given an appropriate MSDS with their initial shipment and an updated MSDS with the first shipment after the update. The producer of the MSDS is responsible for the information being accurate and up to date. Distributors must also ensure that their customer employers are similarly provided with an MSDS.

The MSDS should be in English, but it can be in other languages as well, and include the chemical name and common name of the hazardous chemical. If the

TABLE 9-5
INFORMATION ABOUT THE CHEMICAL SUBSTANCE TO INCLUDE IN MSDS

1 Chemical name and common name of the ingredients
2 Physical and chemical characteristics, for example, vapor pressure and flash point
3 Physical hazards, for example, potential for fire, explosion, and reactivity
4 Health hazards, for example, signs and symptoms of exposure
5 Primary routes of entry
6 OSHA PEL and ACGIH TLV
7 Carcinogen or potential carcinogen designation
8 Safe handling and use precautions, for example, hygienic practices
9 Control measures, for example, administrative controls, engineering controls, and personal protective equipment
10 Emergency and first aid procedures
11 Date of MSDS preparation or latest revision
12 Name, address, and telephone number of the party responsible for preparation or distribution of the MSDS

Source: 29 CFR 1910.1200.

hazardous chemical is a mixture, the chemical and common name of the ingredients that contribute to the hazard should be given, as well as the common name for the mixture. The MSDS should contain information for the hazardous chemical, as shown in Table 9-5.

The MSDS should be readily accessible to all employees who are working in the vicinity of the hazardous chemicals during each workshift. Alternatives to paper copies of MSDSs, such as microfiche and electronic access, are acceptable so long as there are no barriers to MSDS access in each workshift and workplace. For multiple locations, MSDSs may be kept at a primary workplace facility, with the information immediately available in an emergency.

No hazardous chemical should be used in the workplace unless an MSDS is available. Whenever new and significant health hazard information is developed for a material in the workplce, a revised MSDS should be made available as soon as practicable and employees should be notified of changes in work procedures or personal protective equipment to protect health and safety. The MSDSs should be available upon request to OSHA, NIOSH, and employee representatives.

UNDERSTANDING THE MSDS

The MSDS is the cornerstone of hazard communication because it provides employers and employees with the information needed to establish proper work procedures, handle chemicals safely, and maintain regulatory compliance. OSHA requires chemical manufacturers to provide correct data and information on an MSDS. The manufacturer has no such OSHA requirement for a specific MSDS format, although there are specific information requirements. OSHA has a nonmandatory format, Form 174 (Fig. 9-1), which may be used by manufacturers and importers for information compliance. [4, 8] OSHA Form 174 follows the requirements of 29 CFR 1910.1200. Many MSDSs follow the general format of Form 174.

The two-page OSHA MSDS in Figure 9-1 has eight sections, which focus on these principal areas:

- Manufacturer
- Hazardous constituents
- Physical and chemical characteristics
- Fire, explosion, reactivity, and health hazards
- Handling and use precautions
- Hazard control

This OSHA Hazard Communication Standard was welcomed in 1985 by chemical producers and users, who prefer a single federal regulation to diverse state and local regulations, all aimed at the same worker right-to-know goal. Many of these producers and users did not want OSHA to mandate a rigid MSDS format because the companies already had their own formats and information systems.

OSHA established a performance-oriented standard that permitted the MSDS to be presented in any format as long as the OSHA information requirements were achieved. Industry has considerable latitude in the manner of compliance. [3, 19] Some companies use the OSHA two-page format, but most companies have chosen to expand the basic information requirements into more than two pages. This has resulted in some confusion in accessing chemical hazards, particularly in an emergency situation. In subsequent Superfund legislation, Congress mandated that MSDSs be made available to state emergency response commissions, local emergency planning committees, fire departments, and local citizens to aid in planning and control of chemical hazards in communities. [12] As a result, the MSDSs, which were originally developed by and for health and safety professionals, are now being used by workers and citizens, who found much of their language to be user-unfriendly.

The chemical industry has had similar problems with multiple MSDS formats, depth of information, and sometimes subjective interpretations, as both a user and producer of MSDSs. There is usually a need for additional hazard information beyond the OSHA requirements, depending upon the situation of the MSDS user. Chemical users want adequate information from suppliers so that the user can independently assess the degree of hazard for all workplace chemicals, hazardous or not.

Periodically, the MSDS producer reviews and updates the hazard information based upon user experience, regulatory requirements, and other changing needs. The updated MSDS is then made available to the chemical users and other interested parties.

Understanding the MSDS involves training with MSDS examples, guides, and a glossary of terms. This combination of training tools should provide employers

Material Safety Data Sheet
May be used to comply with
OSHA's Hazard Communication Standard.
29 CFR 1910.1200 Standard must be
consulted for specific requirements.

U.S. Department of Labor
Occupational Safety and Health Administration
(Non-Mandatory Form)
Form Approved
OMB No. 1218-0072

IDENTITY *(As used on Label and List)*	*Note: Blank spaces are not permitted. If any item is not applicable, or no information is available, the space must be marked to indicate that.*

Section I

Manufacturer's Name	Emergency Telephone Number
Address *(Number, Street, City, State, and Zip Code)*	Telephone Number for Information
	Date Prepared
	Signature of Preparer *(optional)*

Section II — Hazardous Ingredients/Identity Information

Hazardous Components (Specific Chemical Identity; Common Name(s))	OSHA PEL	ACGIH TLV	Other Limits Recommended	% *(optional)*

Section III — Physical/Chemical Characteristics

Boiling Point		Specific Gravity (H$_2$O = 1)	
Vapor Pressure (mm Hg.)		Melting Point	
Vapor Density (AIR = 1)		Evaporation Rate (Butyl Acetate = 1)	
Solubility in Water			
Appearance and Odor			

Section IV — Fire and Explosion Hazard Data

Flash Point (Method Used)		Flammable Limits	LEL	UEL
Extinguishing Media				
Special Fire Fighting Procedures				
Unusual Fire and Explosion Hazards				

(Reproduce locally)

OSHA 174, Sept. 1985

Section V — Reactivity Data

Stability	Unstable		Conditions to Avoid	
	Stable			

Incompatibility *(Materials to Avoid)*

Hazardous Decomposition or Byproducts

Hazardous Polymerization	May Occur		Conditions to Avoid	
	Will Not Occur			

Section VI — Health Hazard Data

Route(s) of Entry:	Inhalation?	Skin?	Ingestion?

Health Hazards (Acute and Chronic)

Carcinogenicity	NTP?	IARC Monographs?	OSHA Regulated?

Signs and Symptoms of Exposure

Medical Conditions Generally Aggravated by Exposure

Emergency and First Aid Procedures

Section VII — Precautions for Safe Handling and Use
Steps to Be Taken in Case Material is Released or Spilled

Waste Disposal Method

Precautions to Be Taken in Handling and Storing

Other Precautions

Step VIII — Control Measures

Respiratory Protection *(Specify Type)*			
Ventilation	Local Exhaust		Special
	Mechanical (General)		Other
Protective Gloves		Eye Protection	
Other Protective Clothing or Equipment			
Work/Hygienic Practices			

Page 2

* U S G P O 1994-491-529/45775

FIGURE 9-1
OSHA-SUGGESTED FORMAT FOR A MATERIAL SAFETY DATA SHEET.

and employees with a level of understanding that enables them to properly interpret relevant safety and health protection information. [19]

PREPARATION OF THE MSDS

In the preparation of MSDSs, many recognized physical, chemical, and health hazards should be identified on the MSDS. These hazards, shown in Tables 9-6 and 9-7, are recognized by OSHA as potentially harmful to safety and health in the workplace. It is generally desirable on MSDSs to provide sufficient information based on Table 9-8 (page 250) to allow the reader to readily identify the degree of hazard present. There are many scientific terms used in the preparation of MSDSs. The glossary provided in Table 9-8 on pages 250–253 contains the meanings of the terminology found in typical MSDSs.

In the OSHA nonmandatory form (Fig. 9-1), the chemical is identified by chemical name, code name, chemical number, or trade name.

Section I: Information about Chemical Manufacturer

Manufacturer   The manufacturer, importer, employer, or other responsible party preparing or distributing the material safety data sheet should be a source for additional information on the hazardous chemical and appropriate emergency procedures. Include the mailing address, telephone number for general information, and emergency telephone number.

MSDS number (optional)   A space could be made available for cross-referencing an MSDS file. A list of chemicals in the workplace is required by OSHA.

CAS number (optional)   Chemical Abstract Service numbers provide an additional reference for information concerning specific chemicals. The number identi-

TABLE 9-6
PHYSICAL AND CHEMICAL HAZARDS FOR MSDS INCLUSION

Explosive
Flammable
Combustible
Pressure
Pyrophoric
Oxidizer
Initiator
Polymerization
Stability
Water-reactive
Incompatibility

TABLE 9-7
HEALTH HAZARDS FOR MSDS INCLUSION

Carcinogen
Reproductive toxin
Blood toxin
Irritant
Nephrotoxin
Hepatoxin
Neurotoxin

fies the specific compound and allows identification regardless of the name or naming system used.

Date prepared   Gives the date the MSDS was prepared and also provides a reference when updated MSDSs have been prepared.

Prepared by (optional)   Shows the name of the responsible person who prepared the data sheet.

Section II: Hazardous Components

The chemical and common names of all ingredients which have been determined to be reportable health hazards are listed.

If the hazardous chemical is a single substance, its chemical name and common names are listed. The common names listed should be those ordinarily in use for that product.

If the hazardous chemical is a mixture that has been tested as a whole to determine its hazardous properties, the chemical and common names of the ingredients that contribute to those known hazards and the common names for the mixture are listed.

If the hazardous chemical is a mixture that has not been tested as a whole, the chemical and common names are listed for all ingredients that are

1 Determined to be health hazards and comprise 1 percent or more of the mixture
2 Identified as carcinogens and present at 0.1 percent or greater
3 Determined to present a physical hazard when present in the mixture

Inclusion of the percentage composition is optional.

OSHA PEL values can be found in the General Industry Standards, specifically OSHA Safety and Health Standards, 29 CFR 1910.

ACGIH TLV values are found in *Threshold Limit Values and Biological Exposure Indices,* which is updated by ACGIH on an annual basis.

Where other sources have recommended exposure limits, these are to be included in the section "Other Limits Recommended."

TABLE 9-8
SCIENTIFIC TERMINOLOGY USED IN THE PREPARATION OF MATERIAL SAFETY DATA
SHEETS

Acute   A short-term period of action measured in seconds, minutes, hours, or days.

Acute effects of overexposure   The adverse effects that are normally evident immediately or shortly after the exposure to a hazardous material without implying a degree of severity.

Asphyxiant   A vapor or gas that can cause injury by reducing the amount of oxygen available for breathing.

Carcinogen   A chemical that has been demonstrated to cause cancer in humans or animals and is therefore considered capable of causing cancer in humans. A chemical is considered to be a carcinogen if (1) it has been evaluated by the Internal Agency for Research on Cancer (IARC) and found to be a carcinogen or potential carcinogen, (2) it is listed as a carcinogen or potential carcinogen in the *Annual Report on Carcinogens,* published by the National Toxicology Program (NTP) (latest edition); or (3) it is regulated by OSHA as a carcinogen.

Chronic   A long time period of action in weeks, months, or years.

Chronic effects of overexposure   The adverse effects that develop slowly over a long period of time or upon repeated prolonged exposure to a hazardous material without implying a degree of severity.

Combustible liquid   A liquid having a flash point at or above 100°F (37.8°C) but below 200°F (93.3°C). This term does not include any liquid mixture that has one or more components with a flash point above 200°F (93.3°C) that make up 99% or more of the total volume of the mixture. [For test method, see definition of flash point, 1910.106(a) (18); 49 CFR 173.115(b).]

Compressed gas   (1) A gas or mixture of gases having in a container an absolute pressure exceeding 40 lb/in² at 70°F (21.1°C); or (2) a gas or mixture of gases having in a container an absolute pressure exceeding 104 lb/in² at 130°F (54.4°C), regardless of the pressure at 70°F (21.1°C); or (3) a flammable liquid having a vapor pressure exceeding 40 lb/in² absolute pressure at 100°F (37.8°C), as determined by the American National Standard Method of Test for Vapor Pressure of Petroleum Products (Reid Method) Z11.44-1973 (ASTM D 323-72).

Corrosive material   A chemical liquid or solid that causes visible destructrion or irreversible alteration in human skin tissue at the site of contact or, in the case of leakage from its packaging, a liquid that has a severe corrosion rate on steel.

(1)   A material is considered to be *destructive* or to *cause irreversible alteration in skin tissue* if, when tested on the intact skin of the albino rabbit by the method described in Appendix A of 49 CFR Part 173, the structure of the tissue at the site of contact is destroyed or changed irreversibly after an exposure period of 4 h or less.

(2)   A liquid is considered to have a *severe corrosion rate* if its corrosion rate exceeds 0.250 inch per year (IPY) on steel (SAE 1020) at a test temperature of 130°F. An acceptable test is described in NACE Standard TM-01-69.

Explosive   A chemical that causes a sudden, almost instantaneous release of pressure, gas, and heat when subjected to sudden shock, pressure, or high temperature.

Exposure   An employee is subjected to a hazardous chemical in the course of employment through any route of entry (inhalation, ingestion, skin contact, or absorption, etc.); includes potential (e.g., accidental or possible) exposure.

Flammable material   A chemical substance that falls within any of the following categories:

(1)   *Flammable aerosol:* A chemical substance or mixture dispensed from its container as a mist, spray, or foam by a propellant under pressure, which, when tested by the method

TABLE 9-8
SCIENTIFIC TERMINOLOGY USED IN THE PREPARATION OF MATERIAL SAFETY DATA
SHEETS *(CONTINUED)*

---

described in 16 CFR 1500.45, yields a flame projection exceeding 18 in at full valve opening or a flash-back (a flame extending back to the valve) at any valve opening.

(2) *Flammable gas:* A gas which, at atmospheric temperature and pressure, forms a flammable mixture with air when present at a concentration of 13% or less by volume, or that forms a range of flammable mixtures with air wider than 12%, regardless of lower limit; see also *flammable aerosol.*

(3) *Flammable liquid:* A liquid having a flash point below 100°F (37.8°C). This does not include any liquid mixture having one or more components with a flash point at or above 100°F (37.8°C) which make up 99% or more of the total volume of the mixture. [For test method, see definition of flash point, 1910.106(a) (19); 49 CFR 173.115 (a).]

(4) *Flammable solid:* A solid other than an explosive that can cause fire through friction, absorption of mixture, spontaneous chemical change, or retained heat from manufacturing or processing, or that can be readily ignited, and when ignited, will continue to burn or be consumed after removal of the source of ignition (49 CFR 173.150).

Flash point   The minimum temperature at which a liquid gives off a vapor in sufficient concentration to ignite when tested as follows: [1910.106(a) (14); 49 CFR 173.115(d).]

(1) *Tagliabue Closed Tester:* See American National Standard Method of Test for Flash Point by Tag Closed Tester, Z11.24 1971(ASTM D 56-77). For liquids with a viscosity of less than 45 Saybolt Universal Seconds (SUS) at 100°F (37.8°C) that do not contain suspended solids or that have a tendency to form a surface film under test.

(2) *Pensky-Martens Closed Tester:* See American National Standard Method of Test for Flash Point by Pensky-Martens Closed Tester, Z11.7-1974 (ASTM D 93-79). For liquids with a viscosity equal to or greater than 45 SUS at 100°F (37.8°C) or that contain suspended solids or have a tendency to form a surface film under test.

(3) *Setaflash Closed Tester:* See American National Standard Method of Test for Flash Point by Setaflash Closed Tester (ASTM D 3278-78). For mixtures, if the result of any test method is above 100°F (37.8°C), a fresh sample shall be evaporated to 90% of the original volume and retested. The lower of the two values shall be taken as the flash point.

Foreseeable emergency   Any potential occurrence such as but not limited to equipment failure, rupture of containers, or failure of control of equipment that could result in an uncontrolled release of a hazardous chemical into the workplace.

Hazardous chemical substance or mixture   A substance that is one or more of the following: an extremely toxic material, a highly toxic material, a toxic material, a corrosive material, an irritant, a strong sensitizer, a dangerously reactive material, an extremely flammable material, a combustible liquid, a pyrophoric material, a strong oxidizer, a pressure-generating material, or a compressed gas.

Health hazard   A chemical for which there is statistically significant evidence based on at least one study conducted in accordance with established scientific principles that acute or chronic health effects may occur in exposed employees. The term *health hazard* includes chemicals that are carcinogens, toxic or highly toxic agents, reproductive toxins, irritants, corrosives, sensitizers, hepatotoxins; nephrotoxins, neurotoxins, agents that act on the hematopoietic system, and agents that damage the lungs, skin, eyes, or mucous membranes.

TABLE 9-8
SCIENTIFIC TERMINOLOGY USED IN THE PREPARATION OF MATERIAL SAFETY DATA
SHEETS *(CONTINUED)*

Highly toxic   A chemical falling within any of the following categories:

(1)   A chemical that has a median lethal dose (LD_{50}) of ≤50 mg/kg of body weight when administered orally to albino rats weighing between 200 and 300 g each.

(2)   A chemical that has a median lethal dose (LD_{50}) of ≤200 mg/kg of body weight when administered by continuous contact for 24 h (or less if death occurs within 24 h) with the bare skin of albino rabbits weighing between 2 and 3 kg each.

(3)   A chemical that has a median lethal concentration (LC_{50}) in air of ≤200 ppm by volume of gas or vapor, or ≤2 mg/L of mist, fume, or dust, when administered by continuous inhalation for 1 h (or less if death occurs within 1 h) to albino rats weighing between 200 and 300 g each.

Irritant   A chemical substance or mixture, not a corrosive, which on immediate, prolonged, or repeated contact with normal living tissues induces a local inflammatory response in the skin, eyes, or mucous membranes (16 CFR 1500.41).

Lower explosive limit (LEL)   The lowest concentration of gas or vapor (% vol in air) which will burn or explode if an ignition source is present.

Material safety data sheet (MSDS)   A document that contains information and instructions on the chemical and physical characteristics of a substance, its hazards and risks, the safe handling requirements and actions to be taken in the event of fire, spill, or overexposure, etc.

Median lethal concentration (LC_{50})   The concentration in air of gas, vapor, mist, fume, or dust for a given period of time that is most likely to kill one-half of a group of test animals using a specific test procedure. Inhalation is route of exposure and the value LC_{50} is usually expressed as ppm or mg/m³.

Median lethal dose (LC_{50})   The dosage of a substance or mixture that is most likely to kill one-half of a group of test animals using a specified test procedure. The dose is expressed as the amount per unit of body weight, the most common expression being milligrams of material per kilogram of body weight (mg/kg of body weight). Usually refers to oral or skin exposure.

Mutagen   Those chemicals or physical effects that can alter genetic material in an organism and results in physical or functional changes in all subsequent generations.

Oxidizer   A chemical other than a blasting agent or explosive that initiates or promotes combustion in other materials, thereby causing fire either of itself or through the release of oxygen or other gases.

Physical hazard   A chemical for which there is scientifically valid evidence that it is a combustible liquid, a compressed gas, explosive, flammable, an organic peroxide, an oxidizer, pyrophoric, unstable (reactive), or water-reactive.

Pyrophoric material   A chemical substance or mixture that will ignite spontaneously in dry or moist air at or below 130°F (54.4°C).

Reactive material   A chemical substance or mixture that may vigorously polymerize, decompose, condense, or become self-reactive under conditions of shock, pressure, or temperature and includes a chemical substance or mixture that falls within any of the following categories:

(1)   *Explosive material:* A chemical substance or mixture that causes sudden, almost instantaneous release of pressure, gas, and heat when subjected to sudden shock, pressure, or high temperature.

TABLE 9-8
SCIENTIFIC TERMINOLOGY USED IN THE PREPARATION OF MATERIAL SAFETY DATA
SHEETS *(CONTINUED)*

(2) *Organic perioxide:* An organic compound that contains the bivalent -0-0- structure
which may be considered a structural derivative of hydrogen peroxide, in which one or
both of the hydrogen atoms has been replaced by an organic radical.

(3) *Pressure-generating material:* A chemical substance or mixture that may spontaneously
polymerize, with an increase in pressure, unless protected by the addition of an
inhibitor, or by refrigeration or other thermal control; may decompose to release gas in
its container; or comprises the contents of a self-pressurized container.

(4) *Water-reactive material:* A chemical substance or mixture that reacts with water to
release heat or gas which is flammable, highly toxic, or toxic.

Sensitizer   A chemical substance or mixture that causes a substantial number of persons
to develop a hypersensitive reaction in normal tissue upon reapplication of the chemical or
mixture through an allergic or photodynamic reaction.

Strong oxidizer   A chemical substance or mixture that initiates or promotes combustion in
other materials, thereby causing fire either of itself or through the release of oxygen or other
gases.

Teratogen   A chemical that has been demonstrated to cause physical defects in the
developing embryo.

Threshold limit values (TLV)   The airborne concentration of the substance that represents
conditions under which it is believed nearly all workers may be repeatedly exposed day after
day without adverse effect. There are three categories of threshold limit values (TLVs).

(1) *Time-weighted average (TWA):* The time-weighted average concentration for a normal
8-h work day or 40-h work week, to which nearly all workers may be exposed, day after
day, without adverse effect.

(2) *Short-term exposure limit (STEL):* The maximum concentration to which workers can
be exposed for a period up to 15 min continuously without suffering from irritation,
chronic or irreversible tissue change, or narcosis of sufficient degree to increase
accidental injury, impair self-rescue, or materially reduce work efficiency, provided that
no more than four excursions per day are permitted, with at least 60 min between
exposure periods, and provided that the daily TWA also is not exceeded.

(3) *Ceiling (C):* The concentration that should not be exceeded, even instantaneously.

Toxic   A chemical falling within any of the following toxic categories:

(1) A chemical that has a median lethal dose (LD_{50}) of more than 30 mg/kg but not more
than 500 mg/kg of body weight when administered orally to albino rats weighing
between 200 and 300 g each.

(2) A chemical that has a median lethal dose (LD_{50}) of more than 200 mg/kg but not more
than 1000 mg/kg of body weight when administered by continuous contact for 24 h (or
less if death occurs within 24 h) with the bare skin of albino rabbits weighing between 2
and 3 kg each.

(3) A chemical that has a median lethal concentration (LC_{50}) in air of more than 200 ppm
but not more than 2000 ppm by volume of gas or vapor, or more than 2 mg/L but not
more than 20 mg/L of mist, fume, or dust, when administered by continuous inhalation
for 1 h (or less if death occurs within 1 h) to albino rats weighing between 200 and 300
g each.

Unstable   A chemical that in the pure state, or as produced or transported, will vigorously
polymerize, decompose, condense, or become self-reactive under conditions of shock,
pressure, or temperature.

Units used for these measurements are usually milligrams per cubic meter (mg/m³) or parts per million (ppm).

Section III: Physical and Chemical Characteristics

Section III tells what the material or mixture is like and how it behaves. The conditions of testing, including the temperature scale used (°C or °F), is shown for each entry. This information is useful for the design of ventilation systems and for providing adequate fire and spill containment equipment and procedures.

Boiling point   Refers to the temperature at which a material boils, in °F, under ordinary atmospheric pressure (1 atm = 760 mmHg). If the material is a mixture, a boiling range may be given.

Vapor pressure   Indicates how much vapor the material may produce. It refers to the pressure of saturated vapor above the liquid and the temperature where measured. A high vapor pressure indicates that a liquid will evaporate easily.

Vapor density   Tells how heavy the pure gaseous form of the material is in relation to air. The weight of a given volume of a vapor or gas (with no air present) is compared with the weight of an equal volume of air. Values should be given in the ambient temperature range of 60 to 90°F to facilitate field usage. High vapor densities pose a particular problem because these vapors will collect in the bottom of tanks or buildings.

Water solubility   Indicates the solubility of the material in distilled water at 50°F. Solubility may be given in weight percent, or the following terms may be used instead of numbers.

Negligible	<0.1 percent solubility
Slight	0.1 to 1 percent solubility
Moderate	1 to 10 percent solubility
Appreciable	>10 percent solubility
Complete	soluble in all proportions

Specific gravity   Shows how heavy the material is compared to water and tells whether it will float or sink. The weight of a given volume of material is compared to the weight of an equal volume of water at 39.2°F (4°C).

Melting point   The temperature where a solid becomes a liquid.

Evaporation rate   When butyl acetate is used for less volatile solvents, weights may be recorded for equal times of evaporation. In this case, values greater than 1 indicate evaporation rates greater than butyl acetate.

Water reactive   Indicates if the chemical reacts with water to release a gas that is flammable or presents a health hazard.

Section IV: Fire and Explosion Hazard Data

Flash point and method   Indicates the temperature at which a liquid will give off enough flammable vapor to ignite. The closed-cup values should be given but open-cup methods are also used. The results of the two methods can vary by several degrees. The method used to determine the flash point should be stated.

Autoignition temperature   Refers to the minimum temperature needed to cause self-sustained combustion in the absence of a spark or flame. The temperature and temperature scale are given.

Flammable limits in air   Reports the range of gas or vapor concentrations (percent by volume of air) that will burn or explode if an ignition source is present. Lower and upper limits are noted. Knowledge of the lower limit will aid in determining the volume of ventilation needed for an enclosed space to prevent fires and explosions. If the material tested was in the form of a dust in air, this fact is also noted.

Extinguishing media   Gives the fire-fighting extinguishing media suitable for use on the burning material. These should be indicated by generic name. The standard fire-fighting agents are water fog, foam, alcohol foam, carbon dioxide, and dry chemical.

Special fire-fighting procedures   If water is unsuitable, the fire-fighting medium to be used is specified. Also listed are any necessary personal protective equipment, including respirator selection, protective clothing, eye protection and self-contained breathing apparatus (SCBA).

Unusual fire and explosion hazards   Indicates if the material presents an unusual hazard and any special conditions that might affect it. If evacuation is necessary, that fact should be indicated.

Section V: Reactivity Data

Section V deals with chemical reactivity and associated hazards. It will aid in safe storage and handling of hazardous or unstable substances. Instability or incompatibility to common substances such as water, direct sun, metals used in piping or containers, acid, or alkalies should be listed.

Stability   "Unstable" means that a chemical in the pure state or as produced or transported will vigorously polymerize, decompose, condense or become self-reactive under conditions of shock, pressure, or temperature. A check mark shows whether the material is stable or unstable under "reasonable, foreseeable conditions." If the material is unstable, the conditions under which a dangerous reaction may occur are given.

Incompatibility   Provides information on common materials and contaminants with which the material may reasonably come in contact to produce a reaction that would release large amounts of energy. If no such incompatibility exists, enter "none." For example, oxidizing agents will react with reducing agents; acids react with bases.

Hazardous decomposition products   Hazardous materials produced in dangerous amounts by burning, oxidation, or heating are listed. Thermal decomposition products might include CO, CO_2, SO_3, NH_3, oxides of nitrogen, phosgene, and hydrogen chloride. For example, carbon tetrachloride undergoes thermal oxidation to produce phosgene and hydrogen chloride.

Hazardous polymerization   Hazardous polymerization releases large amounts of energy. A check mark indicates whether or not hazardous polymerization can occur. If it can occur, the reasonably foreseeable storage conditions that would start the polymerization are listed. Included is the expected time period in which the inhibitors may be used up. Conditions to avoid are listed. They might include catalysts that cause polymerization, heat or temperature, sunlight, and so on.

Section VI: Health Hazard Data

Primary route of entry   Indicate by check mark the potential routes of exposure for the hazardous chemical during the course of normal usage or a foreseeable emergency. A "foreseeable emergency" is one that would normally be planned for as a presumed potential occurrence determined by the nature of the work. Equipment failure and rupture of containers should be considered. If the chemical is not hazardous, this is also indicated.

Health hazards   Indicate acute and chronic hazards that result from exposure to the hazardous chemical. *Acute hazards* are quickly apparent effects of the chemical as a result of short-term exposure and are of short duration. Tissue damage or irritation sensations and lethal dose are among those things considered. *Chronic effects* generally result from long-term exposure. The effects may not be immediately apparent and are likely to be of long duration. Long-term changes in the body should be included. These are some of these characteristics of the chemicals:

Carcinogen (cancer-causing)   Chronic bronchitis
Teratogen (tumor-causing)   Liver degeneration
Mutagen (genetic changes)   Kidney damage
Blood abnormality

Carcinogen listed in   Shows by check mark whether the hazardous chemical

1 Is listed in the National Toxicology Program (NTP) *Annual Report on Carcinogens* (latest edition).

2 Has been found to be a potential carcinogen in the International Agency for Research on Cancer (IARC) monographs

3 Has been found to be a potential carcinogen by OSHA.

If the chemical is not listed, this is also indicated.

Signs and symptoms of exposure   Provide descriptions of the most common sensations that an exposed person will feel and their appearance. Symptoms of exposure can be varied and may depend on individual susceptibility, concentration, and the type of material. Attention should be given to effects caused by eye contact, skin contact, inhalation, and ingestion.

Medical conditions generally aggravated by exposure   Those medical conditions that are recognized as being aggravated by exposure should be listed.

Emergency and first aid procedures   Provide the immediate temporary steps to be taken in case of eye contact, skin contact, inhalation, and ingestion. These are emergency procedures only. The victim should be examined by a doctor as soon as possible after exposure. Procedures for removing contamination from skin and eyes, neutralization if recommended, treatment for inhalation including use of oxygen or artificial respiration, and what to do in case of ingestion are given.

Section VII: Precautions for Safe Handling and Use

Steps to be taken if material is released or spilled   Give precautions to be taken in the event of spills or leaks. These would include such things as avoiding breathing gases and vapor, avoiding skin contact with liquid or solid, and removing sources of ignition. Special equipment used for cleanup such as glass or plastic scoops and types of containers are listed. Also given are specific absorbents, neutralization materials, decontamination materials, whether evacuation is necessary or safety personnel are required, and so on.

Waste disposal method   Gives methods for disposal of spilled solids or liquids. Methods must always follow federal, state, and local regulations. They may include incineration with or without scrubbing of the waste gases, landfill burial, licensed waste disposal firm, scrap recovery, flushing with water, return of material to original container, etc. The manufacturer or supplier may provide specific recommendations. Cautions concerning disposal such as "do not flush to sewer" or "do not incinerate" may be included in this section.

Precautions to be taken in handling and storage   This section gives any special precautions to be taken in storage and handling such as avoiding reaction hazards with oxidizing agents, reducing agents, acids, etc. Conditions for storage such as temperature, ventilation, no smoking or other sources of ignition are also given. When applicable, the safe storage life is indicated.

Other precautions   Lists any other general precautions to be taken that have not previously been mentioned.

NFPA rating (optional)   The National Fire Protection Association (NFPA) has developed a rating system for indicating the health, flammability, and reactivity hazards of chemicals) (listed below). Health hazard describes short-term contact or inhalation hazard only. In addition, a special precaution symbol may be used where necessary.

Health

4 *Danger:* May be fatal on short exposure. Specialized protective equipment required.

3 *Warning:* Corrosive or toxic. Avoid skin contact or inhalation.

2 *Warning:* May be harmful if inhaled or absorbed.

1 *Caution:* May be irritating.

0 No unusual hazard.

Flammability

4 *Danger:* Flammable gas or extremely flammable liquid.

3 *Warning:* Flammable liquid flash point below 100°F.

2 *Caution:* Combustible liquid flash point of 100 to 200°F.

1   *Caution:* Combustible if heated.

0   Not combustible.

Reactivity

4   *Danger:* Explosive material at room temperature.

3   *Danger:* May be explosive if shocked, heated under confinement, or mixed with water.

2   *Warning:* Unstable or may react violently if mixed with water.

1   *Caution:* May react if heated or mixed with water but not violently.

0   *Stable:* Not reactive when mixed with water.

HMIS ratings (optional)   The National Paint and Coatings Association has proposed the Hazardous Material Identification System. Ratings are given for health, flammability, reactivity, and personal protection (see below). These ratings are similar to the NFPA ratings.

Health

0   *Minimal hazard:* No significant risk to health.

1   *Slight hazard:* Irritation or minor reversible injury possible.

2   *Moderate hazard:* Temporary or minor injury may occur.

3   *Serious hazard:* Major injury likely unless prompt action is taken and medical treatment is given.

4   *Severe hazard:* Life-threatening, major, or permanent damage may result from single or repeated exposures.

Flammability

0   *Minimal hazard:* Materials that are normally stable and will not burn unless heated.

1   *Slight hazard:* Materials that must be preheated before ignition will occur. Flammable liquids in this category will have flash points ≥200°F (OSHA Class IIIB).

2   *Moderate hazard:* Material that must be moderately heated before ignition will occur, including flammable liquids with flash points of 100 to 200°F (OSHA Class II and Class IIIA).

3   *Serious hazard:* Materials capable of ignition under almost all normal temperature conditions, including flammable liquids with flash points <73°F and boiling points ≥100°F as well as liquids with flash points of 73 to 100°F (OSHA Classes IB and IC).

4   *Severe hazard:* Very flammable gases or very volatile flammable liquids with flash points <73°F and boiling points <100°F (OSHA Class IA).

Reactivity

0   *Minimal hazard:* Materials that are normally stable, even under fire conditions; will not react with water.

1   *Slight hazard:* Materials that are normally stable, but can become unstable at high temperatures and pressures. These materials may react with water, but will not release energy violently.

2   *Moderate hazard:* Materials that in themselves are normally unstable and will readily undergo violent chemical change, but will not detonate. These materials may also react violently with water.

3 *Serious hazard:* Materials that are readily capable of detonation or explosive decomposition or reaction but require a strong initiating source or must be heated under confinement before initiation. These materials may react explosively with water without requiring heat or confinement.

4 *Severe hazard:* Materials that are readily capable of detonation or explosive decomposition or explosive reaction at normal temperature and pressure.

Personal Protection   Indicated by letter designation, as follows:

A   Safety glasses
B   Safety glasses, gloves
C   Safety glasses, gloves, apron
D   Face shield, gloves, apron
E   Safety glasses, gloves, dust respirator
F   Safety glasses, gloves, apron, dust respirator
G   Safety glasses, gloves, vapor respirator
H   Splash goggles, gloves, apron, vapor respirator
I   Safety glasses, gloves, dust and vapor respirator
J   Splash goggles, gloves, apron, dust and vapor respirator
K   Airline, hood or mask, gloves, full suit, boots
X   Ask your supervisor for guidance.

Section VIII: Control Measures

This section has the safety and health control measures for chemical hazards. The type of personal protective equipment, the type of ventilation to be used and the precautions to be taken when using the material for its intended purpose are given.

Respirator protection   The type of respirator to be used is specified.

Ventilation   The type of ventilation required in work areas and under what conditions it is suitable are given. For very volatile and low-TLV materials, local exhaust that captures fumes at their source is probably the most effective control.

Protective gloves   Gives the type of glove, including the materials of construction.

Eye protection   Indicates the type of eye protection, such as full-face shield or safety goggles.

Other protective clothing and equipment   When special units or other clothing, clothing of special material or construction, or other special handling is required for personal protection, it should be indicated here.

Work hygienic practices   Indicates personal hygienic steps to be taken when handling the chemical. Washing hands after use, not smoking, or disposal or laundering of contaminated clothing may be indicated here.

PROSPECTS FOR HAZARD COMMUNICATION

Some chemical manufacturers, in an effort to assure completeness, have created a problem for MSDS users. The MSDS should contain user-friendly, easy-to-understand information to serve as a quick reference for employers and employees in an emergency situation. [2] Many MSDSs are too complex for most users to understand, and the numerous different MSDS formats can be confusing and difficult to use. The NFPA 704 system is more user-friendly than MSDSs, particularly in an emergency. [7]

In an effort to make MSDSs more user-friendly to all parties, the American National Standards Institute and the Chemical Manufacturers Association have worked together to develop a voluntary consensus standard, ANSI Z400.1-1993, for MSDSs. It recommends a data sheet structure of 16 sections that always appear in a particular sequence.

The first ten sections should meet all of the U.S. OSHA requirements. The other sections are useful for compliance with international standards under the International Standards Organization (ISO).

Several examples of well-conceived and prepared MSDSs that use the ANSI Z400.1-1993 format are shown in Figures 9-2 and 9-3 on page 265. Both of these examples have been adapted from the ANSI voluntary standard Z400.1-1993. The considerable detail in the ANSI format contains more information than required by OSHA, which can be verified by comparing these ANSI formatted examples with the OSHA format in Figure 9-1.

Two examples of industrial MSDSs are shown in Figures 9-4 on pages 268–271 and 9-5 on pages 274–280. MSDSs for sulfuric acid and hydrogen peroxide, two widely used industrial chemicals, were furnished by DuPont Specialty Chemicals. They follow the ANSI format.

Users of these ANSI/CMA data sheets should always be able to locate the needed information in the same location in these uniform MSDSs. The structure of these data sheets is based upon four fundamental questions listed below as Questions A through D. [13, 21]

There are several other OSHA standards that relate to hazard communication for specific workplaces. This includes 29 CFR 1910.120 because of the health and safety risk of hazardous waste operations and emergency response (HAZWOPER). [9, 16, 18] Occupational exposure to hazardous chemicals in laboratories is covered by 29 CFR 1910.1450. [5, 6, 14] Pilot plants represent some unique health and safety problems that require special consideration by industry. [15]

Question A: What is the material and what do I need to know immediately in an emergency?

Section 1: Chemical Product and Company Identification
- Links the MSDS to the material.
- Identifies the supplier of the MSDS.
- Identifies a source for more information.

Section 2: Composition and Information on Ingredients
- Lists the OSHA hazardous components.
- May also list significant nonhazardous components.
- Lists corresponding Chemical Abstracts Service registry numbers, where appropriate, for each component.
- May include additional information, such as exposure guidelines, about components.

Section 3: Hazards Identification

MATERIAL SAFETY DATA SHEET
Effective Date: 02/28/96 Date Printed: 03/05/96 MSDS: 54321

1. CHEMICAL PRODUCT AND COMPANY IDENTIFICATION
     Product Code: 12345
     Product Name: BETA SOLVENT
     Synthetic Chemical Company, 100 Roosevelt Road, Wheaton, IL 60187,
     24-hour emergency phone: 708-690-0800

2. COMPOSITION/INFORMATION ON INGREDIENTS

Component	CAS#	Exposure Limits	% by Wt.
Bulanone	13567-89-9	ACGIH TLV 180 ppm, (TWA), 450 ppm (STEL) OSHA PEL 180 ppm (TWA), 450 ppm (STEL)	100%

3. HAZARDS IDENTIFICATION

———————— EMERGENCY OVERVIEW ————————

Extremely Flammable. Causes eye irritation. Can cause severe respiratory irritation. Can cause severe central nervous system depression (including unconsciousness). Highly toxic to fish. *Clear* free-flowing liquid *with sweet odor.*

Potential Health Effects
EYE: May cause severe irritation (tears, blurred vision and redness). May result in damage to cornea, vision impairment and even blindness.
SKIN: Prolonged exposure may cause skin irritation. May cause drying of skin. A single prolonged exposure is not likely to result in the material being absorbed through skin in harmful amounts.
INGESTION: Small amounts (tablespoonful) swallowed are not likely to cause injury. Swallowing larger amounts may cause injury.
INHALATION: A single brief (minutes) inhalation exposure may cause serious effects. Overexposure may cause respiratory irritation. Symptoms include tightness in the chest, difficult breathing and coughing. Prolonged overexposure may cause central nervous system depression with anesthetic effects (numbing) or narcotic effects (headaches, dizziness, sleepiness, loss of coordination, and unconsciousness).
CHRONIC (CANCER) INFORMATION: Does not cause cancer, based on animal studies.
TERATOLOGY (BIRTH DEFECT) INFORMATION: Does not cause birth defects, based on animal studies.
REPRODUCTION INFORMATION: No data available.

4. FIRST AID MEASURES
     EYES: Immediately flush eyes with plenty of water for at least 15 minutes. Get medical attention.
     SKIN: Wash with soap and water. Get medical attention if irritation develops or persists.
     INGESTION: If swallowed, get medical attention.
     INHALATION: Remove to fresh air. If not breathing, give artificial respiration. Get immediate medical attention.
     NOTE TO PHYSICIANS: Supportive care. May aggravate pre-existing respiratory conditions.

5. FIRE FIGHTING MEASURES
     FLAMMABLE PROPERTIES:
          FLASH POINT: −8°F (−22°C)
          Extremely flammable. Material will readily ignite at ambient temperatures. Colorless vapors may travel considerable distance to ignition source and cause flash fires or explosions.

(continued)

FIGURE 9-2
EXAMPLE OF AN ANSI FORMAT FOR AN MSDS FOR A LIQUID SOLVENT.

FLAMMABLE LIMITS:
Lower flammable limit: 2.4% FLAMMABLE
Upper flammable limit: 15.9%
AUTOIGNITION TEMPERATURE: 865°F (463°C)
HAZARDOUS COMBUSTION PRODUCTS: Carbon dioxide and possibly carbon monoxide.
EXTINGUISHING MEDIA: Alcohol resistant foam, carbon dioxide, and dry chemical.
FIREFIGHTING INSTRUCTIONS: Water can be used to cool fire-exposed containers, to protect personnel and to disperse vapors and spills. Water runoff can cause environmental damage. Dike and collect water used to fight fires.

Fire-fighters should wear normal protective equipment and positive-pressure self-contained breathing apparatus.

6. ACCIDENTAL RELEASE MEASURES
SMALL SPILL:
Absorb spill with an inert material, then place in a chemical waste container.
LARGE SPILL:
Contain spilled liquid with sand or earth. DO NOT use combustible materials such as sawdust. Eliminate all ignition sources. Use plastic or aluminum shovel to transfer absorbed waste material into drums. Prevent runoff from entering into storm sewers and ditches which lead to natural waterways.

7. HANDLING AND STORAGE
HANDLING: Avoid contact with eyes. Keep container closed when not in use. Use only in a well-ventilated area. Wash thoroughly after handling. Avoid prolonged or repeated breathing of vapor. Avoid prolonged or repeated contact with skin. Use grounding and bonding connection when transferring material to prevent static discharge, fire or explosion. Use sparkproof tools and explosion proof equipment.
STORAGE: Store in well-ventilated area. Keep away from heat, sparks, and flame. Precautions apply to emptied containers.

8. EXPOSURE CONTROLS/PERSONAL PROTECTION
ENGINEERING CONTROLS: Control airborne concentrations below the exposure limits. Use only with adequate ventilation. Local exhaust ventilation may be necessary. Lethal concentrations may exist in areas with poor ventilation.
RESPIRATORY PROTECTION: When respiratory protection is required, or concentrations are unknown use a NIOSH/MSHA approved air-purifying respirator equipped with organic vapor cartridges or canisters. For emergency and other conditions where exposure limit may be greatly exceeded, use an approved positive-pressure, self-contained breathing apparatus or positive-pressure airline with auxiliary self-contained air supply.
SKIN PROTECTION: For brief contact, no precautions other than clean body-covering clothing should be needed. Use impervious gloves such as Neoprene.
EYE PROTECTION: Use chemical goggles. If vapor exposure causes eye discomfort, use a full-face, supplied-air respirator.

9. PHYSICAL AND CHEMICAL PROPERTIES.
BOILING POINT: 123°F (51°C)
MELTING POINT: −110°F (-79°C)
VAPOR PRESSURE: 185 mmHg at 68°F (20°C)
VAPOR DENSITY: 1.99 (air = 1.0)
SOLUBILITY IN WATER: Completely miscible
SPECIFIC GRAVITY: 0.64 at 77°F- (25°C)
pH: No data available
ODOR: Sweet
APPEARANCE: Clear colorless liquid

FIGURE 9-2 *(continued)*

10. STABILITY AND REACTIVITY

CHEMICAL STABILITY: (CONDITIONS TO AVOID) Keep away from flames and spark-producing equipment.

INCOMPATIBILITY: Nitric plus acetic acid and nitric plus sulfuric acid.

HAZARDOUS DECOMPOSITION PRODUCTS: None.

HAZARDOUS POLYMERIZATION: Will not occur.

11. TOXICOLOGICAL INFORMATION

EYE: Concentrations over 200 ppm can cause irritation.

SKIN: Prolonged or repeated contact may defat the skin causing dermatitis. Bulanone is not absorbed through the skin in acutely toxic amounts. The dermal LD_{50} for rabbits is >3500 mg/kg.

INGESTION: The oral LD_{50} for rats is > 5000 mg/kg.

INHALATION: The minimum lethal concentration for laboratory animals is in the range of 15,000 ppm for 1-4 hour exposures. In humans, irritation begins at 210 ppm. Bulanone is a central nervous system (CNS) depressant with effect occurring at concentrations >10,000 ppm; however, motor skills may be impaired at 950 ppm.

SUBCHRONIC: Rats exposed to 7000 ppm for 6 hours per day, 5 days per week for 10 weeks showed no adverse effects other than slight evidence of CNS depression during exposure.

CHRONIC/CARCINOGENICITY: Administration of bulanone to rats for 2 years in drinking water at 950 ppm and lower caused no carcinogenic response.

TERATOLOGY: A study in rats exposed to 300, 1800, or 9000 ppm showed no teratogenic effect, though maternal toxicity was evidenced by CNS depression at the top dose. No fetotoxicity was noted.

REPRODUCTION: No data available.

MUTAGENICITY: Negative in the Ames test.

12. ECOLOGICAL INFORMATION

ECOTOXICOLOGICAL INFORMATION: Static acute LC_{50} for fathead minnow is 2.4 mg/l.

Static acute LC50 for rainbow trout is reported to be 0.15 mg/l.

Material is highly toxic to fish on an acute basic.

Static acute LC50 for daphnids is 6.0 mg/l.

Material is moderately toxic to aquatic invertebrates on a static acute basis.

CHEMICAL FATE INFORMATION: Appreciable evaporation from water to air is expected in the environment.

Partitioning from water to oil is low.

Potential for mobility in soil is very high.

Biodegradation is expected in a wastewater treatment plant.

13. DISPOSAL CONSIDERATIONS:

RCRA hazardous waste if discarded. ID #D001. Recycle or incinerate at an EPA approved facility or dispose in compliance with Federal, State and local regulations.

14. TRANSPORT INFORMATION (Not meant to be all inclusive):

Bulanone/FLAMMABLE/UN1090

15. REGULATORY INFORMATION (Not meant to be all inclusive—selective regulations represented)

U.S. FEDERAL REGULATIONS:

OSHA: Hazardous by definition of Hazard Communication Standard (29 CFR 1910.1200).

CERCLA: SARA HAZARD CATEGORY: This product has been reviewed according to the EPA 'Hazard Categories' promulgated under Sections 311 and 312 of the Superfund Amendment and Reauthorization Act of 1986 (SARA Title III) and is considered, under applicable definitions, to meet the following categories:

An immediate health hazard

A fire hazard

FIGURE 9-2 (continued)

SECTION 313: This product contains NONE of the substances subject to the reporting requirements of Section 313 of Title III of the Superfund Amendments and Reauthorization Act of 1986 and 40 CFR Part 372.

INTERNATIONAL REGULATIONS:

CANADIAN WHMIS: Controlled Product Hazard Class B2, D2. This MSDS has been prepared in compliance with Controlled Product Regulations. Product is used primarily as an industrial solvent in reactions.

CANADIAN ENVIRONMENTAL PROTECTION ACT (CEPA): All components of this product are on the Domestic Substances List (DSL), and acceptable for use under the provisions of CEPA.

EINECS: All components of this product are on the European Inventory of Existing Commercial Chemical Substances.

STATE REGULATIONS:

There are no known additional requirements necessary for compliance with state Right-To-Know regulations.

VOLATILE ORGANIC COMPOUNDS (VOC): Contents of containers are 100% VOC's. Material is photochemically reactive.

16. OTHER INFORMATION

MSDS Status: Revised Sections 5, 8, and 15.

The information Herein Is Given in Good Faith, But No Warranty, Express Or Implied, Is Made.

Consult The Synthetic Chemical Company For Further Information.

FIGURE 9-2 *(continued)*

• Provides information on the potential adverse health effects and symptoms that might result from reasonably foreseeable use and misuse of the material.

• May provide an emergency overview that describes the material's appearance and severe, immediate health, physical, and environmental hazards associated with emergency response situations.

Question B: What should I do if a hazardous situation occurs?

Section 4: First Aid Measures

• Provides easily understandable instructions on what to do when results of exposure require immediate treatment and when simple measures may be taken before professional medical assistance is available. Instructions provide for each route of exposure.

• May also include instructions to medical professionals.

Section 5: Fire-Fighting Measures

• Provides basic fire-fighting guidance, including appropriate extinguishing media.

• Describes other fire and explosive properties useful for fighting fires involving the material such as flash points, explosive limits.

Section 6: Accidental Release Measures

• Describes action to be taken to minimize the adverse effects of an accidental spill, leak, or release of the material.

MATERIAL SAFETY DATA SHEET

DATE PREPARED: July 9, 1996

MSDS No.: 2345

1. CHEMICAL PRODUCT AND COMPANY IDENTIFICATION

Product Identifier: PRODUCT NAME
General Use: Demulsifier
Product Description: Surfactants in hydrocarbon solvents.

MANUFACTURER: EMERGENCY TELEPHONE NUMBERS:

Company Name	Health	(123) 123-1234 8 am-6 pm EST M-F
321 Main St.	Transportation	(123) 123-2345 8 am-6 pm EST M-F
Town, USA 12345-1234	CHEMTREC	(800) 424-9300 24 hours Everyday
(123) 123-4567 (Sales Office)		

2. COMPOSITION INFORMATION ON INGREDIENTS

	wt%	CAS Registry #
Component A	75	12345-67-8
Component B	10	1234-56-7
Component C	10	54321-12-1
Component D	5	(Trade Secret)

OSHA HAZARDOUS COMPONENTS (29 CFR 1910.1200):

	EXPOSURE LIMITS OSHA PEL	8 hrs. TWA (ppm) ACGIH TLV
Component A Flammable Liquid	110	110
Component B Skin Sensitizer	170	170
Component C Eye Irritant	400	400

3. HAZARDS IDENTIFICATION

EMERGENCY OVERVIEW:

Clear, amber, flammable liquid with a solvent odor.
Irritating to eyes, skin and respiratory tract.

POTENTIAL HEALTH EFFECTS:

INHALATION:

Vapor may be irritating to the nose, throat, and lungs. May cause unconsciousness or central nervous system depression.

EYE CONTACT:

Irritating.

SKIN CONTACT:

Frequent or prolonged contact may irritate the skin and cause a skin rash (dermatitis). May cause skin sensitization.

INGESTION:

Small amounts may be drawn into the lungs. This may cause severe health effects such as inflammation of the lungs and infection of the bronchi. Ingestion may cause irritation of digestive tract and diarrhea.

CHRONIC:

Test with similar materials have produced skin cancer in experimental animals. The relationship to humans has not been fully established.

4. FIRST AID MEASURES

INHALATION:

Use positive pressure air supplied respiratory protection when removing the affected victim from area. Keep the victim at rest. Give artificial respiration if necessary. Call for prompt medical attention.

EYE CONTACT:

Flush eyes with large amounts of water until irritation subsides. If irritation persists, get medical attention.

(continued)

FIGURE 9-3
EXAMPLE OF AN ANSI FORMAT FOR AN MSDS FOR A HYDROCARBON
DEMULSIFIER.

SKIN CONTACT:
Flush with large amounts of water. Use soap if available. Remove severely contaminated clothing. If irritation persists, seek medical attention.
INGESTION:
If swallowed, DO NOT induce vomiting. Keep at rest. Get prompt medical attention.

5. FIRE FIGHTING MEASURES

Flashpoint:	50°F/10°C
Flammable Limits:	1.0 to 6.7% by volume
Autoignition Temperature:	Not Available

GENERAL HAZARD:
Flammable liquid; may release vapors that form flammable mixtures when temperatures are at or above the flash point. Toxic gases will form upon combustion.
FIRE FIGHTING INSTRUCTIONS:
Either allow fire to burn out under controlled conditions or extinguish with foam or dry chemical. Try to cover liquid spills with foam. Use water spray to cool fire exposed surfaces. Avoid spraying water directly into storage containers due to danger of boil over. If a spill has not ignited, use water spray to disperse the vapors.
FIRE FIGHTING EQUIPMENT:
Respiratory and eye protection required for fire fighting personnel. Full protective equipment and a self-contained breathing apparatus should be used for all indoor fires and any significant outdoor fires. For small outdoor fires, which may be extinguished with a portable fire extinguisher, use of an SCBA may not be required.
HAZARDOUS COMBUSTION PRODUCTS:
Smoke, fumes and CO_2 and CO

6. ACCIDENTAL RELEASE MEASURES
LAND SPILL:
Notify the appropriate authorities immediately. Take all additional action necessary to prevent and remedy the adverse effects of the spill. Eliminate ignition source. Isolate hazard area and deny entry to unauthorized or unprotected personnel. Shut off leak if practical. Avoid runoff into storm sewers and ditches which lead to natural waterways. Contain spilled liquid with sand or earth. DO NOT use combustible materials such as sawdust. Absorb spill with an inert material, then place in a chemical waste container.
WATER SPILL:
Remove from water surface by skimming or with suitable absorbents.

7. HANDLING AND STORAGE
STORAGE TEMPERATURE: Ambient   STORAGE PRESSURE: Atmospheric
GENERAL:
Keep container closed when not in use. Loosen closure cautiously. Store in a cool, well ventilated place. Keep away from heat, sparks and flame. Protect material from direct sunlight. Ground and bond containers when transferring materials. Empty containers may retain hazardous properties. Follow all MSDS/label warnings even after container is emptied.

8. EXPOSURE CONTROLS, PERSONAL PROTECTION
ENGINEERING CONTROLS:
Use of local exhaust ventilation to control emissions near the source. Laboratory samples should be handled in a fumehood. Use explosion-proof ventilation equipment.
PERSONAL PROTECTION:
RESPIRATOR:
If concentrations are over the exposure limits use an air supplied respirator.
PROTECTIVE CLOTHING:
People who know they are hypersensitive to Component B should avoid contact with this product. Wear Neoprene or butyl rubber gloves, rubber boots, a chemical worker's suit, chemical goggles and a face shield, as appropriate.

FIGURE 9-3 *(continued)*

9. PHYSICAL AND CHEMICAL PROPERTIES
   Vapor Pressure   : 55 mmHg at 100°F
   Specific Gravity   : 0.85 at 60°F (16°C)
   Solubility in Water: Insoluble
   pH   : Not available
   Boiling Point   : 325°F

   Vapor Density: 4.2
   (Air=1)
   Evaporation Rate: 0.8
        (n-Butyl Acetate=1)
   Freezing Point: Not Available
   Odor: Solvent
   Appearance: Clear, amber
   Physical State: Liquid

10. STABILITY AND REACTIVITY
    GENERAL:
       This product is stable.
    INCOMPATIBLE MATERIALS AND CONDITIONS TO AVOID:
       Strong oxidizing agents.
    HAZARDOUS DECOMPOSITION:
       None

11. TOXICOLOGICAL INFORMATION
    TOXICITY DATA IS AVAILABLE, Call (123) 123-1234

12. ECOLOGICAL INFORMATION
    No data available.

13. DISPOSAL CONSIDERATIONS
    RCRA Hazard Class: D001
       Ensure disposal in compliance with government requirements and ensure confor-
       mity to local disposal regulations.

14. TRANSPORT INFORMATION
    DOT (Department of Transportation):
       PROPER SHIPPING NAME: Flammable Liquid, n.o.s. (Component A), UN 1996
       HAZARD CLASS: Flammable Liquid
       IDENTIFICATION NUMBER: UN 1996

15. REGULATORY INFORMATION
    TSCA (Toxic Substance Control Act):
       Components of this product are listed on the TSCA Inventory.
    CERCLA (Comprehensive Response Compensation, and Liability Act):
       No reportable quantity substances. We recommend you contact your local authori-
       ties to determine if there may be other local reporting requirements.
    SARA TITLE III (Superfund Amendments and Reauthorization Act):
       311/312 Hazard Categories:
          Immediate Health, Delayed Health, Fire.
       313 Reportable Ingredients:
          None

16. OTHER INFORMATION
    No specific notes.
    REVISION SUMMARY:
    Since Mar. 13, 1995, this MSDS has been revised in the following sections:
    Emergency Overview
    First Aid, Inhalation

    THE INNFORMATION RELATES TO THIS SPECIFIC MATERIAL. IT MAY NOT BE
    VALID FOR THIS MATERIAL IF USED IN COMBINATION WITH ANY OTHER MA-
    TERIALS OR IN ANY PROCESS. IT IS THE USER'S RESPONSIBILITY TO SAT-
    ISFY ONESELF AS TO THE SUITABILITY AND COMPLETENESS OF THIS INFOR-
    MATION FOR HIS OWN PARTICULAR USE.

FIGURE 9-3 *(continued)*

MATERIAL SAFETY DATA SHEET

DuPont Chemicals

4950CR	Revised 31-MAY-1995	Printed 12-SEP-1995
	Sulfuric Acid, 77 to 100%	

CHEMICAL PRODUCT/COMPANY IDENTIFICATION
   Material Identification
      Corporate MSDS NumberDU00051
      CAS Number   7664-93-9
      Formula H_2SO_4
      Molecular Weight   98.08
      CAS Name   SULFURIC ACID
      Grade   77 to 100% TECHNICAL
   Tradenames and Synonyms
      CC0036
   Company Identification
      MANUFACTURER/DISTRIBUTOR
            DuPont
            1007 Market Street
            Wilmington, DE 19898
      PHONE NUMBERS
         Product Information   1-800-441-7515
         Transport Emergency   CHEMTREC: 1-800-424-9300
         Medical Emergency   1-800-441-3637

COMPOSITION/INFORMATION ON INGREDIENTS
   Components

Material	CAS Number	%
*SULFURIC ACID	7664-93-9	
60 DEG TECHNICAL		77.7
66 DEG TECHNICAL		93.2
1.835 ELECTROLYTE		93.2
98% TECHNICAL		98
99% TECHNICAL		99
100% TECHNICAL		100
WATER	7732-18-5	0-22

*Disclosure as a toxic chemical is required under Section 313 of Title III of the Superfund Amendments and Reauthorization Act of 1986 and 40 CFR part 372.

HAZARDS IDENTIFICATION
   Potential Health Effects
      INHALATION
         Exposure to mists may cause: Irritation of the nose and throat with sneezing, sore throat or runny nose. Non-specific effects such as headache, nausea and weakness. Gross overexposure may cause: Irritation of nose, throat, and lungs with cough, difficulty breathing or shortness of breath. Pulmonary edema (body fluid in the lungs) with cough, wheezing, abnormal lung sounds, possibly progressing to severe shortness of breath and bluish discoloration of the skin; symptoms may be delayed. Repeated and/or prolonged exposure to mists may cause: Corrosion of teeth.
      SKIN CONTACT
         Contact with liquid may cause: Skin corrosion, burns or ulcers. Contact with a 1% solution may cause slight irritation with itching, redness or swelling. Repeated and/or prolonged exposure to mists may cause: Irritation with itching, burning, redness, swelling or rash.
      EYE CONTACT
         Contact with liquid may cause: Eye corrosion or ulceration; blindness may result. Re-

(continued)

FIGURE 9-4
SULFURIC ACID MSDS FROM DUPONT SPECIALTY CHEMICALS.

peated and/or prolonged exposure to mists may cause: Eye irritation with tearing, pain or blurred vision.

INGESTION

Immediate effects of overexposure may include: Burns of the mouth, throat, esophagus and stomach, with severe pain, bleeding, vomiting, diarrhea and collapse of blood pressure—damage may appear days after exposure.

ADDITIONAL HEALTH EFFECTS

The International Agency for Research on Cancer (IARC) classified "strong inorganic acid mists containing sulfuric acid" as a Category 1 carcinogen, a substance that is "carcinogenic to humans." This classification is for strong inorganic acid mists only and does not apply to sulfuric acid or sulfuric acid solutions. The basis for the IARC classification rests on several epidemiology studies which have several deficiencies. These studies did not account for exposure to other substances, some known to be animal or potential human carcinogens, social influences (smoking, etc.) and included small numbers of subjects. Based on the overall weight of evidence from all human and chronic animal studies, no definitive causal relationship between sulfuric acid mist exposure and respiratory tract tumors has been shown.

Increased susceptibility to the effects of this material may be observed in persons with pre-existing disease of the lungs.

Carcinogenicity Information

None of the components present in this material at concentrations equal to or greater than 0.1% are listed by IARC, NTP, OSHA or ACGIH as a carcinogen.

FIRST AID MEASURES

First Aid

INHALATION

If inhaled, immediately remove to fresh air. If not breathing, give artificial respiration. If breathing is difficult, give oxygen. Please note: Symptoms may be delayed; prompt medical attention may be required. Call a physician.

SKIN CONTACT

In case of contact, immediately flush skin with plenty of water for at least 15 minutes, while removing contaminated clothing and shoes. Call a physician. Wash contaminated clothing before reuse.

While the patient is being transported to a medical facility, continue the application of cold, wet compresses. If medical treatment must be delayed, repeat the flushing with cold water or soak the affected area with cold water to help remove the last traces of sulfuric acid. Creams or ointments should not be applied before or during the washing phase of treatment.

EYE CONTACT

In case of contact, immediately flush eyes with plenty of water for at least 15 minutes. Call a physician.

INGESTION

If swallowed, do not induce vomiting. Give large quantity of water. Call a physician immediately. Never give anything by mouth to an unconscious person.

Notes to Physicians

Continued washing of the affected area with cold or iced water will be helpful in removing the last traces of sulfuric acid. Creams or ointments should not be applied before or during the washing phase of the treatment.

FIRE FIGHTING MEASURES

Flammable Properties

Will not burn.

Fire and Explosion Hazards:

Reacts with most metals, especially when dilute, to give flammable, potentially explosive hydrogen gas. Follow appropriate National Fire Protection Association (NFPA) codes.

FIGURE 9-4 *(continued)*

Extinguishing Media

Use media appropriate for surrounding material.

Use water spray to cool containers exposed to fire; do not get water inside containers.

Fire Fighting Instructions

Evacuate personnel to a safe area. Keep personnel removed and upwind of fire. Generates heat upon addition of water, with possible spattering. Wear full protective clothing. Runoff from fire control may cause pollution. Neutralize run-off with lime, soda ash, etc., to prevent corrosion of metals and formation of hydrogen gas. Wear self-contained breathing apparatus if fumes or mists are present.

ACCIDENTAL RELEASE MEASURES

Safeguards (Personnel)

NOTE: Review FIRE FIGHTING MEASURES and HANDLING (PERSONNEL) sections before proceeding with clean-up. Use appropriate PERSONAL PROTECTIVE EQUIPMENT during clean-up.

Accidental Release Measures

Stop flow if possible. Review "Fire and Explosion Hazards" and "Safety Precautions" before proceeding with clean-up. Use appropriate protective equipment during clean-up. Soak up small spills with dry sand, clay or diatomaceous earth. Dike large spills, and cautiously dilute and neutralize with lime or soda ash, and transfer to wastewater treatment system. Prevent liquid from entering sewers, waterways, or low areas.

If this product is spilled and not recovered, or is recovered as a waste for treatment or disposal, the Reportable Quantity is 1000 lb. (based on the sulfuric acid content of the solution spilled). Comply with Federal, State, and local regulations on reporting releases.

DuPont Emergency Exposure Limits (EEL) are established to facilitate site or plant emergency evacuation and specify airborne concentrations of brief durations which should not result in permanent adverse health effects or interfere with escape. EEL's are expressed as airborne concentration multiplied by time (C;tsT) for up to a maximum of 60 minutes and as a ceiling airborne concentration. These limits are used in conjunction with engineering controls/monitoring and as an aid in planning for episodic releases and spills. For more information on the applicability of EEL's, contact DuPont.

The DuPont Emergency Exposure Limit (EEL) for Sulfuric Acid is 10 mg/m^3 for 15 to 60 min and 20 mg/m^3 for up to 15 min with a not-to-exceed ceiling of 20 mg/m^3.

HANDLING AND STORAGE

Handling (Personnel)

Do not get in eyes, on skin, or on clothing. Avoid breathing vapors or mist. Wash thoroughly after handling.

Keep containers closed. Do not add water to contents while in container because of violent reaction.

Storage

Keep out of sun and away from heat, sparks, and flame. Keep container tightly closed and (drum) closure up to prevent leakage. Loosen closure carefully. Relieve internal pressure when received and at least weekly thereafter. Do not use pressure to empty. Be sure closure is securely fastened before moving container. Do not wash out container or use it for other purposes; replace closure after each withdrawal and return it with empty container.

EXPOSURE CONTROLS/PERSONAL PROTECTION

Engineering Controls

Good general ventilation should be provided to keep vapor and mist concentrations below the exposure limits.

FIGURE 9-4 *(continued)*

Personal Protective Equipment

Have available and wear as appropriate for exposure conditions when handling containers or operating equipment containing sulfuric acid: chemical splash goggles; full-length face shield/chemical splash goggles combination; acid-proof gauntlet gloves, apron, and boots; long sleeve wool, acrylic, or polyester clothing; acid proof suit and hood; and appropriate NIOSH/MSHA respiratory protection. In case of emergency or where there is a strong possibility of considerable exposure, wear a complete acid suit with hood, boots, and gloves. If acid vapor or mist are present and exposure limits may be exceeded, wear appropriate NIOSH/MSHA respiratory protection.

Exposure Guidelines

Exposure Limits
Sulfuric Acid, 77 to 100%

PEL	(OSHA)	1 mg/m3, 8 Hr. TWA
TLV	(ACGIH)	1 mg/m3, 8 Hr. TWA
		STEL 3 mg/m3
AEL*	(DuPont)	1 mg/m3, 8 & 12 Hr. TWA
		3 mg/m3, 15 minute TWA

*AEL is DuPont's Acceptable Exposure Limit. Where governmentally imposed occupational exposure limits which are lower than the AEL are in effect, such limits shall take precedence.

PHYSICAL AND CHEMICAL PROPERTIES

Physical Data

Boiling Point	193-327 C (379-621 F) @ 760 mmHg
Vapor Pressure	<0.3 mmHg @ 25 C (77 F)
	<0.6 mmHg @ 38 C (100 F)
Vapor Density	3.4
Melting Point	−35 to 11 C (−31 to 52 F)
Evaporation Rate	<1 (Butyl Acetate=1.0)
Solubility in Water	100 WT%
pH	<1
Odor	Odorless.
Form	Oily; clear to turbid liquid
Color	Colorless to light gray

GRADE	BOILING PT. DEG C	BOILING PT. DEG F	MELTING PT. DEG C	MELTING PT. DEG F	SPECIFIC GRAVITY
60 DEG TECHNICAL	193	380	−12	10	1.706
66 DEG TECHNICAL	279	535	−35	−31	1.835
1.835 ELECTROLYTE	279	535	−35	−31	1.835
98% TECHNICAL	327	621	−2	29	1.844
99% TECHNICAL	310	590	4	40	1.842
100% TECHNICAL	274	526	11	51	1.839

STABILITY AND REACTIVITY

Chemical Stability

Stable, but reacts violently with water and organic materials with evolution of heat.

Incompatibility with Other Materials

Vigorous reactions with water; alkaline solutions; metals, metal powder; carbides; chlorates; fuminates; nitrates; picrates; strong oxidizing, reducing, or combustible organic materials. Hazardous gases are evolved on contact with chemicals such as cyanides, sulfides, and carbides.

FIGURE 9-4 *(continued)*

Decomposition
Releases sulfur dioxide at extremely high temperatures.
Polymerization
Polymerization will not occur.
TOXICOLOGICAL INFORMATION
Animal Data
EYE:
Animal testing indicates this material is corrosive to the eye, when tested undiluted. Animal testing indicates this material is a moderate eye irritant, when tested as 10% solution.
SKIN:
The concentrated compound is corrosive. Animal testing indicates this material is a slight skin irritant, when tested as 10% solution.
INGESTION:
LD_{50}, rat: 2140 mg/kg.
INHALATION:
8 hour, LC_{50}, guinea pigs: 30 mg/m^3.

Single and repeated exposure caused: Irritation of the respiratory tract. Corrosion of the respiratory tract. Lung damage. Labored breathing. Altered respiratory rate. Pulmonary edema. Repeated exposure caused: Altered red blood cell count.

CARCINOGENIC, DEVELOPMENTAL, REPRODUCTIVE, MUTAGENIC EFFECTS:

No adequate animal data are available to define the carcinogenic potential of this material. Limited studies do not suggest effects. In animal testing this material has not caused developmental toxicity. No animal data are available to define the following effects of this material: reproductive toxicity. This material has not produced genetic damage in bacterial cultures. It has not been tested for genetic toxicity in mammalian cell cultures or in animals.

ECOLOGICAL INFORMATION
Ecotoxicological Information
AQUATIC TOXICITY:
Slightly to moderately toxic.
96 hour LC_{50}—Bluegill sunfish: 10.5 ppm.
48 hour TLm—Flounder: 100-300 ppm
DISPOSAL CONSIDERATIONS
Waste Disposal
Cleaned-up material may be an RCRA Hazardous Waste on disposal due to the corrosivity characteristic. Do not flush to surface water or sanitary sewer system. Comply with Federal, State, and local regulations. If approved, neutralize and transfer to waste treatment system.
TRANSPORTATION INFORMATION
Shipping Information
DOT/IMO

Proper Shipping Name	SULFURIC ACID
Hazard Class	8
UN No.	1830
DOT/IMO Label	CORROSIVE
Packing Group	II
Reportable Quantity	1000 lb (454 kg)

Shipping Containers
Tank Cars.
Tank Trucks.
Barge.

# Indicates updated section.

FIGURE 9-4 *(continued)*

REGULATORY INFORMATION
U.S. Federal Regulations
  TSCA Inventory Status   Reported/Included.
  TITLE III HAZARD CLASSIFICATIONS SECTIONS 311, 312
  Acute: Yes
  Chronic: Yes
  Fire: No
  Reactivity: Yes
  Pressure: No
  HAZARDOUS CHEMICAL LISTS
  SARA Extremely Hazardous Substance: Yes
  CERCLA Hazardous Substance: Yes
  SARA Toxic Chemical: Yes

OTHER INFORMATION
  NFPA, NPCA-HMIS
    NFPA Rating
    Health   3
    Flammability   0
    Reactivity   2

    Water Reactive.

    NPCA-HMIS Rating
    Health   3
    Flammability   0
    Reactivity   2
    Personal Protection rating to be supplied by user depending on use conditions.

Additional Information
  For further information, see DuPont Sulfuric Acid "Storage and Handling Bulletin."

The data in this Material Safety Data Sheet relates only to the specific material designated herein and does not relate to use in combination with any other material or in any process.
  Responsibility for MSDS: DuPont Chemicals
  Address: Engineering & Product Safety
           P.O. Box 80709, Chestnut Run
           Wilmington, DE 19880-0709
  Telephone: (302) 999-4946

FIGURE 9-4 *(continued)*

Question C: How can I prevent hazardous situations from occurring?

Section 7: Handling and Storage
• Provides information on appropriate practices for safe handling and storage of the material.

Section 8: Exposure Controls and Personal Protection
• Provides information on practices and equipment useful for minimizing worker exposure.
• Provides guidance on personal protection equipment.
• May also include exposure guidelines.

Section 9: Physical and Chemical Properties
• Identifies the physical and chemical properties that characterize the material and how to design safe work practices.

MATERIAL SAFETY DATA SHEET

DuPont Chemicals

8251CR   **Revised 29-NOV-1993**   **Printed 13-AUG-1995**
                     HYDROGEN PEROXIDE (20 TO 60%)

CHEMICAL PRODUCT/COMPANY IDENTIFICATION
  Material Identification
    Corporate MSDS Number   DU000114
    Formula   H202
    Molecular Weight   34.02
  Tradenames and Synonyms
    "ALBONE"
    "PERONE"
    "TYSUL"
    "ALBONE," "PERONE" AND "TYSUL" ARE REGISTERED TRADEMARKS OF DUPONT.
    CC0007
    HYDROGEN PEROXIDE SOLUTIONS
  Company Identification
    MANUFACTURER/DISTRIBUTOR
              DuPont
              1007 Market Street
              Wilmington, DE 19898

    PHONE NUMBERS
      Product Information   1-800-441-9442
      Transport Emergency   CHEMTREC: 1-800-424-9300
      Medical Emergency   1-800-441-3637

COMPOSITION/INFORMATION ON INGREDIENTS
  Components

Material	CAS Number	%
HYDROGEN PEROXIDE	7722-84-1	20-60
WATER	7732-18-5	40-80
"ALBONE" 50M ALSO CONTAINS		
ADIPIC ACID		124-04-9
SUCCINIC ACID		110-15-6

  Components (Remarks)
    Strength is expressed in weight percent (WT%)

HAZARDS IDENTIFICATION
  Potential Health Effects
    Hydrogen peroxide may cause severe irritation or burns of the skin, eyes and mucous membranes. Splashes in the eye can cause severe eye damage with ulceration of the cornea, and may cause irreversible eye damage, including blindness. Skin exposure can result in bleaching of the skin and hair.

    Inhalation of concentrated vapors can cause irritation of the nose and throat with chest discomfort, cough, difficulty in breathing and shortness of breath.

    Ingestion can cause irritation of the upper gastrointestinal tract with pain and distention of the stomach and esophagus due to liberation of oxygen.

    Gross overexposure by ingestion may be fatal.

(continued)

FIGURE 9-5
HYDROGEN PEROXIDE MSDS FROM DUPONT SPECIALTY CHEMICALS.

HUMAN HEALTH EFFECTS:

Skin contact with aqueous solutions of less than 50% may cause irritation with discomfort or rash. Higher or prolonged exposure may result in skin burns or ulceration. Evidence suggests that skin permeation can occur in amounts capable of producing systemic toxicity. Effects of eye contact with aqueous solutions of less than 5% may include eye irritation with discomfort, tearing, or blurring of vision. Higher or prolonged exposure may result in eye corrosion with corneal or conjunctival ulceration. Contact with aqueous concentrations of greater than 10% may result in eye corrosion with corneal or conjunctival ulceration with possible irreversible eye damage, including blindness.

Overexposure by inhalation may cause irritation of the upper respiratory passages or nonspecific discomfort such as nausea, headache, or weakness. Higher inhalation exposures may lead to temporary lung irritation effects with cough, discomfort, difficulty breathing, or shortness of breath; or fatality from gross overexposure. Ingestion may cause irritation of the gastrointestinal tract with upper abdominal pain, "heartburn," nausea, vomiting, and diarrhea. "Coffee grounds" vomitus and black tarry stools may occur as a result of gastrointestinal tract bleeding. Additional effects from overexposure include red blood cell destruction, or gas embolism. When used as colonic lavage, hydrogen peroxide has caused gas embolism and gangrene of the intestine at concentrations down to 0.75%. Gross overexposure by ingestion may be fatal.

Individuals with preexisting diseases of the skin, eyes, or lungs may have increased susceptibility to the toxicity of excessive exposures.

Carcinogenicity Information

None of the components present in this material at concentrations equal to or greater than 0.1% are listed by IARC, NTP, OSHA or ACGIH as a carcinogen.

FIRST AID MEASURES
First Aid

INHALATION

If inhaled, immediately remove to fresh air. If not breathing, give artificial respiration. If breathing is difficult, give oxygen. Call a physician.

SKIN CONTACT

In case of contact, immediately flush skin with plenty of water for at least 15 minutes while removing contaminated clothing and shoes. Call a physician. Wash contaminated clothing and shoes promptly and thoroughly.

EYE CONTACT

In case of contact, immediately flush eyes with plenty of water for at least 15 minutes. Call a physician.

INGESTION

If swallowed, do not induce vomiting. Give large quantities of water. Never give anything by mouth to an unconscious person. Call a physician.

Notes to Physicians

If swallowed, large amounts of oxygen may be released quickly. The distention of the stomach or esophagus may be injurious. Insertion of a gastric tube may be advisable.

FIRE FIGHTING MEASURES
Flammable Properties

Will not burn, but decomposition, which may be caused by heat or contamination will release oxygen which will increase the explosive limit range and burning rate of flammable vapors.

FIGURE 9-5 *(continued)*

Fire and Explosion Hazards:

Strong oxidizer. Contact with clothing or combustibles may cause fire. Effect may be delayed. Contact with organic liquids or vapors may cause immediate fire or explosion, especially if heated. Under certain circumstances, detonation may be delayed. Oxygen release from hydrogen peroxide may force organic or hydrogen vapors into an explosive range. Follow appropriate National Fire Protection Association (NFPA) codes.

Extinguishing Media
Use only water.

Fire Fighting Instructions
Flood with water. Cool tank/container with water spray.

Wear full protective clothing (rubber suit and boots) including chemical splash goggles or hood and self-contained breathing apparatus.

ACCIDENTAL RELEASE MEASURES
Safeguards (Personnel)
NOTE: Review FIRE FIGHTING MEASURES and HANDLING (PERSONNEL) sections before proceeding with clean-up. Use appropriate PERSONAL PROTECTIVE EQUIPMENT during clean-up.

Accidental Release Measures
Comply with Federal, State, and local regulations on reporting releases of wastes. Flood area with water and drain to an approved chemical sewer or wastewater treatment system, including municipal sewers if approved. May be destroyed with sodium metabisulfite or sodium sulfite (1.9 lbs. SO_2 equivalent per lb. of peroxide) after diluting to 5–10% peroxide.

The Extremely Hazardous Substance List Reportable Quantity for >52% Hydrogen Peroxide is 1 lb.

If Hydrogen Peroxide (20 to 60%) is spilled and not recovered, or is recovered as a waste for treatment or disposal, the CERCLA Reportable Quantity is 100 lbs. (release of an unlisted Hazardous Waste characteristic of ignitibility).

HANDLING AND STORAGE
Handling (Personnel)
Use extreme care when attempting any reactions because of fire and explosion potential (immediate or delayed). Conduct all initial experiments on a small scale and protect personnel with adequate shielding as the reactions are unpredictable and may be delayed, and may be affected by impurities, contaminants, temperature, etc. Do not get in eyes. Avoid contact with skin and clothing. Wash thoroughly after handling. Avoid contact with flammable or combustible materials. Avoid contamination from any source including metals, dust, and organic materials. Never use pressure to empty drums; container is not a pressure vessel. In the event of an accident where large volumes of hydrogen peroxide might come into contact with external fires or with incompatible chemicals, a one-half mile area from the incident should be evacuated.

Storage
Store in a properly vented container or in approved bulk storage facilities. Do not block vent. Do not store on wooden pallets. Do not store where contact with incompatible materials could occur, even with a spill. (See "Hazardous Reactivity.") Have water source available for diluting. Do not add any other product to container. Never return used or unused peroxide to container, instead dilute with plenty of water and discard. Rinse empty containers thoroughly with clean water before discarding. (See "Waste Disposal.")

EXPOSURE CONTROLS/PERSONAL PROTECTION
Engineering Controls
Use sufficient ventilation to keep employee exposure below recommended exposure limits.

FIGURE 9-5 *(continued)*

Personal Protective Equipment

EYE/FACE PROTECTION

Wear coverall chemical splash goggles. In addition, where the possibility exists for eye or face contact due to splashing or spraying of material, wear chemical splash goggles/full-length face shield combination.

RESPIRATORS

Where there is potential for airborne exposure in excess of applicable limits, wear NIOSH/MSHA approved respiratory protection.

PROTECTIVE CLOTHING

Where there is potential for skin contact, have available and wear as appropriate: impervious gloves, apron, pants, jacket, hood, and boots; or totally encapsulating chemical suit with breathing air supply. Permeation data supplied by vendors indicate that impervious materials such as natural rubber, natural rubber plus Neoprene, nitrile, or polyvinylchoride afford adequate protection.

Do not wear leather gloves or leather shoes (uppers or soles) because they can ignite following contact with peroxide. Cotton clothing can also ignite. This effect may be within minutes, or delayed. Clothing fires and skin damage occur less quickly with 50% or lower hydrogen peroxide than with 70% material, but adequate personal protection is essential for all industrial concentrations. Protective skin creams offer no protection from hydrogen peroxide and should not be used.

#Exposure Guidelines

Applicable Exposure Limits

HYDROGEN PEROXIDE

PEL	(OSHA)	1 ppm, 1.4 mg/m^3 (90%) - 8 Hr TWA
TLV	(ACGIH)	1 ppm, 1.4 mg/m^3, 8 Hr. TWA
AEL*	(DuPont)	None Established

ADIPIC ACID

PEL	(OSHA)	None Established
TLV	(ACGIH)	5 mg/m^3, 8 Hr. TWA
AEL*	(DuPont)	10 mg/m^3, 8 Hr. TWA
WEEL	(AIHA)	5 mg/m^3, 15 minute TWA

*AEL is DuPont's Acceptable Exposure Limit. Where governmentally imposed occupational exposure limits which are lower than the AEL are in effect, such limits shall take precedence.

PHYSICAL AND CHEMICAL PROPERTIES

Physical Data

Evaporation Rate: Greater than 1
Solubility in Water: 100 WT%
Form: Clear liquid
Color: Colorless
Odor: Slightly pungent, irritating

	HYDROGEN PEROXIDE CONCENTRATION			
WT%	20%	35%	50%	60%
Boiling Point C	103	104	114	119
F	217	226	237	246
Melting Point C	-14.6	-33.0	-52.2	-55.5
F	5.7	-27.4	-62.0	-67.9
Specific Gravity 25 C (77 F)	1.07	1.13	1.19	1.24
Vapor Pressure-mmHg 25 C (77 F)	20.6	17.4	13.5	10.7

# Indicates updated section.

FIGURE 9-5 *(continued)*

STABILITY AND REACTIVITY

Chemical Stability
Unstable with heat or contamination; liberation of oxygen gas may result in dangerous pressures. (See "Decomposition," below.)

Incompatibility with Other Materials
Incompatible with most flammables/combustibles (See "Fire and Explosion Hazards") as well as cyanides, nitric acid, potassium permanganate, and many other oxidizing and reducing agents. Mixtures with both organics and some acids may be especially reactive.

Decomposition
Contamination or heat may cause self-accelerating exothermic decomposition with oxygen gas and steam release that can cause dangerous pressures. May react dangerously with rust, dust, dirt, iron, copper, heavy metals or their salts (such as mercuric oxide or chloride), alkalis, and with organic materials (especially vinyl monomers).

Polymerization
Polymerization will not occur.

TOXICOLOGICAL INFORMATION

Animal Data
Inhalation 8-hour LC_{50}:   >2000 ppm in rats (90% H_2O_2)
Skin absorption LD_{50}:   >2000 mg/kg in rabbits (35% H_2O_2)
Oral LD_{50}:   1232 mg/kg in rats (35% H_2O_2)

At aqueous concentrations of less than 50% hydrogen peroxide skin irritation occurs, but at greater concentrations hydrogen peroxide is corrosive to the skin. Concentrations less than 5% in aqueous solutions are eye irritants; solutions between 5% and 10% range from severe eye irritants to being corrosive; concentrations greater than 10% are corrosive to the eye. The compound is not a skin sensitizer in animals.

Repeated inhalation exposures produced nasal discharge, bleached hair, and respiratory tract congestion with some deaths occurring in rats and mice exposed to concentrations greater than 67 ppm. Dogs exposed by inhalation to 7 ppm for 6 months had lung and skin irritation.

The effects from single high oral doses include convulsions. Repeated administration of the compound in the diet of animals resulted in growth inhibition, reduced weight gain, abnormal liver function, ulcers, and discoloration of the stomach lining with swelling. Long-term administration to mice in the drinking water resulted in gastric erosion and duodenal hyperplasia.

One study by skin application suggested no carcinogenic activity. Results of an ingestion study with mice suggested that hydrogen peroxide might be carcinogenic. However, the FDA and other organizations have reviewed this study and concluded there is insufficient evidence that hydrogen peroxide is carcinogenic. An unpublished, long-term study with rats revealed no evidence of carcinogenicity. Female rats treated with 10% hydrogen peroxide produced offspring of lower body weight and some structural abnormalities, but these changes were attributed to maternal toxicity. Hydrogen peroxide produced genetic damage to bacterial and mammalian cells in culture, but one study in animals indicated it did not produce genetic damage. Limited tests in animals demonstrate no reproductive toxicity.

ECOLOGICAL INFORMATION

Ecotoxicological Information
Aquatic Toxicity

96-hour LC_{50}, catfish: 37.4 mg/L

DISPOSAL CONSIDERATIONS

Waste Disposal
Comply with Federal, State, and local regulations. If approved, may be diluted and drained to a municipal sewer or waste treatment plant. May be diluted and drained

FIGURE 9-5 *(continued)*

through a scrap metal pit (iron, copper, etc.) to reduce peroxide concentration. Hydrogen peroxide may be an RCRA regulated hazardous waste upon disposal due to the oxidizing characteristic under the ignitibility category.

TRANSPORTATION INFORMATION
Shipping Information
DOT/IMO
Proper Shipping Name   HYDROGEN PEROXIDE, AQUEOUS SOLUTIONS
Hazard Class   5.1
UN No.   2014
DOT/IMO Label   OXIDIZER, CORROSIVE
Subsidiary Hazard Class   8
Packing Group   II
Shipping Containers
Tank Cars
Tank Trucks
ISO (Sea) Tanks
Drums
Bottles

REGULATORY INFORMATION
U.S. Federal Regulations
TSCA Inventory Status   Reported/Included.
TITLE III HAZARD CLASSIFICATIONS SECTIONS 311, 312
Acute: Yes
Chronic: No
Fire: No
Reactivity: Yes
Pressure Yes
LISTS:
SARA Extremely Hazardous Substance   (Yes)*
CERCLA Hazardous Material   (**)
SARA Toxic Chemical   No

  *For greater than 52% material.
 **See Disposal Information.

CANADIAN WHMIS CLASSIFICATIONS:

  C; E; F

HYDROGEN PEROXIDE >52% is a flammable liquid as defined by OSHA in 29 CFR 1910.1200(c). Use of this product may require compliance with 29 CFR 1910.119, Process Safety Management of Highly Hazardous Chemicals.

OTHER INFORMATION
NFPA, NPCA-HMIS
NFPA Rating
Health   2
Flammability   0
Reactivity   1

Oxidizer.
NPCA-HMIS Rating
Health   3
Flammability   0
Reactivity   1

Personal Protection rating to be supplied by user depending on use conditions.

FIGURE 9-5 *(continued)*

Additional Information
For further information, see DuPont HYDROGEN PEROXIDE Storage and Handling Bulletin.

The data in this Material Safety Data Sheet relates only to the specific material designated herein and does not relate to use in combination with any other material or in any process.

Responsibility for MSDS	DuPont Chemicals
Address	Engineering & Product Safety
	P.O. Box 80709, Chestnut Run
	Wilmington, DE 19880-0709
Telephone	302-999-4946

FIGURE 9-5 *(continued)*

Section 10: Stability and Reactivity
• Describes the conditions that could result in a potentially hazardous chemical reaction.

Question D: Is there any other useful information about this material?

Section 11: Toxicological Information
• May be used to provide information on toxicity testing of the material and its components for medical professionals, occupational safety and health professionals, and toxicologists.
Section 12: Ecological Information
• May be used to provide information on the effects the material may have on plants and animals and its environmental fate.
Section 13: Disposal Considerations
• May provide information useful to determine appropriate disposal measures.
Section 14: Transport Information
• May provide basic shipping classification information.
Section 15: Regulatory Information
• May be used to provide additional information on state, federal, and international regulations affecting the material or its components.
Section 16: Other Information

BIBLIOGRAPHY

1 *Accident Prevention Manual for Business and Industry: Administration and Programs,* National Safety Council, Itasca, Ill., 1992.
2 Altvater, T. S., "Material Safety Data Sheets," *Professional Safety,* October 1990.
3 Anton, T. J., *Occupational Safety and Health Management,* 2d ed., McGraw-Hill, New York, 1989.
4 Crowl, D. A., and J. F. Louvar, *Chemical Process Safety and Fundamentals with Applications,* Prentice-Hall, Englewood Cliffs, N.J., 1990.
5 Ducommun, J. C., *How to Use Sample Bombs Safely, Fire Protection Manual* vol. 1; Gulf Publishing, Houston, 1985.

6 Ferrante, L. M., "Hazardous Materials and the Laboratory," *Pollution Engineering,* September 1990.

7 *Fire Protection Guide on Hazardous Materials,* 8th ed., NFPA, Quincy, Mass., 1984.

8 29 CFR 1910.1200, OSHA, Washington, D.C., 1994.

9 19 CFR 1910.120, OSHA, Washington, D.C., 1994.

10 Kamp, J., "Worker Psychology," *Professional Safety,* May 1994.

11 Kavianian, H. R., and C. A. Wentz, *Occupational and Environmental Safety Engineering and Management,* Van Nostrand Reinhold, New York, 1990.

12 Lepkowski, W., "Chemical Companies Make Public Worst Case Accident Scenarios," *C&E News,* June 20, 1994.

13 Long, J. R., "Standard for Material Safety Data Sheets in the Offing," *C&E News,* May 18, 1992.

14 29 CFR 1910.1450, OSHA, Washington, D.C., 1994.

15 Palluzi, R. P., "Develop an R&D Safety Standards Program," *Chemical Engineering Progress,* February 1995.

16 Pina, J. J., "Preventing the IFS," *Professional Safety,* September 1995.

17 Plog, B. A., et al., *Fundamentals of Industrial Hygiene,* 3d ed., National Safety Council, Itasca, Ill., 1988.

18 Roughton, J., "Protecting the Hazardous Waste Worker," *Pollution Engineering,* June 1995.

19 Slote, L., *Handbook of Occupational Safety and Health,* Wiley, New York, 1987.

20 *Hazardous Industrial Chemicals: Precautionary Labeling,* Z129.1-1994, ANSI, New York, 1994.

21 *Hazardous Industrial Chemicals: Material Safety Data Sheets, Preparation,* Z400.1-1993, ANSI, New York, 1993.

PROBLEMS

1 What is the intent of the Hazard Communication Standard?

2 What type of items are included in Hazard Evaluation and Communication?

3 What are the most commonly cited areas of the OSHA Hazard Communication Standard?

4 What are the Hazard Communication Standard areas that employers should include in employee training?

5 Why is the Hazard Communication Standard the most often violated OSHA standard for industry?

6 What is a hazardous chemical for purposes of the Hazard Communication Standard?

7 Under 29 CFR 1910.1200, what chemicals are carcinogens?

8 What are the criteria for determining if a chemical mixture is hazardous?

9 When a container of a hazardous chemical is being shipped, what is required to be on the container label?

10 What information about the chemical substance should be included on the MSDS?

11 What protective measures would you suggest to protect employees at a rail-car loading dock for the liquid solvent listed in Figure 9-2?

12 What safety and health hazards could result if a fork truck dropped several 55-gal drums of the hydrocarbon demulsifier listed in Figure 9-3 in an enclosed, heated material warehouse and the drums ruptured?

13 What safety and health hazards result if an employee is exposed to an outdoor 400-gal spill of the hydrocarbon solvent listed in Figure 9-2?

14 Design your own generic MSDS—one that would be universally acceptable to chemical users.

15 An industrial warehouse contains 55-gal drums of gasoline, hydrochloric acid, sodium chloride, benzene, calcium hydroxide, and vinyl chloride. What are the potential hazards in this warehouse?

16 Using the ANSI MSDS standard format, place the OSHA-required data elements into the appropriate sections of the ANSI Z400.1-1993 form.

17 A chemical process area uses several 55-gal drums of sulfuric acid per day. What potential hazards could result from a spill of one drum in the workplace?

18 What precautions should be taken in storing ten 55-gal drums of hydrogen peroxide in an enclosed warehouse?

PROCESS AND PLANT
SAFETY HAZARDS

Process and plant safety involves the application of scientific, engineering, and management principles in the identification, characterization, and control of process and plant hazards so as to prevent injuries and incidents that can cause harm to either the employees at the plant facility or the public. [2, 8, 33]

A number of professional organizations have been actively involved in this field for many years. The American Society of Safety Engineers and the National Safety Council consider the safety, health, and welfare of workers and the public to be of paramount importance. Whenever their safety, health, and welfare are in question, there is an obligation to advise employers or appropriate authorities.

The Center for Chemical Process Safety (CCPS) of the American Institute of Chemical Engineers was established in 1985 to develop and disseminate technical information for use in the prevention of major chemical accidents. The CCPS has prepared publications and programs for facilities that handle or use hazardous chemicals. Relative to other industries, the chemical industry has a fairly low accident rate, but more can be done to reduce the risks even further and protect employees and the general public against the effects of acute and chronic exposure to chemicals and catastrophic events. [30]

Chemical safety management requires an integrated approach, as shown in Table 10-1 on page 284. [9]

GOALS AND OBJECTIVES OF THE SAFETY PROGRAM

The most important goal of process and plant design must be the safe development, design, construction, start-up, and operation of the plant. This must be a pri-

TABLE 10-1
AN INTEGRATED APPROACH TO PROCESS AND PLANT SAFETY MANAGEMENT

Goals and objectives of the safety program
Risk analysis and management
Industrial process design and operation
Mechanical integrity of process equipment
Safety regulations, codes, and standards
Project safety and health review
Hazard evaluation of process design and operation
Personnel motivation and training
Process and plant modification and change
Incident investigation and safety audits

---

ority for all technical and business matters. It will reduce the occurrence of major industrial accidents, such as occurred with Union Carbide in Bhopal, India, Phillips Petroleum Company in Houston, Texas, and ARCO in Channelview, Texas. These tragedies might have been avoided if process and plant safety had been the number one priority. [22, 23]

Safety risks should be reduced to the lowest practical level through a clear understanding of the process design, operation, and mechanical integrity of equipment. Employees must be made aware of safety regulations, codes, and standards, and these regulatory matters should never be knowingly violated in the design and operation of the facility. The management of projects should include a safety and health review coupled with a hazard evaluation of process and plant design. Personnel involved in the development, design, construction, and operation of the facility should be given adequate training and be properly motivated to accomplish the goals of the safety program. All process and plant modifications and any subsequent future changes in design or operation of the facility should be reassessed through this integrated approach to process and plant safety management. Any incident involving the process or plant should cause immediate concern and result in a thorough, objective critique and investigation to determine the cause of the incident. In a similar fashion, near-miss incidents should also be investigated to determine their cause. Periodically the process and plant design and operation should be given a thorough safety audit by an experienced, qualified team to reduce or eliminate the risk of potential safety hazards.

Management should dedicate sufficient resources (personnel, equipment, time, and money) to reduce the risk of hazards. The development of process technology, the achievement of plant production objectives, the availability of products to supply customers, and the motivation of personnel all have the potential to conflict with safety goals. A realistic approach to safety is best achieved by a management commitment and organizational understanding that resolves these conflicts before they occur. In practical terms, if a process or plant design and operation cannot be done safely, it will be in the best interests of all concerned parties to develop another safer alternative that will also satisfy the additional needs of the organization.

The investment of resources to develop safe processes and plant facilities usually adds to the capital expenditures required and the operating costs. The long-term return on this short-term investment in safety should result in improved on-stream factors and product quality, better product availability for the marketplace, lower insurance costs, higher morale of personnel, more objective project evaluation, and improved public and personnel relations.

The safety objectives should be understood and accepted by all technical and business functions of the organization and have the full support and commitment of management. Top management must be readily accessible in order to demonstrate its support of the safety commitment. This can best be achieved through a direct reporting relationship of the safety function to the top management of the organization. Under no circumstances should the safety function of an organization report to production, marketing, or other line business functions that might have conflicting goals in carrying out the safety program. Safety personnel must communicate with the entire organization to promote safety awareness and knowledge that will lead to improved safety performance and greater confidence in the public and regulatory communities.

All practical techniques and alternatives should be used to enhance safety as much as is practical. [1, 12] In process and plant design and operation this can be done by identifying first the risk and then those alternative actions that can be used. This activity can be used for acute or chronic exposure to chemical hazards listed on material safety data sheets, process chemistry hazards, process equipment and instrumentation, materials of construction, and any other hazards. Chemicals, the process design, the plant layout, and the process operating conditions can all play important roles in creating hazards. [18, 21, 24]

RISK ANALYSIS AND MANAGEMENT

Risk management begins with hazard identification. [7, 10, 27, 29, 37, 38] Material safety data sheets are excellent sources for chemical hazard identification. Other hazard evaluation techniques will be discussed later in this chapter. The identified hazards need to be assessed to determine the probability and severity of the impact of the hazard on the safety and health of the public, the surrounding community, the plant employees, the environment, and plant property. The Delphi method, which ranks hazards according to their probable occurrence and severity, is a recommended technique for hazard assessment. Next, all practical steps should be taken to reduce both the probability and severity of the risks. A thorough knowledge of the related hazards for the chemicals, equipment, and processes in the plant facility should be in the background and experience of those personnel doing the risk analysis and management. [14] By integrating risk management into the process and plant design, the remaining risk can be managed in a practical and effective manner for both normal and upset plant operating conditions. [11]

INDUSTRIAL PROCESS DESIGN AND OPERATION

A desirable process and plant safety program begins with a well-conceived process and plant design that will be maintained throughout the entire operation of the facility, unless it is modified. The design of all industrial processes that use hazardous chemicals should begin with the available information about those chemical hazards. Material safety data sheets should be reviewed to identify the type of hazard each chemical represents and the methods of control (see Table 10-2).

Information Gathering

Process chemistry involves the conversion of raw materials into desirable products. Process chemistry can also create or reduce chemical hazards in process and plant design and may therefore influence or impact on process and plant design as shown in Table 10-3.

As the design proceeds, information about the chemical hazards is integrated into the development of process technology and operating equipment. Technology-based information is needed for each category listed in Table 10-4. Necessary operational information is shown in Table 10-5 on page 288. The better the planning and organizing of the items in Tables 10-4 and 10-5, the greater the probable success of the operating facility. During the development of process technology and equipment the focus should be on accurate and reliable process flow diagrams and piping and instrumentation diagrams. All of the other items in Tables 10-4 and 10-5 are necessary information and components for these two diagrams, which will be used in the final design and construction of the operating facility.

Plant Operation

Normal plant operations will be based upon controlling the process operating parameters within desirable operating ranges. Significant deviations may result in an upset condition. As shown in Table 10-6 on page 288, provisions must be made in the plant and process design to return these upset conditions to normal operations

TABLE 10-2
DATA AND INFORMATION FOR CHEMICAL HAZARD ASSESSMENT

Name and chemical composition

Regulatory requirements

Physical and chemical property data

Fire and explosion data

Reactivity data

Health information

Spill, leak, and disposal procedures

Special protective equipment and precautions

TABLE 10-3
PROCESS CHEMISTRY IN PROCESS AND PLANT DESIGN

Chemical raw materials and catalysts
Chemical reactions and kinetic data
Process operating parameters and ranges
Preliminary process flow diagram
Material, product, and waste inventories

or institute emergency procedures. A critical situation in the safe operation of any process or operating facility is the capability to control deviations and to restore upset conditions to normal operations without an incident.

Before reaching normal operations an operating facility needs to go through a start-up period. During the start-up phase, major deviations from normal operating conditions can be expected. Smooth trouble-free plant start-ups are the exception rather than the norm. The probability for a successful start-up is greatly enhanced through operator training and process design that anticipates start-up problems before they actually occur.

It will be necessary to shut down the operating facility on occasion for routine maintenance, adjustments in product inventories, or emergency situations. During the shutdown phase, deviations outside of normal operating ranges can also be expected. It is important that operating personnel and plant designs anticipate operating problems connected with plant shutdown. A smooth shutdown can greatly assist a subsequent successful start-up of the operating facility. Shutdown procedures should be well-conceived; otherwise, plugged lines, damage to instrumentation,

TABLE 10-4
TECHNOLOGY INFORMATION FOR PROCESS AND PLANT DESIGN

Products, by-products, and wastes produced
Process technology to be used
Plant capacity
Process flow diagrams
Process operating variables
Selection of the plant site
Hazard identification and safety control
Waste identification and environmental control
Process control techniques
Equipment specifications
Government permit requirements
Piping and instrumentation diagrams

TABLE 10-5
OPERATIONAL INFORMATION FOR PROCESS AND PLANT DESIGN

Construction timetable
Inspection and testing of equipment and instrumentation
Personnel and training requirements
Startup and troubleshooting of potential problems
Debottlenecking opportunities
Operations, maintenance, and emergency procedures
Procedures for upset operating conditions
Procedures for safety and environmental audits

stress and corrosion of construction materials, poor documentation of procedures followed, and other safety hazards may result. Smooth plant shutdowns could also be thought of as preventive maintenance prior to the subsequent plant start-up.

During the process and plant design some thought should be given to future expansion or termination of the facility. Sometimes modest initial expenditures during the original process and plant design can greatly facilitate and enhance future plant expansions. Good planning practices can allow for the ultimate termination of operations without creating unnecessary safety or environmental hazards.

MECHANICAL INTEGRITY OF PROCESS EQUIPMENT

The design, fabrication, installation, and maintenance of process equipment will greatly affect the safety and reliability of both the process and plant operation. Only the construction materials specified should be used in fabrication and maintenance. All process vessels, pumps, valves, piping, and other equipment used in the operating facility should be fabricated and installed in accordance with the design specifications. When design criteria and specifications are compromised, the prob-

TABLE 10-6
SAFETY EQUIPMENT AND SYSTEMS
FOR
PROCESS AND PLANT DESIGN

Fire protection
Gas and vapor detection
Alarm and interlock
Pressure relief and vent
Isolation of equipment and plant
Emergency relief and vent
Emergency and backup services

ability of an accident is greatly increased. Even small items, such as O-rings, gaskets, and welds improperly selected or installed, can mean the difference between safe operations and operating hazards.

Welding procedures and equipment dimensions and capacities need to be verified. All of the resources that have focused on the process and plant design to produce a safe and reliable operating facility could be of little value unless they have been inspected and tested to confirm their reliability. The methods and frequencies for testing critical instrumentation in a process and plant design should be custom-tailored to meet the requirements for each plant. An initial testing of these systems should be carried out prior to plant start-up and periodically thereafter once the plant has commenced operation.

During the start-up and normal operating phases of the new facility, procedures should be in place to maintain the integrity of process equipment, particularly where hazardous materials are involved. During maintenance operations the facility is vulnerable to the creation of a hazardous condition because of the nature of the hazardous materials that could be exposed during the maintenance work. This is particularly true of processes with flammable hazardous materials, which have the potential to be exposed to ignition sources. A fire can easily start in a plant operation that has been shut down for maintenance unless a well-conceived safety program reduces the risk of an incident. Welding, flame cutting, the use of equipment that could produce sparks, or static electricity could become the ignition source in hot-work areas involving flammable or combustible materials. During the process and plant design consideration should be given for the need for such hot work during the operation of the facility. Hot-work permits are normally issued only after a thorough evaluation and assessment of the hazards present in the area. These hot-work permits are typically valid only for one 8-h shift at a time.

Safety relief valves and rupture disks with emergency venting capability are widely used in emergency planning. This is particularly true when the process system involves high pressures well above ambient conditions. Safety relief devices are designed to operate during responses to emergencies at the operating facility. [35] The test and inspection program for safety relief devices must assure that they are maintained in a reliable condition to adequately respond to plant emergencies. Testing programs should require the relief devices to be tested prior to plant start-up, with subsequent testing to be performed at least on an annual basis. [6]

PROCESS HAZARD ANALYSIS

The Occupational Safety and Health Administration (OSHA) is responsible for regulating safety hazards in the workplace. A number of OSHA regulations affect process and plant design and operation, including 29 CFR 1910.1200, the Hazard Communication Standard, which regulates the way chemicals are managed in the workplace. Another OSHA standard, 29 CFR 1910.120, involves safety standards for chemical hazards associated with hazardous waste operations and environmental cleanup. [16]

The most significant OSHA standard is 29 CFR 1910.119, Process Safety Management of Highly Hazardous Chemicals. [3, 5, 25] It requires employers to manage hazards associated with processes using materials identified as highly hazardous to prevent or reduce the consequences of catastrophic releases or other incidents. Manufacturers of explosives and pyrotechnics and firms having processes involving flammable gases or liquids in quantities of at least 10,000 lb at one location are subject to this OSHA rule. These employers must perform a process hazard analysis using material safety data sheets, process chemistry, block-flow diagrams (process flow diagrams), and piping and instrumentation diagrams. The employer's hazard analysis must use at least one of the following methodologies:

What-if analysis
Checklist
What-if/checklist
Hazard and operability study
Failure mode and effects analysis
Fault tree analysis
An appropriate equivalent methodology

These process hazard analysis techniques will be discussed in more detail later in this chapter. The exceptions to 29 CFR 1910.119 are hydrocarbons used only as fuels and flammable liquids that are below their normal boiling point at ambient conditions. OSHA regulations do not apply to retail facilities, oil and gas Exploration and Production (E & P) operations, or unoccupied remote facilities.

Employers must develop and implement written operating procedures for each operating phase that includes operating limits, safety and health considerations, and safety systems. Personnel who are involved in these operating procedures must be given appropriate training in this area. Employers must inform contractors about potential hazards related to their work and be certain that the contractor has in turn informed their employees. Employers must further assure that mechanical integrity of equipment, prestart-up safety reviews, hot-work permits, modifications and process changes, investigation of incidents, emergency planning and response, and safety compliance audits are all done in accordance with 29 CFR 1910.119. [31]

The goal of this OSHA regulation is to assure that both technology and management controls are in place and working together to avoid catastrophic incidents in the future. Between 1987 and 1991, the chemical process industry, for example, has incurred property losses in the range of $400 million annually. In 1990, 44 U.S. employees were killed because of plant explosions. This new OSHA standard ensures that hazardous chemical industries will be more closely monitored and that future process and plant design and operation will be forced to comply with OSHA regulations as well as with industry and company standards.

The American Petroleum Institute published an industry standard for the management of process hazards. [26] The Chemical Manufacturers Association's resource guide is intended to prevent fires, explosions, or accidental chemical releases. The American Institute of Chemical Engineers Center for Chemical Process

Safety describes technical elements of a model chemical process safety program, as well as guidelines for hazard evaluation procedures that can be used in connection with the OSHA-mandated process hazard analysis. All of these regulations, codes, and standards should be incorporated, when applicable, into the design of safe and reliable processes and operating facilities.

A project safety and health review should be performed on both the process and the operating facility. [4] This review will depend upon the number and complexity of the hazards, the potential risks in the plant siting and layout, the process design, the plant operation, and the associated chemical inventory management. Chemical hazards should be addressed under the guidelines of the OSHA Hazard Communication Standard, 29 CFR 1910.1200. Table 10-2 has described the data and information that need to be addressed for this chemical hazard assessment. The amount of each chemical that constitutes a hazard is important in the assessment process. The larger the quantity, the greater the potential hazard. This chemical hazard assessment process should apply to raw material inventories, chemicals in process, product inventories, and inventories of wastes that are being managed prior to final disposal. [28, 34]

The siting of the operating facility can play an important role in the safety and health review. Generally, communities do not want chemical processing facilities located in densely populated areas. Communities would prefer such operating facilities to be sited in more remote locations so as to lessen the potential hazard to the community. Related transportation of hazardous chemicals and traffic patterns should minimize the probability of a catastrophic incident. Safety should greatly influence the layout of the plot plan and ensure proper receiving, storage, shipping procedures, and inventory reduction, which are all interdependent in the inherent safety at the facility.

The indentification of hazards early in the review process will aid in either removing the hazard or reducing it to a more manageable level of risk. If plant logistics are considered early in the design, the need for extensive plant layout or process revisions may be eliminated. Project safety and health review using process flow diagrams and piping and instrumentation diagrams should begin well before the design package is finalized.

Adequate spacing needs to be provided for the process plant, various storage tanks, and traffic systems. Hazardous chemical loading and unloading facilities should not be in the process area or in storage facilities for hazardous chemicals. By the same token, hazardous material storage facilities should be located at a safe distance from the process area. Storage of hazardous chemicals has a much lower risk of accidents than loading, unloading, and processing of the same chemical hazards. This logistical problem should be managed in the process and plant layout with the objective of keeping chemical hazardous storage tanks as far as possible from property lines and yet maintaining adequate distances between loading and unloading for storage and process areas. The primary reason that thousands of people were killed in the incident at the Union Carbide Bhopal plant was that residential areas were allowed to develop near the process and storage areas at the plant site.

Safety considerations that could reduce the risk of an incident in the storage of hazardous materials include the following:

Inventory minimization of hazardous materials in storage
Pollution prevention and waste minimization process practices
Liquified gases stored only at low temperatures and pressures
Liquid storage vessels designed so that leaks and spills will not accumulate under the storage tanks or process equipment

Many industrial processes involve reactive chemicals and chemical reactions. One of the primary safety objectives is to control these reactions and chemicals so as to prevent incidents and mishaps. Exothermic reactions generate heat, and the resulting temperature increase may influence the rate of the chemical reaction. The thermostability of reactive chemicals should also be reviewed. Other reactive chemical properties may include shock sensitivity, flashpoint, flammability limits, autoignition temperature, and dust. The minimization of reactive chemical inventories should reduce the risk of that chemical hazard to the processing facility. Substitution of less hazardous materials may improve process and plant safety by eliminating or reducing the risk of a reactive chemical incident.

The sudden decrease in pressure of a vapor confined at a temperature above its boiling point results in an explosive vaporization of the liquid to form a vapor cloud. This boiling-liquid expanding-vapor explosion (BLEVE) is an unusual physical phenomena that has been observed in incidents involving propane, butane, chlorine, and ammonia and has resulted in thousands of deaths. [32]

Materials of construction at the operating facility are usually based on mechanical properties and corrosion and chemical resistance. They should also be reviewed for their potential to catalyze undesirable and unsafe chemical reactions. Corrosion products of common steel materials have been shown to catalyze the exothermic hydrogenation of ethylene. The well-intended selection of corrosion-resistant alloys may introduce an undesirable catalyst that promotes chemical side reactions.

Management should require project safety and health reviews for all process and plant designs. These reviews are necessary to evaluate the potential hazards and associated risk that may affect the reliability and safety of the operating facility. They help protect the safety and health of employees and the public by avoiding bad business decisions and violations of regulations and standards in process and plant design.

A hazard analysis is a study to identify, analyze, and evaluate the risk of hazards associated with a process or operating facility. [13] Process and plant hazard analysis is used to identify limitations in the siting, design, layout, and operation of facilities that could result in safety, health, and environmental hazards to employees and the public. The objective of process and plant hazard analysis is to improve the safety and management of risk at operating facilities that use industrial processes. The need for safety equipment and safety systems in the process and plant design, as shown in Table 10-6, will be identified during this analysis.

The occurrence of major industrial incidents, and exposure to acute and chronic health hazards, has resulted in legislation and regulations concerning process and plant hazards. The principal federal agencies involved in the regulation of process and plant hazards are the Occupational Safety and Health Administration, the U.S. Environmental Protection Agency, and the U.S. Department of Transportation. These agencies have promulgated numerous regulations concerning safety and health hazards, and this in turn has resulted in more formalized process and plant safety management programs by industry. Each of these safety programs addresses the probability, severity, and consequences of process and plant hazards based on an understanding of the following risk-related questions:

1 *Severity:* What can go wrong? How bad can it get?
2 *Probability:* How likely is it to occur?
3 *Consequences:* What are the impacts?

Hazardous materials under certain physical and chemical conditions have the ability to create incidents, accidents and exposures, which are sequenced in Table 10-7 on page 294. These can have serious consequences for employees, the community, and the environment. There is usually an initiating event and possibly additional intermediate events that act in sequence to bring the process hazard to the point where it becomes an incident phenomena. The initiating event may be a process upset, a management control failure, a human error, or some external event beyond the control of the operating facility such as sabotage or an earthquake. What follows next in this accident sequence are propagating factors and risk-management failures. Ignition sources and human error are two examples of propagating factors that may be enhanced by adverse weather conditions resulting in a chain of events that cause an incident. If the safety system or the mitigation system does not perform effectively and personnel are inadequately trained, an incident may occur. If there is a discharge leak or release of hazardous material, a fire, explosion, or toxic chemical exposure may result. Plant personnel and the public in the surrounding community may suffer adverse health effects or even death from this occurrence. Property damage and harm to the environment may also result from such an occurrence.

Planning and preparation are important to the ultimate success of any process and plant hazard analysis. Management must be supportive of this analysis and have input into the definition of the objectives for the hazard analysis being undertaken. A hazard analysis can be performed at any phase in a project life cycle, as shown in Table 10-8 on page 295.

A team of knowledgeable personnel should perform the hazard analysis. Table 10-9 on page 296 lists the various disciplines and backgrounds that should be represented on the hazard analysis team. An interdisciplinary approach is the most effective. Since every hazard analysis will be unique to the situation, not all disciplines may be represented, though there should be no compromise in the most strategic areas identified. All team members should be experienced in their discipline, possess good communication skills, be able to work well with others, and have an understanding of the process and plant facility under analysis.

The team leader must stress the main objectives: to identify process and plant hazards, improve the safety of the operating facility, and make the plant more productive. Criticism should be constructive. The team should limit its efforts to hazard identification and possible consequences, rather than engineering detailed solutions.

TABLE 10-7
THE SEQUENCE OF HAZARDOUS MATERIAL INCIDENTS, ACCIDENTS, AND
EXPOSURES

Presence of hazardous materials
    Physical, chemical, and toxic properties
    Quantities and type of containment

Physical and chemical conditions present

Initiating events
    Process deviations and upsets
    Equipment failure
    Loss of utilities
    Management control failure
    Human error
    External events

Propagating factors
    Chain of events
    Ignition source
    Management control failure
    Human error
    Weather conditions

Risk management failure
    Safety system
    Mitigation system
    Emergency plan
    Human error
    Training

Occurrence
    Discharge, leak, release
    Fire, explosion, toxic chemical exposure

Consequences
    Employee and public health, injuries, and death
    Property damage
    Environmental damage

CONSEQUENCES OF HAZARD IDENTIFICATION

Many adverse consequences can result from process and plant hazards, as shown in Table 10-10 on page 296. The team should discriminate between minor hazards and major hazards, which may be critical to the overall safety of the process and plant.

Material safety data sheets are excellent sources of information about potential hazards. Consensus standards of the National Fire Protection Association and the American National Standards Institute, the Chemical Hazard Response Information System (CHRIS) of the U.S. Coast Guard, and DOT guidelines from the U.S. Dept. of Transportation can also point to potential hazards.

Hazard identification should begin with MSDS information and its relation to the basic process chemistry. Laboratory experiments in research and development, followed by pilot plant operations, will help define process conditions to achieve

TABLE 10-8
TYPICAL PROCESS AND PLANT HAZARD ANALYSIS NEEDS FROM PROJECT
INCEPTION

Project phase	Information needs
Research and development	Material safety data sheets Process chemistry Chemical interactions
Pilot plant	Design safety opportunities Hazardous operations interfaces Waste reduction and pollution prevention
Facility siting and layout	Comparison of sites Employee and community risk Transportation and storage of hazardous materials
Detailed design	Safety systems specifications Materials of construction Inventory, process and spill control
Construction	Inspection and testing of equipment Safety systems verification Housekeeping
Start-up	Anticipate problem situations Provide training Verify operating and safety parameters
Operation	Develop operating manual and procedures Integrity of equipment Maintenance of safety systems
Expansion, modification, change	Identify safety compromises Provide training Revalidate hazard analysis
Decommission	Revalidate hazard analysis

optimal yields and performance. These R & D activities are important because they may reveal necessary process changes and unexpected side reactions and by-products and reinforce ideas about process contaminates and potential materials of construction. Some commonly used industrial chemicals, shown in Table 10-11 on page 296, are potentially hazardous and need to be assessed in relation to other chemicals being used.

A matrix can be constructed to study the interaction of chemicals, operating conditions, materials of construction and contaminants, and associated health and environmental effects (see Table 10-12 on page 297). OSHA and EPA regulatory limits and ceilings should also be included in this hazard identification matrix for every chemical that may become involved in the process at the operating facility. Mixtures of these chemicals should also be considered because of their possible synergistic effect in producing hazards.

Hazard identification will typically result in lists of materials and conditions that could create potential hazards. The material hazard lists may include flamma-

TABLE 10-9
DISCIPLINES AND BACKGROUNDS OF
HAZARD ANALYSIS TEAM MEMBERS

Chemistry

Construction

Engineering

Environmental

Industrial hygiene

Instrumentation

Maintenance

Operations

Process control

Research and development

Safety

Transportation

Other specialists or consultants

TABLE 10-10
POTENTIAL CONSEQUENCES FROM
PROCESS AND PLANT HAZARDS

Human consequences
   Employee injury and death
   Public injury and death
   Loss of employment
   Psychological effects
   Personnel and public relations

Environmental consequences
   Air pollution
   Water pollution
   Land pollution
   Ecological damage
   Wildlife injury and death

Economic consequences
   Property damage
   Loss of employment
   Lost production and inventories
   Reduction in sales
   Legal liability

TABLE 10-11
POTENTIALLY HAZARDOUS INDUSTRIAL
CHEMICALS IN COMMON USE

Acids

Aldehydes

Alkaline metals

Alkyl metals

Amines

Bases

Carbonyls

Chlorates

Cresols

Cyanides

Epoxies

Ethers

Halogens

Hydrocarbons

Mercaptans

Organonitro compounds

Organophosphates

Peroxides

Phenols

Silanes

TABLE 10-12
HAZARD IDENTIFICATION MATRIX

	Chemicals				
	1	2	3	Mixtures	Comments
Chemicals					
1					
2					
3					
Conditions					
Process conditions					
Ambient conditions					
Materials of construction					
Contaminates					
Health effects					
Environmental effects					
Regulatory limits					

ble, explosive, toxic, and reactive materials; the conditions list may include process operating parameters, flow rates and quantities of chemicals, process system hazards, hazardous contaminants, incompatible chemicals, and special hazardous situations.

HAZARD ANALYSIS TECHNIQUES

The OSHA process safety management of highly hazardous chemical regulation requires that employers perform hazard analyses (PHA) to identify and control process hazards. Employers must use one of the six techniques (see Table 10-13). Using an "appropriate equivalent methodology" could involve additional expense and time and may only add to the uncertainty of the results from a regulatory perspective.

TABLE 10-13
OSHA-SPECIFIED TECHNIQUES FOR PROCESS HAZARD ANALYSIS

What-if
Checklist
What-if/checklist
Hazard and operability study (HAZOP)
Failure mode and effects analysis (FMEA)
Fault tree analysis (FTA)
Appropriate equivalent methodology

Source: 29 CFR 1910.119

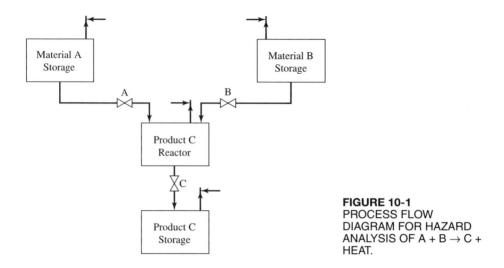

FIGURE 10-1
PROCESS FLOW
DIAGRAM FOR HAZARD
ANALYSIS OF A + B → C +
HEAT.

Figure 10-1 shows a process flow diagram for the following exothermic reaction:

$$A + B \rightarrow C + heat$$

This chemical reaction involves two liquids, A and B, which flow from their respective storage tanks into the stirred product C reactor. The flow rate of materials A and B is regulated by control valves A and B, respectively. The liquid-phase chemical reaction that results is highly exothermic and generates considerable quantities of heat, which must be controlled by the cooling water on the jacket of the product C reactor. The reaction of materials A and B to produce product C proceeds rapidly, and product C is drawn off from the bottom of the reactor. Product C flows through valve C into the product C storage vessel to await shipment. The reactor and each of the storage tanks have pressure-operated relief valves in case of pressure buildup during the process operation. Temperature, pressure, and level indicators are provided in the product C reactor for use in controlling the process chemistry.

What-if analysis should be used for relatively uncomplicated processes. The application of what-if analysis to the process in Figure 10-1 is shown in Table 10-14. The number of what-if questions may be many and varied depending on the brainstorming session of the analysis team.

In *checklist analyses* a standardized checklist is developed and used for each step of the process. While considerable time and effort is involved in developing the standardized checklist, once in place it can be used regularly to ensure compliance with a standard set of procedures even for complex processes with similar hazards.

The application of checklist hazard analysis to the example process in Figure 10-1 is shown in Table 10-15. Each of the questions on the checklist are given a

TABLE 10-14
WHAT-IF ANALYSIS OF SAMPLE PROCESS

What-if	Consequence	Comments
Material A does not flow to the reactor?	Unreacted material B will contaminate product C	Alarm and shutoff valve B on low flow through valve A
Reactor temperature exceeds operating limit?	Reactor may be damaged if pressure relief fails	Supply more cooling water to reactor; alarm and shut off valves A and B when reactor temperature limit exceeded
Material B is contaminated?	Product may be off-specification; undesirable reactions may occur	May need improved supplier B quality control; verify material B receiving and procedures

yes or no answer, which is then amplified under the comments. The checklist is a much more rigid procedure than the what-if brainstorming session. The number and type of checklist questions are limited only by the knowledge, experience, background, and creativeness of the preparer. Only experienced personnel should be involved in its preparation.

The *combined what-if and checklist analyses* produce a more broad-based hazard assessment technique. The initial phase of this analysis involves a brainstorming what-if session followed then by a standardized checklist analysis. When these two sessions are treated as separate distinct analyses, the answers that are developed are much broader and more comprehensive than if the analysis is performed with only one of these methods. The combined results become highly effective for PHA.

Hazard and operability analysis (HAZOP) is a systematic study of each process element and requires a thorough examination of process-flow diagrams and piping

TABLE 10-15
CHECKLIST ANALYSIS OF SAMPLE PROCESS

Checklist	Answer	Comments
Is each delivery of material B checked for contamination?	Yes	Supplier of material B has been reliable
Has the pressure relief on the reactor been checked in the past 6 months, as required?	Yes	The inspection scheduled valve has been followed
Are the operators receiving the required training?	No	Several new hires have not completed the training program as scheduled

and instrumentation diagrams as well as an analysis of deviations of process operating parameters. [15] HAZOP analysis is particularly well suited to complex processes and facilities where potential exists for making operability improvements. [19] This systematic and thorough review of the operating facilities design and operation also evaluates the consequences of operator errors. It provides a better understanding about plant operations and should lead to improved plant efficiency. HAZOP analysis should be done only by an experienced analysis team that is familiar with HAZOP analyses. The HAZOP team leader should have a strong technical background and be intimate with the HAZOP analysis technique. HAZOP analysis uses guide words that are shown in Table 10-16. These guide words are applied to HAZOP process parameters, as shown in Table 10-17, to arrive at potential deviations for each item under study. Table 10-18 on page 302 shows the application of the HAZOP hazard analysis to the example process shown in Figure 10-1.

The *failure mode and effects analysis (FMEA)* begins with a listing of all of the equipment and process components of the system under study. These individual components are analyzed methodically to identify potential failure modes, consequences, operating safeguards, and recommended actions to mitigate the hazard. The FMEA hazard analysis requires that a detailed piping and instrumentation diagram be made available to the team members for study. This hazard analysis would be most applicable to projects that are well into the design phase. The application of FMEA hazard analysis to the example process in Figure 10-1 is shown in Table 10-19 on page 302.

Fault tree analysis begins with a graphic diagram of all sequences of events that could result in an incident, accident, or exposure. This is a highly specialized hazard analysis technique that uses logic and event symbols (see Table 10-20 on page

TABLE 10-16
HAZOP GUIDE WORDS

Guide words	Meaning
OSHA-required	
No	Never, none
Less	Quantitative decrease, low, too short
More	Quantitative increase, high, too long
Part of	Qualitative decrease, too little
As well as	Qualitative increase, contaminates, too much
Reverse	Opposite of forward or intent
Other than	Complete substitution, another
Other possible guide words	
Yes	Always
Same as	Constant
Forward	Opposite of reverse
Begin	Start
End	Completion
Reached	Achieved

TABLE 10-17
TYPICAL HAZOP PROCESS PARAMETERS

Pressure	Addition
Temperature	Data
Flow	Information
Level	Separation
Time	Viscosity
Composition	Voltage
pH	Frequency
Reaction	Speed
Heating	Density
Cooling	Solubility
Mixing	

303). As shown in Figure 10-2 on page 303, this is a fairly complicated analysis that requires an in-depth knowledge of the interactions of the information in Tables 10-2 through 10-6. Fault tree analysis requires specially trained and skilled team members and should be used on specialized and individual components of the process plant design. It requires significantly more time and effort than other more broad-based approaches.

The what-if/checklist analysis, the HAZOP analysis, and the FMEA analysis are generally best for assessing hazards during the design, start-up, and normal plant operations. Any checklist must be very reliable. If that is not possible, the HAZOP or FMEA analysis is preferable. Often the final selection of a process and plant hazard analysis technique will come down to the team's familiarity and experience with a particular technique.

Hazard analysis will suggest ways to reduce the risk of the hazardous situation. At the same time, the operating facility should achieve higher productivity and product quality, lower operating costs, and improved employee and public relations through the greater confidence that those identified hazards are now being controlled. [20]

Usually a hazard analysis performed by an experienced team will contain numerous safety improvement recommendations. Management decisions for the implementation of hazard analysis results may be limited by the resources available to implement the recommendations. The advantage of a highly experienced analysis team is that they have the ability to prioritize or rank results of the analysis study. Ranking of safety improvement recommendations from a hazard analysis study should be developed through a risk management process. This process of managing risks involves the probability and severity of the risk that is posed by the potential incident and an estimate of the reduction of that risk that could occur by implementing the team recommendation. The use of HAZOP and FMEA analysis usually provides considerable information that may help in the ranking

TABLE 10-18
HAZOP ANALYSIS OF SAMPLE PROCESS

Item	Deviation	Cause	Consequence	Safeguard	Action
Material A to reactor	No flow	Valve A plugged	Unreacted B will contaminate product C	Periodic maintenance on valve A	Add alarm and shut off the "on" valve B on low flow through valve A
Reactor temperature	High temperature	Excessive reactants in reactor	Reactor may be damaged	Supply more cooling water to reactor jacket; observe reactor temperature closely during operation	Add alarm and shut off "on" valves A and B on excessive reactor temperature
Product C	Low pH	Valve B failed to close	Off-specification product	Periodic maintenance on valve B	Add alarm and shut off "on" valve A on low flow through valve B

TABLE 10-19
FMEA HAZARD ANALYSIS OF SAMPLE PROCESS

Item	Identification	Description	Failure Mode	Consequences	Safeguard	Action
1	Valve A	Motor valve, normally open, Material A service	Fails to open	Excess material A flow to reactor	Periodic maintenance on valve A	Add alarm and shutoff on valve A
2	Valve B	Motor valve, normally open, Material B service	Packing leaks	Small release of material B in the operating area	Periodic maintenance on valve B	Verify packing and valve materials of construction for Material B service
3	Valve C	Motor valve normally open, Material C service	Fails to close	Reactor overflow	Periodic maintenance on valve C and reactor level controller	Add alarm and shutoff on valves A and B

TABLE 10-20
FAULT TREE ANALYSIS LOGIC AND EVENT SYMBOLS

Symbol	Terminology	Description
⌂	OR gate	Output event occurs if any input event occurs
⌂	AND gate	Output event occurs if all input events occur
▭	Top event Intermediate event	Results from interaction of other events
○	Basic event	Component failure: lowest level of resolution
◇	Undeveloped event	Event not examined further
⬠	External or house event	Boundary condition
△-△	Transfer IN-OUT	Used to avoid repetitive logic

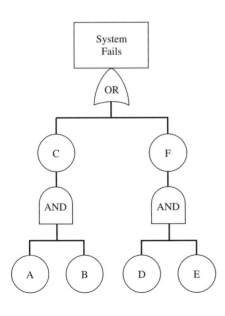

FIGURE 10-2
EXAMPLE OF A FAULT TREE DIAGRAM.
Intermediate events are C and F. Basic events
are A, B, D, and E. The top event occurs if C or
F occurs. Intermediate event C will happen if
both of the basic events A and B occur. Intermediate event F happens when both of the
basic events D and E occur

process. Checklist analysis on the other hand does not provide the type of information that would be useful in prioritizing recommended action items. Certainly any recommendation of the hazard analysis team that would prevent major incidents, accidents, serious injuries, or deaths should require the immediate action of management.

PERSONNEL MOTIVATION AND TRAINING

Operations, maintenance, R & D, engineering design, construction, contractor, and other personnel who are involved in the plant process and design project need to understand the safety and health hazards of the chemicals and processes for the protection of all personnel and the public in nearby communities. The hazard communication standard of OSHA 29 CFR 1910.1200 will assist employees in becoming more knowledgeable about chemical hazards. Process and plant design activities require additional understanding of safety management in the construction, operation, and maintenance of the operating facility.

Operations and maintenance personnel need training in order to possess the knowledge and proficiency to perform their assigned job functions. These personnel are vital to the safe operation of the process and of the operating facility. Operators need to make process and equipment adjustments to safely control the manufacture of products, identify abnormal equipment and process operations, troubleshoot these abnormalities, and identify the need for maintenance when necessary. Both formal and on-the-job training should be required of both operations and maintenance personnel.

Management should recognize that these highly trained and skilled operations and maintenance personnel are an excellent resource for improving both safety and productivity in the operating facility. It is important that management and employees communicate effectively and be open to constructive ideas from all parties that will result in mutually desirable improvements.

Outside contractors are often used to perform work in and around industrial processes that involve chemical hazards. The use of outside contractors in no way releases the employer of the operating facility from the liabilities of their work performance. This is particularly true when outside contractor work is being performed while the plant facility is operating. Employers must communicate with outside contractors about applicable safety management programs, and contractors in turn must be sure their employees are trained to safely perform their contractual commitments. Design engineers need to consider the use of outside contractors in the construction, operation, and maintenance activities associated with the operating facility.

Human error plays an important role in most chemical accidents. [17, 36] The personnel at operating facilities have responsibilities in communications, operations, and decision making. Any errors in judgment or execution can have profound effects upon the safety of an industrial process and plant. Safe and reliable process and plant design can virtually eliminate many failures that might otherwise occur in the equipment and instrumentation hardware at the operating facility. Failures involving human error in equipment, process, and plant operations are more difficult to eliminate because a forgotten or misapplied procedure can result in a catastrophic incident.

Human errors include errors of omission, commission, sequence, and timing. An error of omission is a failure to execute a required task because the employee forgot to do it. In an error of commission the employee performs the required tasks but does it incorrectly. A sequence error involves the employee performing the task correctly but doing so out of order with the required sequence. A timing error is caused by an employee executing a required task either too fast or too slow compared with the specified time frame.

Operator-induced error is caused by an employee who has the knowledge to make the correct decision but acts incorrectly anyway. System-induced errors are created by the integration of incompatible components into a total system. Design-induced errors result from faulty equipment design, fabrication, or installation. Input errors are caused by typographical or other inadvertent errors associated with data, or information for MIS processes. Low- and high-stress errors are human errors that are committed under differing environmental conditions. Low-stress errors are caused by a lapse of memory during an otherwise normal situation, while high-stress errors occur when employees make wrong decisions in life-threatening situations.

The interface between the employee and the process equipment should be compatible. In process control the alarm and information system displays should be user-friendly. An analysis of tasks at the operating facility coupled with ergonomics development should be included in the process and plant design activities.

Well-conceived process control systems should reflect the balance between human and hardware control. Human response to alarms and other instrumentation signals will reflect the human capability at that particular time when the response is required. The hardware response to the same alarm and process instrumentation will be based upon the process hardware design and the mechanical integrity of the hardware at that particular point of demand. From a safety perspective the ideal process control system should have built-in redundancies that balance the attributes associated with both human and hardware control. By the same token, from a safety perspective no process control system should ever rely totally on either human or hardware control response.

A safety program for mechanical integrity of process equipment should be developed and implemented with the following primary goal: to design, construct, install, operate, and maintain equipment and instrumentation to minimize the risk of chemical releases (see Table 10-21 on page 306).

In process and plant design considerable thought should be given to a reliable preventive maintenance system that will maintain the operability of the control hardware. Temperature, pressure and flow instrumentation, materials of construction, valves, packing and gaskets, and relief devices should all be part of the preventive maintenance system that is developed during the design phase of the project.

TABLE 10-21
ENSURING MECHANICAL INTEGRITY OF EQUIPMENT

Equipment design criteria

Develop an overall process flow diagram and detailed piping and instrumentation diagrams.

Compile a list of equipment and instrumentation by categories: e.g., pressure vessels, storage tanks, pumps, reactors, separations equipment, heat exchangers, piping, relief and vent safety systems, fire protection systems, alarms, interlocks, and emergency shutdown systems.

Identify and prioritize critical equipment and instrumentation based on safety considerations.

Design, construct, and install equipment to operate according to specifications and to safely process and contain chemicals.

Operate and maintain equipment in a safe and reliable manner.

Control release of chemicals with backup safety systems.

Maintain emergency safety systems to control or mitigate an unwanted chemical release.

Inspection and testing of equipment and instrumentation

Use mean time to failure to develop frequency and procedure based on codes, standards, manufacturers, and experience.

Provide appropriate training to ensure proper and consistent inspection and testing for inspectors and testing personnel.

Provide appropriate training for maintenance and operations personnel in preventative maintenance, safe work practices, proper use of equipment and tools, and safety hazard identification.

Quality assurance in the field for equipment and instrumentation

Verify materials of construction with specifications.

Verify purchasing, fabrication, and inspection procedures with mechanical drawings and code and standard certifications.

Confirm installation procedures with construction drawings and plot plans.

Verify start-up preparation procedures with operations manual.

PROCESS AND PLANT MODIFICATION AND CHANGE

In the life of the process and plant changes and modifications will be made to the process technology and the equipment for the operating plant. Following plant start-up and the commencement of normal operations additional changes in the process technology, control, safety improvements, and process equipment will be made by maintenance and operating personnel. Further modifications and changes will be needed to satisfy the needs of marketing and other business activities as well as new ideas from research and development. Throughout all of these process and plant modifications and changes it is imperative that they do not result in deviations that could lead to the operating facility being outside safe operating limits. The safety program at the operating facility should contain procedures to properly manage modifications and changes to process chemicals, technology, equipment and facilities other than "replacement in kind." All such changes need to be identified and reviewed prior to the implementation of the modification or change in the operating facility. This requires PHA for all modification or change.

There can be many different types of process equipment and plant modifications or changes, as reflected by the examples in Table 10-22. Prior to the implementation of any such modifications or changes at the operating facility a hazard analysis should be performed. This analysis should be done during the process and plant design for the modification or change. Many catastrophic incidents that have occurred in the past have been the result of unsafe conditions caused by modification and change at the operating facility.

Quality control and assurance programs at the operating facility are often the basis for process and plant modifications and changes. Both quality programs and process safety programs should be compatible because they both stress improvements in design and operations through better communication, training, and teamwork. While achieving improved quality and safety, both of these programs are fo-

TABLE 10-22
EXAMPLE OF PROCESS, EQUIPMENT, AND PLANT MODIFICATIONS

Process changes
    Raw material source or substitution
    Product mix
    Process technology
    Control system
    Operation parameters
    Preventive maintenance
    Operating procedures
    Catalyst system
    Safety alarm settings and interlocks
    Relief valve setting or capacity
    Operating permits

Equipment changes
    Repair or replacement, not in kind
    Maintenance or fabrication procedures
    Inspection and testing procedures
    Addition or deletion of equipment
    Design codes
    Personal protective equipment

Plant changes
    Location of operating process
    Location of raw materials and supplies
    Location of product storage
    Utilities source
    Waste treatment and disposal
    Location of personnel
    Transportation access

cused on increased productivity and profitability for the operating facility. The integration of these two comparable programs offers the opportunity of continuous improvement of both quality and process and plant safety.

INCIDENT INVESTIGATION AND SAFETY AUDITS

A major incident—or even the possibility of one occurring—can have an adverse effect on the future of the operating facility. The loss of public and employee support and confidence, the termination of operating and siting permits, and extraordinary legal and insurance expenses all contribute to the losses incurred by a major fire, an explosion, a runaway reaction, or releases of highly toxic or flammable materials. Major incidents have the potential to kill people and cause tremendous property damage.

It is important during the process and plant operation to study near-miss incidents because a serious consequence could have occurred but did not. Process and plant personnel need to identify and mitigate the causes of incidents and near-misses of similar operations that have been previously constructed. The object of this investigation of incidents is to learn from past experience and thereby avoid repeating the mistakes that caused previous incidents.

Whenever a new operating facility, major plant expansion, or significant process modification is being planned the location of hazardous processes, materials, and products should be a prime consideration. The management of risks associated with this process and plant should involve the health and safety of the employees and the community, the protection of the environment, and the protection of the plant property. The hazard analyses associated with this risk management include a determination of exposure to employees and the community; the maximum release of flammable, explosive, reactive, and toxic materials; an evaluation of property damage and interruption of business activities; and the potential exposure to earthquakes, floods, tornadoes, and the like.

Near-miss occurrences that could have become a major incident should be reviewed whenever possible if they could relate to the process and plant. When these near-miss occurrences are the result of repeated errors in operation or malfunctioning safety devices and alarms, this information can be usefully applied to the improvement of the process and plant safety.

The facts and information developed during a compliance audit are used to verify compliance with regulatory standards and management practices. An effective compliance audit usually includes a review of the relevant documentation and process safety information, an inspection of the operating facilities and interviews with selected plant personnel. Usually an exhaustive preplanned checklist is used to perform a compliance audit by a knowledgeable experienced team. These audits often identify opportunities for improvements in the engineering, operation, maintenance, and regulatory compliance at the facility.

Process and plant safety audits help protect the safety and health of employees, surrounding communities, and the environment. In the process and plant design phase the results of a compliance audit at a similar facility would greatly aid and guide design engineers about situations that could adversely affect the performance of the operating facility. The results from an objective critique of the compliance audit can provide valuable insight in the development of reliable process and plant safety management. These compliance audits and incident investigations coupled with regular systematic hazard analysis revalidations for the operating facility add reliability to the process and plant that cannot be found elsewhere.

BIBLIOGRAPHY

1 Arendt, J. S., et al., *Evaluating Process Safety in the Chemical Industry: A Manager's Guide to Quantitative Risk Assessment,* Chemical Manufacturers Association, Washington, D.C., 1989.

2 Austin, G. T., *Shreve's Chemical Process Industries,* 5th ed., McGraw-Hill, New York, 1984.

3 Bellome, P. J., "OSHA Safety Regulation Calls for Step-by-Step Approach," *Oil & Gas Journal,* June 1, 1992, pp. 49–52.

4 Burk, A. F., "Strengthen Process Hazard Reviews," *Chemical Engineering Progress,* June 1992, pp. 90–94.

5 Chowdhury, J., "OSHA Tightens Its Hold," *Chemical Engineering,* May 1992, pp. 37–42.

6 Crowl, D. A., and J. F. Louvar, *Chemical Process Safety: Fundamentals with Applications,* Prentice-Hall, Englewood Cliffs, N.J., 1989.

7 Dow Chemical, *Fire and Explosion Index Hazard Classification Guide,* 7th ed., AIChE, New York, 1989.

8 Early, W. F., "Process Safety Management," *Hydrocarbon Processing,* February 1991, pp. 105–107.

9 Englund, S. M., "Design and Operate Plants for Inherent Safety," *Chemical Engineering Progress,* March 1991, pp. 85–91.

10 *Extremely Hazardous Substances List,* 40CFR355, U.S. Environmental Protection Agency, Washington, D.C., 1987.

11 Greenberg, H. R., and J. J. Cramer, *Risk Assessment and Risk Management for the Chemical Process Industry,* Van Nostrand Reinhold, New York, 1991.

12 Grose, V. L., *Managing Risk—Systematic Loss Prevention for Executives,* Prentice-Hall, Englewood Cliffs, N.J., 1987.

13 Center for Chemical Process Safety, *Guidelines for Hazard Evaluation Procedures,* 2d ed., AIChE, New York, 1992.

14 Halle, R. T., and M. Vadekar, "Rust-Catalyzed Ethylene Hydrogenation Causes Temperature Runaway," *Oil & Gas Journal,* June 17, 1991, pp. 33–36.

15 Jones, D. W., "Lessons from HAZOP Experiences," *Hydrocarbon Processing,* April 1992, pp. 77–80.

16 Kavianian, H. R., and C. A. Wentz, *Environmental and Occupational Safety Engineering and Management,* Van Nostrand Reinhold, New York, 1990.

17 Kletz, T. A., *An Engineer's View of Human Error,* 2d ed, Institution of Chemical Engineers, London, England, 1991.

18 Kletz, T. A., *Critical Aspects of Safety and Loss Prevention,* Butterworths, Guildford, London, England, 1990.

19 Kletz, T. A., *Hazop and Hazan—Identifying and Assessing Process Industry Hazards,* 3d ed, Institution of Chemical Engineers, London, England, 1991.

20 Kletz, T. A., *Lessons from Disaster: How Organizations Have No Memory and Accidents Recur, London,* Institution of Chemical Engineers, 1992.

21 Kletz, T. A., *Plant Design for Safety: A User-Friendly Approach,* Hemisphere, New York, 1991.

22 Kletz, T. A., *What Went Wrong? Case Histories of Process Plant Disasters,* Gulf Publishing, Houston, 1985.

23 Kletz, T. A., *What Went Wrong? Case Histories of Process Plant Disasters,* 2d ed, Gulf Publishing, Houston, 1988.

24 Kolluru, R. V., "Understand the Basics of Risk Assessment," *Chemical Engineering Progress,* March 1991, pp. 61–67.

25 Kuryla, M. L., and S. C. Yohay, "New Safety Rules Add to Plant Manager's Worries," *Chemical Engineering,* June 1992, pp. 153–160.

26 *Management of Process Hazards: American Petroleum Institute Recommended Practice,* 750, API, Washington, D.C., 1990.

27 *Consensus Standards Related to Fire and Explosion Prevention,* NFPA, Quincy, Mass.

28 *Identification of the Fire Hazards of Materials,* 704, NFPA, Quincy, Mass., 1989.

29 29 CFR 1910.1200, OSHA, Washington, D.C., 1994.

30 Center for Chemical Process Safety, *Plant Guidelines for Technical Management of Chemical Process Safety,* AIChE, New York, 1992.

31 29 CFR 1910.119, OSHA, Washington, D.C., 1994.

32 Prugh, R. W., "Quantify BLEVE Hazards," *Chemical Engineering Progress,* March 1991, pp. 66–67.

33 Sanders, R. E., and J. H. Wood, "Don't Leave Plant Safety to Chance," *Chemical Engineering,* February 1991, pp. 110–118.

34 Sax, N. I., and R. J. Lewis, *Dangerous Properties of Industrial Materials,* 6th ed., Van Nostrand Reinhold, New York, 1984.

35 Simpson, L. L., "Estimate Two-Phase Flow in Safety Devices," *Chemical Engineering,* August 1991, pp. 98–100.

36 Stern, A., and R. R. Keller," Human Error and Equipment Design in the Chemical Industry," *Professional Safety,* May 1991, pp. 37–41.

37 U.S. Coast Guard, *Chemical Hazard Response Information System,* Washington, D.C., 1990.

38 U.S. Department of Transportation, *Guidelines,* Washington, D.C., 1990.

QUESTIONS

1 Describe the components of an integrated approach to process and plant safety management.

2 Why are material safety data sheets useful in assessing chemical hazards?

3 What technology information is needed for process and plant design and operation?

4 List the sequence of hazardous material incidents, accidents, and exposures.

5 What disciplines and backgrounds could possibly be included on a process and plant hazard analysis team?

6 Process and plant hazards create potential human, environmental, and economic consequences. List examples of each of these types of consequences.

7 List six categories of potentially hazardous industrial chemicals, and explain why these chemicals are hazardous.

8 What are the OSHA-specified techniques for process hazard analysis?

9 Add 10 what-if questions to Table 10-14.

10 Based upon the what-if questions derived in the preceding question, list typical consequences and comments to complete your what-if hazard analysis.

11 List six checklist questions in addition to those in Table 10-15, then provide answers and questions to complete the checklist hazard analysis.

12 List the seven OSHA guide words and their meanings that can be used in a HAZOP analysis.

13 List 12 typical HAZOP process parameters that could be used in performing this type of hazard analysis.

14 List five additional items for Table 10-18 having to do with HAZOP hazard analysis.

15 Complete the HAZOP hazard analysis sheet for the items that you have added to Table 10-18. Be sure to include deviations, causes, consequences, safeguards, and actions in your HAZOP.

16 Change the type of failure mode for each of the three valves identified in Table 10-19, and then complete the failure modes and effect analysis for these changed failure modes.

17 In fault tree analysis, explain the difference between an OR gate and an AND gate.

18 List four examples each for process, equipment, and plant changes.

19 According to the American Society of Safety Engineers Code of Ethics, what should their members do if in their professional judgment the safety, health, and welfare of the public is endangered by an industrial operation.

20 When copper and silver come in contact with acetylene, explosive acetylides can be formed. A portion of the steel acetylene line to a process has been taken out of service and replaced with copper tubing. What safety hazard, if any, has been created by this process change?

21 Picric acid (2,4,6-trinitrophenol) is being stored in a jar in the chemical laboratory. Why is this a chemical hazard?

22 Which of the following are perceived by the general public as being high-risk and which are perceived as low-risk?
Cancer
Birth defects
Alcohol
Nuclear waste
Natural radon hazards
Household waste
Hazardous waste
Medical waste
Incinerators
Outdoor air pollution
Indoor air pollution
Cigarette smoking
Secondhand tobacco smoke

23 Why should safety relief valves in an industrial process plant be inspected and tested periodically?

24 What is the right process hazard analysis technique to use on an industrial process?

25 Explain how equipment and piping materials of construction may have a dramatic effect on process chemistry.

26 Explain why a boiling-liquid expanding-vapor explosion differs from other explosions.

27 Describe the what-if technique for process hazard analysis.

28 Describe the checklist process hazard analysis technique.

29 Discuss the combined what-if and checklist process hazard analysis technique.

30 Discuss the hazard and operability (HAZOP) process hazard analysis technique.

31 Discuss the failure mode and effects analysis (FMEA) technique for process hazard analysis.

32 Discuss the fault tree analysis technique for process hazard analysis.

33 Why is it important to identify safety hazards at the early stages of process and plant design?

34 Why is it desirable to minimize the inventory in hazardous materials storage?

35 When constructing dikes around storage tanks of hazardous materials, why is it desirable to design the liquid storage so that leaks and spills do not accumulate under the tanks or process equipment?

PROBLEMS

1 In preparing a reactor for confined-space entry, the operator isolated all process energy sources to the reactor and opened the bottom drain valve to allow a free flow of water to exit the reactor during the cleanout process. After the reactor had been cleaned, the operator exited the reactor, removed all points of isolation for the process energy sources, and restored the reactor to operating condition. The operator then went to the control room and charged the reactor with 1500 gal of a flammable hydrocarbon liquid. In heating the reactor up to process temperature, the operator noticed a rapid temperature increase and a large spill at the bottom of the reactor because the operator had forgotten to close the drain valve at the bottom of the reactor before recharging it. The spilled liquid hydrocarbon ignited, destroying the reactor and severely injuring the operator. What type of human error was involved in this tragic accident?

2 A liquified propane truck tanker was filled and weighed, and the truck driver began to transport the tanker to a customer. While enroute the driver noticed a high-pressure indication in the tanker, stopped the rig, and moved to the rear of the tanker to relieve the excess pressure through a manual vent valve. This was a normal industry procedure. As the driver approached the rear of the tanker a rupture disk blew, releasing a large flammable propane vapor cloud, which ignited. The driver was seriously burned in this accident and later died. A subsequent incident investigation concluded that the tanker had been filled to the incorrect pressure. What type of human error was involved in this incident?

3 A process plant operator disconnected a caustic transfer hose from a circulating pump without first venting the pressure in the line. The operator was sprayed in the face with the caustic that remained in the line and lost the sight of one eye. What type of human error was involved in this incident?

4 An operator was attempting to position a gasket that would seal a reactor lid with his left hand, while closing the reactor lid with his right hand. The heavy lid slipped while the operator was still holding the gasket in place with his left hand and crushed his thumb between the lid and gasket. The operator's thumb had to be amputated. What type of human error was involved in this accident?

5 An operator was adding a shock-sensitive pelletized material to a reactor through a funnel that had been grounded to dissipate any static charge. During the reactor-charging operation the operator poured the pelletized material into the funnel too quickly and caused the funnel to plug. Although the material was shock-sensitive, the operator tried to clear the funnel with a rod. The pelletized material in the funnel exploded, killing the operator and three other people as well as burning the plant to the ground. What type of human errors were involved in this tragic incident?

6 Methyl chloride was being pumped from storage through a preheater into a stainless steel mixing tank at a unit operations laboratory at a major university. An in-line gas pressure regulator had been installed between the preheater and the mixing tank. While this transfer was taking place the students noticed a small amount of white smoke coming from this process and proceeded to investigate its cause. They decided the best way to determine the source of the white smoke was to disassemble the process in a stepwise procedure. First, the feed tank containing the methyl chloride was disconnected from the rest of the system; no smoke was observed. Next, the line was separated at the preheater; again, no smoke was detected. Finally, the in-line gas regulator was removed from the stainless steel mixing tank; no smoke was observed even then. As the students were visually inspecting the in-line gas regulator it burst into flames, severely burning several of the students. Fortunately, one of the students was able to quickly put out the fire with a nearby extinguisher. During the investigation of this accident it was revealed that the students had needed an in-line gas regulator for the process and had found one in an adjacent laboratory drawer. No thought was given at the time as to the compatibility of the regulator's material and construction in relation to the chemicals that were used in the process. The regulator that was installed was constructed of aluminum. Upon contact with methyl chloride, highly pyrophoric aluminum alkyl formed and produced white smoke. When the gas regulator was removed from the system the aluminum alkyl in the regulator contacted oxygen and moisture in the air, thereby creating the fire. What type of human errors were involved in this accident?

7 In designing a distillation column to separate two flammable liquid hydrocarbons the correct number of theoretical plates was put into the column but the column diameter was undersized, thereby not allowing adequate vapor liquid disengagement. Liquid entrainment in the column overheated the line and flooded the overhead heat exchanger, causing a line rupture and a fire that destroyed the distillation column and injured several operators. What type of errors were involved in this accident?

8 In performing a batch chemical reaction the plant operator followed a recipe from a computer database that called for 500 gal of material A, 750 gal of material B, and 900 gal of material C. All of these materials and their respective quantities were charged into a heated stirred chemical reactor that had a 2000-gal capacity. When the operator charged these materials into the reactor, the reactor overflowed, causing a dangerous spill of hazardous corrosive materials. The spill entered the drain system of the plant, overloading the wastewater treatment plant and temporarily putting it out of compliance. Fortunately, no one was injured. What type of error was involved in this accidental spill?

9 Under normal plant operations the operator in the control room pushed the start instead of the stop button on a pump that was being taken out of service. Fortunately, this error was identified and immediately corrected. What type of error was involved in this situation?

10 A group of process engineers are in the administrative building of a polyethylene plant on what appears to be a normal operating day. Suddenly the emergency alarm sounds, requiring all administrative building personnel to immediately leave the plant premises. All of the process engineers immediately leave the administrative building and start toward the plant gate. One engineer realizes she has left her purse on her desk and returns to the administrative building to retrieve it when the plant explodes. She is caught in the shock wave of the plant explosion and suffers severe injury. All of the other process engineers safely escaped the premises before the plant explosion. What type of error was involved in this incident?

11 An acetylene gas cylinder with a maximum pressure of 2000 psig is used to feed a process reactor that has a maximum pressure rating of 300 psig. The acetylene flows

through a control valve and a gas regulator before entering the reactor. The flowing gas pressure is controlled by this gas regulator. In case of reactor or feedline overpressure a pressure relief valve is provided with an appropriate vent. All of the materials of construction of this system have been selected to be compatible with the process chemistry and the operating parameters. The vent gas goes to a flare system.

a Develop a process flow diagram for this situation.

b A flow control valve has been installed between the pressure relief valve and the process system, isolating the relief valve from the process system if the newly installed flow control valve is closed. What would happen in this situation if the reactor pressure exceeded 300 psig?

c The pressure relief valve on the original system above has been set at 325 psig. Describe the safety hazard that has been created and suggest a solution.

d The vent in the original process system is disconnected from the flare line so that when the relief valve functions, the vent gas will remain in the building. How would you correct the hazard created by this situation?

e The piping to the relief valve is 1/2 in in diameter while the vent line piping is 1/4 in in diameter. Describe the hazard that has been created by this situation and offer a solution.

12 An operating plant normally requires city water for a wide variety of uses. Some of these required uses at a plant include a drinking fountain, a safety shower, a chemical reactor, and an eyewash station. Should the chemical reactor overpressure, how could you ensure the integrity of the water supply to the drinking fountain, safety shower, eyewash station, and city water supply main?

13 A promising process to hydrate acetylene using a copper-phosphate catalyst to produce acetaldehyde is under study in research and development. The process chemistry is shown in the following reaction:

$$C_2H_2 + H_2O \xrightarrow[\text{110 kPA}]{175°C} C_2H_4O$$

Based on the research information available, preheated steam will be mixed with acetylene and fed into a fixed-bed catalytic reactor. The effluent from the reactor will be cooled and allowed to flow into a vapor liquid separator with the product acetaldehyde going overhead to a distillation column for cleanup. The bottoms of both the separator and the distillation column will be wastewater. The overhead product acetaldehyde from the distillation column will be cooled before flowing to a product accumulator, which will also provide reflux for the distillation column. Gas from the accumulator will be flared. Using this background information, develop a process flow diagram and lay out an operating facility plot plan, including administrative offices and rail and highway access.

14 Based on the process flow diagram and plot plan that was developed for the acetylene hydration to acetaldehyde in the previous question, perform a process hazard analysis using the what-if technique, the checklist technique, and the HAZOP technique.

15 Research has developed an ethylene and butene polymerization process that is ready for process and plant design. The process chemistry, which uses a catalyst, is shown by the following equation:

$$C_2H_4 + C_4H_6 \xrightarrow[\text{1000 kPA}]{100°C} \text{polyethylene}$$

Ethylene monomer and comonomer butene will be polymerized at elevated temperatures and pressures using a catalyst in a hydrocarbon diluent to produce high-density polyethylene polymer. Ethylene, butene, the liquid diluent, and the catalyst will all be continuously fed into the polymerization reactor. The polymer and residual olefins in the diluent will then flow to a flash separator. The bottom stream from the flash separator will be the product polyethylene. The overhead from the flash separator will be hydrocarbons, which will be returned to the polymerization reactor. Since this will be a highly exothermic reaction, the cooling water requirements for the process will be considerable and some wastewater will be generated. The polyethylene will require further finishing in order to remove residual hydrocarbons and then will be pelletized prior to shipment to the customer. Based on this information, draw a process flow diagram and lay out a plant plot plan for the commercial operating facility, including administrative offices and rail and highway access.

16 Based on the polyethylene process information and plant layout developed in the preceding problem, perform a process hazard analysis using the what-if technique, the checklist technique, and the HAZOP technique.

17 Your company is planning to install an organic chemicals complex that will generate a chlorobenzene hazardous waste stream, which must be incinerated. The resulting hydrogen chloride will be neutralized with liquid caustic according to the following reactions:

$$C_6H_5Cl + 7O_2 \xrightarrow[\text{100 kPA}]{\text{1100°C}} 6CO_2 + 2H_2O + HCl$$

$$HCl + NaOH \xrightarrow[\text{100 kPA}]{\text{110°C}} NaCl + H_2O$$

The chlorobenzene will be incinerated with air and supplemental fuel gas, as needed. It is not anticipated that any significant quantities of ash will result from the incineration process. The effluent gas from the incinerator will be fed into a quench scrubber, where the hydrogen chloride will be neutralized with liquid caustic. The exhaust emissions from the quench scrubber will be vented because they will contain only carbon dioxide and steam. The bottoms from the quench scrubber will be wastewater, which must be treated prior to discharge. It is anticipated that the chlorobenzene waste will be delivered to the operating facility by both rail and truck transport. Based on this information, develop a process flow diagram for the incinerator and quench scrubber, and lay out an operating facility plot plan, including administrative offices and highway and rail access.

18 Based on the information developed in the previous question for the incineration of chlorobenzene waste, perform a process hazard analysis using the what-if technique, the checklist technique, and the HAZOP technique.

19 A company has the following organization functions reporting directly to the top management: research and development, engineering, production, marketing, safety, environmental, and financial. Draw an organization chart to show these reporting relationships.

20 The safety function in the preceding question was moved and is now reporting to the production function, which in turn reports to top management. Discuss how this affects the management awareness of safety hazards in the organization.

21 Using the previous question, the safety function is now put under the marketing function of the organization. Discuss its effect on the awareness of management about safety hazards in the organization.

22 How would you include safety in process and plant design?

23 How would you operate a chemical process plant safely?

24 What are typical equipment safety protection systems at an operating chemical process plant?

25 What types of emergencies might occur at a chemical process plant?

26 Why is a preventive maintenance program important to maintaining safe operations at an operating facility?

27 Why should an explosion hazard check be made before issuing a hot-work permit?

28 Why should block valves that isolate pressure safety relief valves be inspected monthly to ensure that they are sealed in an open position?

29 Why should the installation of a rupture disk be inspected prior to plant start-up?

30 Why should a blowout preventer on an oil well be oriented properly?

31 The nitrogen-inerting system on a maleic anhydride storage tank was turned off by an operator who felt that inerting was not necessary because the electric heating coils in the tank were submerged. Discuss how the storage tank could become an accident waiting to happen.

32 What is an incident?

33 Why is it important to investigate near-miss incidents as well as major incidents? How would you rank these as causes of incidents at chemical process plants?

34 You have just performed a process hazard analysis on an operating plant that has identified two serious safety hazards. One of these safety hazards is strictly an OSHA regulatory compliance problem, while the other hazard is causing product contamination. Either one of these two safety hazards can be corrected in a single day without shutting down the operating plant. The same experienced personnel are needed to correct both of these safety problems. Which safety hazard would you correct first and why?

11

PLANT MAINTENANCE

A well-managed maintenance program is essential for a facility to maintain its operations in a cost-effective and safe manner. A sound maintenance system will result in improved productivity, better safety and health, environmental protection, and desirable public, customer, and employee relations. There are many aspects to well-conceived and organized maintenance. [1]

- Long-term care of equipment, buildings, and grounds
- Normal, routine attention to service and appearance
- Restoration or improvement in service and appearance

Maintenance programs should emphasize preventive maintenance and inspections to discover potential problems. The proper monitoring of operating equipment and structures can alert trained personnel to anticipate breakdowns before they actually occur. Operating personnel and safety, health, and environmental staff are in the best position to assist management in controlling maintenance programs.

The maintenance goals and objectives for an industrial facility should make effective use of appropriate personnel and systematic planning and tracking techniques. True cost savings and high productivity can only be achieved by combining reliability, safety, availability, and maintainability in a cost-effective program that is consistent with management policies. Equipment maintenance should be governed by inherent reliability, failure experience, and the ability to perform effective failure analysis and troubleshooting. [4]

PRINCIPLES OF MAINTENANCE

Maintenance involves many types of problems. The functions of inspection, over-haul, repair, and replacement are common to all maintenance strategies. All of these functions solve problems that are either deterministic or probabilistic.

Deterministic equipment problems are predictable over the life of the equipment. There is low uncertainty associated with the timing or consequence of a maintenance action for a deterministic problem. Many increases in operating costs over time can be attributed to predictable wear and tear. The maintenance costs for such deterministic problems are cyclical, as shown in Figure 11-1. The operating cost per unit time gradually increases, whereupon the maintenance work is performed. The operating cost is predictably reduced and the postmaintenance operating cost trend can be reasonably anticipated. [4] Pump operation, compressor operation, and pipe scale-up are examples of deterministic problems.

Probabilistic equipment problems are indeterminate, making both the timing and cost of maintenance unpredictable. Either the equipment operates, or it fails with little or no warning. The timing and consequence of maintenance depend more on chance, making decisions based upon probability more appropriate. The variations in the failure probability with time elapsed produce a frequency distribution. The elapsed time between the maintenance activity and equipment failure can be estimated and used to generate an optimum maintenance schedule. Vehicle oil change, lightbulb replacement, and replacement of protective coatings are examples of probabilistic equipment problems. The variations in failure probability versus the elapsed time between maintenance and component failure are usually predictable from experience.

Equipment inspection, which is common to all maintenance programs, should be based upon the criteria listed in Table 11-1. The inspection, which determines the equipment condition, should be carried out systematically by qualified, trained personnel. [5]

Equipment overhaul, which is part of preventive maintenance, restores and raises the equipment and components to a much-improved condition, possibly a like-new condition. Minor repairs may also be included under preventive maintenance.

Most repair work fixes equipment or components that have failed or have a defect that is an operating constraint. These limiting characteristics are usually probabilistic, and not predictable within reasonable accuracy.

Replacement of equipment or components is usually based on long-term economics and availability. It is normally done on a case-by-case basis.

Maintenance programs involve much more than the actual maintenance itself. Effective maintenance programs should be well planned, organized, implemented, and controlled to facilitate maintenance activities. The necessary maintenance services must be readily available, when needed, and be reliable and of optimal quality. Consideration should be given to the items in Table 11-2 on page 320 for a sound maintenance program.

The safety of maintenance personnel should be paramount in all maintenance

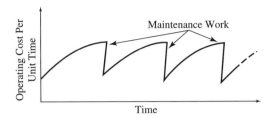

FIGURE 11-1
THE CYCLICAL TREND IN
DETERMINISTIC MAINTENANCE
COSTS. (*Source: Practical Machinery
Management for Process Plants,
Volume 3,* 2nd. ed. by Heinz P. Bloch
and Fred K. Geitner. Copyright © 1990
by Gulf Publishing Company. Used
with permission. All rights reserved.)

activities. These employees may be exposed to electrical, mechanical, toxic, fire, and noise hazards and need to be protected.

Preventive maintenance requires detection of equipment and component defects and associated problems. Whenever periodic inspections are required, the equipment and components should be easily accessible for the inspections. The design and layout of the equipment and the plant should facilitate the inspection and repair to restore the equipment to the proper degree of performance.

The quality of the components, chemicals, lubricants, and cleaning systems should be at least comparable to the original specifications. Equipment manufacturers usually have recommendations that should be followed as closely as is practical. While the maintenance is being performed, good-housekeeping practices will assist in the cleanup efforts after completion.

The lubricants, materials, spare parts, components, and service tools needed during the expected equipment lifetime should be stored in the plant or at the site, as required by a predetermined plan. Those items that are not required to be stored at the facility should be available from vendors on short notice or readily producible from available materials.

Some maintenance activities and materials pose the potential risk of equipment damage. These need to be designed and controlled to reduce the risk to an acceptable level. During maintenance, some components and equipment are dismantled, assembled, or moved at the risk of damage to the components or associated equipment. Equipment protection from the risk of damage from the weather or other en-

TABLE 11-1
CRITERIA FOR EQUIPMENT MAINTENANCE INSPECTIONS

Deterministic failure experience

Probabilistic failure experience

Expected failure frequency

Probable consequences of failure

Cost of inspection

Risk of inspection caused problem

Reliability of the available information

TABLE 11-2
COMPONENTS OF RELIABLE MAINTENANCE PROGRAMS

Safety of maintenance personnel

Ease of inspection and repair

Quality of the maintenance service

Availability of maintenance materials

Low equipment damage potential

Minimal downtime for operations

Protection of employees, facilities, and surroundings

vironmental conditions should be available and used, whenever applicable. The inappropriate or careless application of lubricants and other chemicals to components and equipment can cause harm to other sensitive components and systems.

During maintenance activities, certain equipment or even entire processes may need to be shut down. This should be avoided or minimized to the extent possible through proper design and planning of the maintenance. Plant turnarounds should be planned and scheduled to assure minimal shutdown during normal operations.

Maintenance safety procedures should be strictly followed to protect other employees, the plant property, and surrounding communities from incidents and disasters. Many major fires and toxic releases have been triggered by lax enforcement of sound maintenance safety practices.

MATERIALS AND COMPONENT FAILURES

The identification of the type of material failure is an important step in component failure analysis. There are many possible failure mechanisms, including fracture fatigue, distortion, wear, corrosion, embrittlement, creep, and stress cracking. These failure mechanisms can be applied to many types of materials, such as metals, alloys, and polymers.

Ductile tensile fractures are characterized by tearing of metal accompanied by plastic deformation and expenditure of considerable energy. Ductile fractures are caused usually from an overload of a part that has been underdesigned for the service conditions, improperly fabricated, or made of defective materials. They have a dull, fibrous appearance, may exhibit necking, and have shear lips. An example of a ductile tensile fracture is shown in Figure 11-2. [2]

Brittle tensile fractures exhibit more crack propagation, lower expenditure of energy, and less plastic deformation than ductile fractures (see Figure 11-2). Brittle fractures have a bright, granular appearance with no necking. Brittle fractures are usually caused by product defects either from manufacturing or from service.

Fatigue fractures are caused by the combination of cyclic stress, tensile stress, and plastic strain. All of these fluctuations must be present for fatigue cracks to be initiated and propagated. Cyclical stress and strain initiate a crack, which grows

FIGURE 11-2
EXAMPLE OF DUCTILE AND BRITTLE TENSILE FRACTURES.
(Courtesy of *ASM Handbook,* Vol. 11, *Failure Analysis and Prevention (1986) ASM Int'l., Materials Park, OH, p. 82*).

from tensile stress. The crack grows until the uncracked cross section of the part becomes too weak to handle the imposed load. The remaining cross section is then fractured to complete the fatigue failure of the part. An example of fatigue fracture for the turbine disk of a jet engine is shown in Figure 11-3.

Wear failures are the result of damage to a solid surface caused from material removal or displacement by mechanical action. Usually this surface deterioration occurs gradually over a period of time. Frictional interaction of two surfaces causes wear, increased energy consumption, and heat generation. This may affect the properties of the contacting materials, lubricant performance, and products

FIGURE 11-3
EXAMPLE OF FATIGUE
FRACTURE.
(Courtesy of ASM Handbook,
Vol. 11, *Failure Analysis and
Prevention* (1986) ASM Int'l.,
Materials Park, OH, p. 131).

being processed. Safety hazards from wear can result in mechanical failure, component seizure, or fire. For many applications low friction is desirable, but other situations like braking and belt tension require high friction. This places the safety emphasis on the control of friction.

To reduce friction and wear, lubrication is used between contacting surfaces. When properly maintained a well-conceived lubrication system will inhibit wear and reduce the potential for component failure. The gradual process of wear makes it more difficult to define failure, which then becomes more subjective. For example, equipment performance due to wear may slowly degrade, instead of suddenly dropping off. The selection of more durable materials and the replacement of worn components place more emphasis on economics rather than necessity due to equipment failure. Worn equipment components, such as bearings, seals, gears, piston rings, brakes, and clutches, can be significant factors in productivity, product quality, and process downtime. Even with the best of lubricant systems, some wear occurs, resulting in finely divided debris in the 1- to 2-micron range. This debris becomes suspended in the lubricant as long as the lubricant is in circulation. If the lubricant system fails, the debris buildup, the inability of the lubricant to maintain a continuous wear barrier, and heat dissipation can greatly accelerate component wear and failure. [2]

Corrosion failure of metal parts can occur directly or because corrosion can make components more susceptible to other failure mechanisms. Corrosion is an electrochemical reaction that depends upon the material type, the composition and uniformity of the environment, the temperature, the presence of impurities, the fabrication technique for the part, and the surface imperfections of the part. Corrosion failure is usually determined by the amount of metal thickness removed from the component. Corrosion can be either uniform in nature or localized, as in pitting. This causes corrosion failure to vary with the application, the type of metal, and the critical nature of the component to the overall process. Additional thickness is usually added to the part or component as a safety factor for anticipated corrosion. [12]

Galvanic corrosion is the corrosion rate above normal from the flow of electrical current from an active metal (anode) to a less active metal (cathode) in the same environment. Generally, metals near each other in the galvanic series as shown in Table 11-3 can be bolted or coupled together without causing a substantial increase in the corrosion rate of the more active metal or anode. The galvanic series applies in most environments, including seawater or freshwater. The coupling of dissimilar metals that are far apart in the galvanic series should be avoided because of excessive-corrosion failure potential.

The presence of tensile stress or other forms of mechanical wear or stress in a corrosive environment can greatly accelerate the rate of corrosion of a susceptible metal. Corrosion failure in these situations is usually the result of fine cracks that penetrate deeply into the metal without easily apparent surface corrosion.

Metal embrittlement can occur when a normally ductile metal becomes coated with a thin film of a certain liquid metal and is then stressed in tension. Cata-

TABLE 11-3
GALVANIC SERIES OF METALS OR ALLOYS

Magnesium-anode, corroded end

Aluminum

Zinc

Chromium

Iron

Steel

Nickel

Tin

Lead

Carbon

Copper

Silver

Platinum

Gold-cathode, protected end

strophic brittle failures can result. Ferrous metals can be embrittled by liquid cadmium, copper, gallium, indium, mercury, lead, tellurium, tin, and zinc. Embrittlement of galvanized steel can result from long exposure at elevated temperatures, but below the melting point of the zinc coating. The zinc diffuses into the steel, resulting in a brittle intergranular network of an iron-zinc compound. Mercury embrittles zinc, copper, tantalum, and titanium among the nonferrous alloys. [2] The liquid metal acts to nucleate a crack site on the metal surface, and this is followed by subsequent crack propagation by mechanical means.

Hydrogen embrittlement may occur when small amounts of hydrogen in the ppm range reduce steel tensile ductility to cause premature failure under static load. The hydrogen may be furnished when hydrogen-saturated steel at elevated pressures is cooled, precipitating gaseous hydrogen into microvoids of the steel. Hydrogen could also be introduced into steel voids from hydrogen formed by corrosion, absorption of hydrogen to reduce metal surface energy, or dissolved hydrogen migration to lower the cohesive strength between metal atoms. [2]

Creep metal failures are caused by time-dependent strain under stress, eventually resulting in stress and creep rupture. Creep occurs in a metal or alloy at elevated temperatures. The level of elevated temperature to produce metal creep can be approximated from the melting point of the metal, as shown in Table 11-4 on page 324. This table should be used with caution because the appropriate elevated temperature for each metal or alloy should be determined on an individual basis.

The presence of hydrogen at elevated pressures and temperatures above 230°C (446°F) can cause steel deterioration and failure of steel pressure vessels and pipelines. Atomic hydrogen may permeate and decarburize steel by reducing iron carbide (FeC_3) to form methane. The methane does not diffuse from the steel.

TABLE 11-4
COMMENCEMENT OF CREEP IN METALS AND ALLOYS BASED UPON ABSOLUTE MELTING TEMPERATURE

Material Group	Creep onset temperature		Relation to absolute melting temperature
	°C	°F	
Aluminum alloys	205	400	$0.54\ T_m$
Titanium alloys	315	600	$0.30\ T_m$
Low-alloy steels	370	700	$0.36\ T_m$
High-temperature carbon steel	540	1000	$0.49\ T_m$
High-temperature nickel or cobalt alloys	650	1200	$0.56\ T_m$
Refractory metals or alloys	980–1540	1800–2800	

When the methane pressure exceeds the cohesive strength of the steel, intergranular fissures between grains occur and the steel ductility is greatly lowered. This form of hydrogen embrittlement has been observed in petroleum refinery operations.

Hydrogen sulfide and carbon dioxide in oil and gas production can result in material problems like sulfur stress cracking, hydrogen-induced cracking, stress-corrosion cracking, and weight-loss corrosion. When, H_2S, CO_2, or chlorides are present, the corrosion rate increases. When H_2S at ambient temperature comes in contact with high-strength steels, hydrogen embrittlement failure occurs at well below the normal yield strength of the steel. H_2S at elevated temperatures, for example, above 80°C (176°F), is much less likely to cause hydrogen embrittlement of steel. When H_2S comes in contact with lower-strength steels, hydrogen-induced cracking may occur. The lower-strength steel and H_2S react in the presence of water to form hydrogen gas at internal defects in the steel, particularly at manganese sulfide sites.

Polymers, which are classified as thermoplastic and thermosetting in Table 11-5, are widely used in many industrial applications. The types of failure of polymers have many similarities to metals and alloys. There are many important polymer properties that relate to polymer applications (Table 11-6). When polymer components fail, it is often because of fractures that relate to crazing (a slitlike microcrack), voids, microcracks, shearing, or various combinations. Often this requires laboratory experimental work to reconstruct the probable cause and observed effect. Environmental stress cracking (ESC) is usually caused when the polymer component or application is subjected to severe liquid chemical exposure and measuring component failure for a prolonged period. Usually ESC failure involves a crazing or cracking failure mechanism.

The mechanical polymer properties may be significantly degraded by exposure to ultraviolet light. Heat exposure also produces adverse effects in polymers. Polymers should be protected from sunlight and high-temperature environments to avoid reduction in molecular weight and desirable properties. Stabilizers and pig-

TABLE 11-5
TYPES OF THERMOPLASTIC AND THERMOSETTING POLYMERS

Thermoplastic polymers
  Polyethylene
  Polypropylene
  Polystyrene
  Polyvinyl chloride
  Polyacrylonitrile butadiene styrene
  Polymethyl methacrylate
  Polyethylene terephthalate
  Polybutylene terephthalate
  Polyamide
  Polyacetal
  Polyphenylene sulfide
  Polycarbonate
  Polysulfone
  Polytetrafluoroethylene

Thermosetting polymers
  Phenol-formaldehyde
  Epoxy
  Alkyd
  Melamine-formaldehyde
  Urea-formaldehyde

ments are often added to polymers for outdoor and elevated temperature applications.

CORROSION PROTECTION

The potential for corrosion with accompanying deterioration must be considered in the service life of equipment. Maintenance programs should include an evaluation of corrosion to determine an appropriate service life. [12]

TABLE 11-6
TYPICAL POLYMER PROPERTIES

Density

Tensile strength

Elongation at break

Flexural modulus

Hardness, Rockwell

Linear-mold shrinkage

Environmental stress crack resistance

Notched izod impact

Heat-deflection temperature

The corrosion process is the undesirable and destructive interaction of a metal or other material with its environment. The function of protection systems is to retard or eliminate the deterioration by an economical method. Corrosion occurs because of relatively small physical and chemical differences in the material and its environment. Material variations are from minor impurities, localized composition, variations, scratches and abrasions, breaks in the outer protective layer, built-in thermal stresses, stray electrical currents, galvanic coupling of dissimilar metals, and bacterial action. Variations in environmental conditions may occur because of temperature changes, oxygen concentration, presence of corrosive agents, water, or humidity. When these conditions happen, they create dissimilar material surface sites, and electrochemical corrosion occurs. There are two major types of corrosion concern. Uniform corrosion is the general roughening of a metal surface and its gradual thinning. Pitting corrosion is a concentrated and localized form of attack. Pitting corrosion rates may be 3 to 10 times the uniform corrosion rate. [9]

There are many methods and techniques that can be used to prevent or retard corrosion, as shown in Table 11-7. Corrosion inhibitors, which are absorbed on the metal surface, and protective coatings, which are thicker and form adhesive films, reduce electrical current flow. They separate the metal from the electrolyte, thereby reducing electrical current flow.

The selection of a suitable coating or inhibitor system, proper surface preparation, and careful application are essential to the performance of corrosion inhibitors and protective coatings. Inorganic anionic inhibitors like chromates, phosphates, molybdates, and silicates are used to control dissolved oxygen corrosion. Organic cationic inhibitors like long chained amines are used to control H_2S and CO_2 corrosion. A wide range of protective coatings, including epoxies, urethanes, polyesters, extruded polyolefins, asphalt and inorganic zinc, may be used depending upon the application.

The failure of coating and inhibitor systems is usually because of improper and incomplete surface preparation, rather than poor-quality coatings. The surface must be clean and free of dirt, rust, or scale, and no moisture should be present. The application should be closely monitored to ensure that the application follows directions with the proper film thickness or concentration.

Cathodic protection is based on an outside source of electrical energy that constantly supplies electrons to the noncorroding cathode in a corrosion cell. As

TABLE 11-7
CORROSION PREVENTION METHODS AND TECHNIQUES

Corrosion inhibitors

Protective coatings

Cathodic protection

Moisture control

Removal of corrosive environment

Material substitution

shown in the example in Figure 11-4 on page 328, a continuous electrolyte or soil path must always connect the anode and the cathode. Usually sacrificial galvanic anodes of zinc, magnesium, or aluminum are used because these metals are much higher in the galvanic series of Table 11-3 than steel. When an electrical rectifier is used to drive current through the cell the steel structure to be protected should be connected to the negative terminal of the direct current power supply. The material of the impressed current anodes is usually selected for long life, like graphite or a lead alloy. [9,10]

Often corrosion is enhanced by the presence of corrosive gases dissolved in water. Oxygen, carbon dioxide, and hydrogen sulfide are such examples. Deaeration or stripping of gases and chemical treatment are among the most commonly used techniques for dissolved-gas removal from water.

All of this discussion is based on the premise that the most economical metal, alloy, or polymer has been selected for the application. If the corrosion problems are too severe or become uneconomical for safe operation, an alternative material may need to be selected. Material substitution may solve these difficult corrosion problems in a reliable, economic manner.

INSPECTION AND EVALUATION

There are many inspection techniques to detect equipment and component nonconformance to specifications and operating standards. These techniques cover a wide range of technologies and degrees of sophistication. Nondestructive testing and observation of operations by trained personnel should be part of all quality assurance operations. The detection of flaws, imperfections, and other potential equipment problems is best determined before failures occur so that maintenance can be performed safely and efficiently. The aircraft and nuclear industries have developed many nondestructive testing techniques for component inspections in the field. Both of these industries have to emphasize the protection of life in their safety maintenance programs. Many of their inspection and evaluation techniques have found far-reaching applications in numerous other industrial operations.

Maintenance and repair work is normally performed in the field, in shops, at service centers, and at the locations of manufacture. The methods and techniques in Table 11-8 on page 329 are commonly used for nondestructive inspections, testing, and evaluation of equipment and components. [16]

Visual inspection is the most common inspection method and is based on the judgment of appearance by trained personnel. Operations and maintenance personnel are most likely to detect equipment problems through routine visual inspection during their normal work at an operating facility. Surface flaws like finish, cracks, corrosion, and contamination can usually be detected and examined by visual inspection. A wide variety of equipment may be used to assist the naked eye in performing visual inspections. These may include magnifying systems for evaluating surface finish and shapes, borescopes for illuminating and observing inaccessible areas, or image sensors for remote sensing and visual records. A borescope is a long tubular optical device that illuminates for inspection narrow, difficult-to-reach

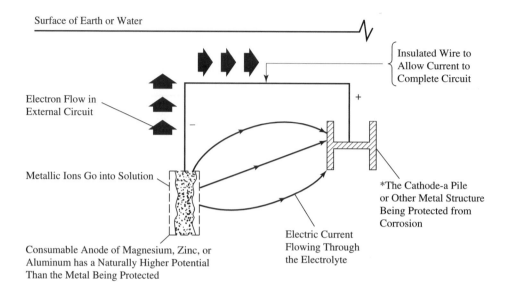

Surface of Earth or Water

Insulated Wire to Allow Current to Complete Circuit

Electron Flow in External Circuit

+

−

Metallic Ions Go into Solution

*The Cathode-a Pile or Other Metal Structure Being Protected from Corrosion

Electric Current Flowing Through the Electrolyte

Consumable Anode of Magnesium, Zinc, or Aluminum has a Naturally Higher Potential Than the Metal Being Protected

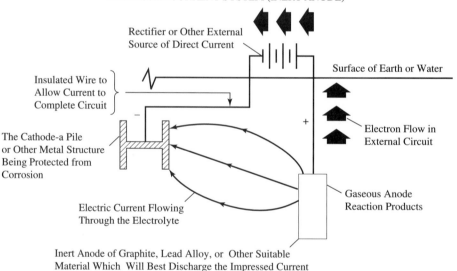

IMPRESSED CURRENT SYSTEM (INERT ANODE)

Rectifier or Other External Source of Direct Current

Surface of Earth or Water

Insulated Wire to Allow Current to Complete Circuit

−

+

Electron Flow in External Circuit

The Cathode-a Pile or Other Metal Structure Being Protected from Corrosion

Gaseous Anode Reaction Products

Electric Current Flowing Through the Electrolyte

Inert Anode of Graphite, Lead Alloy, or Other Suitable Material Which Will Best Discharge the Impressed Current

FIGURE 11-4
CATHODIC PROTECTION OF STEEL.

TABLE 11-8
METHODS FOR NONDESTRUCTIVE INSPECTION AND TESTING OF EQUIPMENT

Visual inspection
Liquid-penetrant inspection
Ultrasonic inspection
X-ray inspection
Magnetic-particle inspection
Eddy-current inspection

areas. Typical examples of borescopes are shown in Figure 11-5 on page 330. Borescopes can range from rigid, for straight-line paths between the observer and the observed area, to flexible, with video monitors for nonlinear pathways. Borescopes can inspect inside piping, walls, and tanks without costly teardowns. [3]

Microscopes and magnifying glasses that may range in magnification from 3 to 500 times actual size can be used to visually examine microstructure in the metallurgy and accurately measure tolerances.

Image sensors for visual inspections can be used with television cameras to focus the probing beam on the surface to be inspected. This application can be most useful for real-time control of industrial processes.

Liquid-penetrant inspection is used to detect discontinuities that are open to the surface of solid materials. A fluorescent or visible liquid is spread on the cleaned surface to seep into cracks and flow by capillary action. Excess liquid is wiped or washed off the surface; then a developer or blotter is used to enhance the seepage and color of the penetrant out of the surface opening for visual penetrant indication. Inspection of visible penetrant is performed in white light. Fluorescent penetrants are inspected in a darkened area with an ultraviolet light. Fluorescent penetrants are usually green, while visible penetrants are usually red in color. This technique is only applicable to cracks or flaws that are open to the surface. It will not detect flaws that are closed to the surface. [2]

Ultrasonic inspection is based upon the deflection of sound waves that travel through a test component or sample. It can be used to detect internal flaws and measure thickness and the extent of corrosion. Ultrasonic test equipment is portable, highly sensitive, reliable, and versatile. Experienced technicians are required to calibrate and accurately interpret the testing equipment.

X-ray testing provides a recording of the condition of the component being tested. The absorption of the penetrating gamma radiation will record differences in density, thickness, or composition as shades of gray. X-rays are most useful in detecting internal flaws in castings, fabrications, and weld seams. Gamma rays are hazardous, requiring strict safety procedures in testing.

Magnetic-particle inspection is useful in locating surface and subsurface discontinuities in ferromagnetic materials. When the component to be tested is electri-

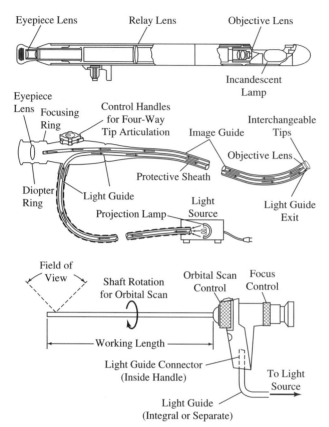

FIGURE 11-5
TYPICAL EXAMPLES OF INSPECTION BORESCOPES. (*Source: ASM Handbook, Vol. 17, Nondestructive Evaluation and Quality Control,* (1989), p. 3, fig. 1. ASM International, Materials Park, OH 44073-0002.)

cally magnetized, internal discontinuities cause a magnetic leakage field to be formed at the surface of the component. Once the leakage field is formed, it can be detected by distributing powdered ferromagnetic particles over the surface. The magnetically held particles form a sharp outline on the surface, indicating the size, shape, and extent of the internal discontinuity. After the inspection, the electrical current is shut off and the part is demagnetized. This is a simple, inexpensive test that requires no precleaning or removal of protective coating. It requires iron, nickel, or cobalt ferromagnetic alloys for test parts. Small and shallow surface cracks are most readily detected in final inspection by this technique.

Eddy-current inspection can be used on any metallic material that conducts electricity. It will detect cracks, seams, voids, or inclusions and measure metal thickness and nonconductive coatings on conductive surfaces. A probe or coil is placed adjacent to the part. An alternating current in the coil produces electromagnetic induction in the part, causing eddy-current flow within closed loops. Discontinuities change the eddy-current flow, which can be monitored by observing the change in induced voltage or current or electrical impedance. This technique requires careful interpretation by skilled inspectors.

Mechanical testing of components usually involves hardness, tensile strength, fatigue, elongation, stress crack, or impact tests. It often involves test specimens that must be carefully prepared and analyzed. [2]

Corrosion detection and monitoring may be done with sensors like test coupons, measurement of actual penetration rates on the equipment, or analysis of process fluids for dissolved materials from corrosion reactions. Usually several duplicate test coupons and holders are mounted in flow-line tees for easy retrieval in 1 to 3 months and should be positioned to be representative of the corrosive liquid phase (Fig. 11-6). The test coupons should be cleaned and weighed before and after exposure. The weight loss is converted to an average corrosion rate usually expressed as mils per year. [10]

$$CR = 2831 \frac{W}{AT}$$

where CR = corrosion rate, mils per year (1 mil = 0.001 in)
$\quad$ W = coupon weight loss, g
$\quad$ A = coupon surface area, in²
$\quad$ T = coupon exposure time, days

There are about 25.4 microns per year for a rate of 1 mil per year.

PREVENTIVE MAINTENANCE

Preventive maintenance is the performance of periodic scheduled maintenance to assure optimum, cost-effective, and safe operation throughout the life expectancy of the equipment and the facility. It usually includes periodic inspections, over-

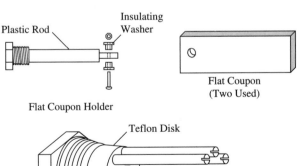

Insulating Washer

Plastic Rod

Flat Coupon Holder

Flat Coupon
(Two Used)

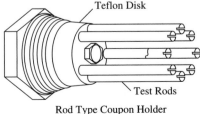

Teflon Disk

Test Rods

Rod Type Coupon Holder

FIGURE 11-6
CORROSION TEST COUPONS AND HOLDERS. (*Source:* Jones, Loyd W., *Corrosion and Water Technologies for Petroleum Producers,* 1988, p. 89.)

hauls, repairs, and replacements to extend the expected life of the system. The overall facility operating time is increased by reducing unscheduled shutdowns and minimizing the system damage when equipment and components fail. [14]

Routine walk-around inspections by personnel of operating facilities help ensure that equipment is properly functioning and being maintained. These walk-arounds should be regular and systematic, for example, hourly, per shift, daily, weekly. The detection of any abnormal or obvious problem or operating change, indicating incipient failure, allows adjustments and minor repairs to be made easily. Observation of lubricant, fuel, or gas leaks, unusual noise and vibration levels, changes in air, wastewater, or cooling-water quality, unusual odors, and changes in operating temperature, pressure, flow, and level are typical examples of potential maintenance problems. These problems require further investigation and possible maintenance. [14]

Scheduled routine maintenance is a necessary component of a sound preventive-maintenance program. Systematic lubrication, housekeeping, equipment cleaning, and corrosion control reenforcement are examples of scheduled routine maintenance activities that restore and improve operating performance. More detailed and thorough inspections should be planned periodically in addition to the walk-around inspections previously discussed.

Usually the time, supplies, personnel, and money spent on preventive maintenance is a much greater value to the operating facility than all of the necessary resources spent on avoidable repairs and cleanup. [1]

Planning plays an important role in preventive maintenance. It should begin with the selection of what is to be prevented from occurring, that is, deterioration, defects, or failure of equipment or facilities. Then it should be determined whether the deterioration, defect, or failure can be prevented or retarded by periodic actions. If prevention is not practical, then the consequence of the undesirable situation could be reduced by continuous or at least periodic surveillance. Based upon this information, the preventive-maintenance procedure should be selected. This procedure should be implemented during periodic intervals within the normal life span of the equipment and facility. Finally, the procedure or task should be assigned to a specific organizational function, for example, operations or maintenance crews, who will be responsible for the performance of the preventive maintenance. [4]

PREDICTIVE MAINTENANCE

Predictive maintenance assumes that most equipment failures are probabilistic and indeterminate. The consequences of unexpected defects or failures are controlled by checking equipment at predetermined intervals, with the maintenance action to be carried out only if the inspection shows the specific need for it. This is in contrast to preventive maintenance, where the maintenance action is performed when a predetermined time interval has elapsed.

Predictive maintenance requires accurate, reliable data and information that will be used as the basis for maintenance decisions. The monitoring methods and associated instrumentation should be carefully selected to be user-friendly for the skilled, trained personnel who will make the observations. The maintenance personnel should be flexible so that they can be available whenever required by predictive-maintenance observations. [4, 14]

UNSCHEDULED MAINTENANCE

Even with the best of plans, which generally focus on scheduled maintenance activities, there will be times when equipment requires unscheduled maintenance. When equipment fails to operate in an acceptable manner, is automatically shut down by safety systems or to avoid unsafe conditions, there is usually associated unscheduled maintenance work. This type of maintenance is often costly and time-consuming. If it does occur, the cause should be determined and corrected through either future preventive or predictive scheduled maintenance, if practical. Recurrences of similar unscheduled-maintenance activities should be avoided and should generate the management attention it deserves.

TURNAROUND MAINTENANCE AND REPAIR

The profitability and safety of a plant is affected by both onstream factors and maintenance costs. All maintenance work should be planned and scheduled, unless an unforeseen equipment breakdown forces unscheduled maintenance. Plant turn-arounds, which may occur annually in many process industries, involve considerable planning and teamwork among personnel from operations, maintenance, procurement, safety, and technical support services. The plant turnaround is the ideal time to overhaul major equipment that is vital to the plant onstream factor. Qualified personnel need to be assigned well in advance to develop and coordinate the complex and considerable work list for the turnaround. All critical spare parts and materials and other necessary items must be on hand when they are needed. A pre-work safety checklist should be developed to include the shutdown of process electrical power, have appropriate blinds in place, satisfactorily purge vessels and lines, keep desirable instrumentation on line, and adequately monitor for hazardous conditions and environments. [4, 11]

As equipment is carefully disassembled, photographs, written notes, and even a tape or video recorder can be useful to the accurate reconstruction of maintenance events and observations. As equipment is made ready for maintenance work, particular attention should be given to visible deposits. These deposits may be telltale signs of operational problems or malfunctions of equipment components like seals, rotors, bearings, or corrosion of materials of construction.

Once the equipment has been cleaned, and inspected, and the appropriate components have been replaced, the reassembly can begin. It is important that all parts and components be well protected from weather, contaminates, abuse, and transportation damage prior to reassembly. The considerable expense of equipment overhaul can be wasted if the reassembly is done in a careless or haphazard manner. Since turnaround activities take place 24 hours per day, there is a great deal of reliance placed on all of the maintenance personnel involved to perform their assigned tasks in a conscientious manner without close supervision. The start-up that follows the overhaul should be monitored to verify satisfactory performance. All significant events and findings throughout the entire turnaround should be well documented. This allows future discussions with equipment vendors and manufacturers to be on a more factual basis, improves maintenance activities, and will be useful in the preparation of future turnarounds. [4, 15]

INTEGRATION OF SAFETY PROCEDURES

Maintenance personnel should have good mechanical ability and a well-rounded background in their assigned work areas. These employees should have more in-depth safety training than other employees because the hazards of their work conditions change from day to day. They must know how to use many specialty tools, machines, and personal protective equipment. [1]

Proper training must be provided to make maintenance crews aware of job hazards and control. Maintenance employees need to know the properties and hazard control of all irritating, toxic, or corrosive materials present in the workplace.

Whenever a new or nonroutine maintenance activity is necessary, hazards and safe methods of that job should be thoroughly discussed with the affected maintenance crew. Safety and health professional staff should be available for maintenance planning and implementation, particularly for complex or hazardous work.

By doing many maintenance activities throughout the plant, these personnel become familiar with all of the processes of normal plant operation and the related equipment in the plant. This unique exposure allows maintenance employees the ability to identify, analyze, and correct many unsafe conditions in the plant and its process units. [1]

CONFINED SPACES

Many fatal and serious injuries to maintenance employees and rescuers have been caused by the hazards of confined spaces. Insufficient preparation for unforeseen hazards is usually the reason for these deadly accidents. These confined spaces have limited access and ventilation. In fact, they never were designed and constructed for long-term human occupancy. These spaces may be susceptible to entry of hazardous materials while workers are inside. Mechanical internals in confined spaces could crush, shred, or otherwise injure workers if inadvertent motion should commence. Oxygen deficiency, explosion, fire, toxic chemicals, electrical hazards, and falls are some of the other hazards encountered.

OSHA has developed a safety system requiring a permit to enter a confined space (see Table 11-9). A confined space is defined as being large enough that an employee can bodily enter and perform limited assigned work and has limited or restricted means for entry or exit. A confined space could include tanks, silos, boilers, manholes, pipelines, process vessels, or trenches. When any of the hazards in Table 11-10 are present in an identified confined space and cannot be eliminated, a

TABLE 11-9
PRINCIPAL AREAS OF THE OSHA PERMIT–REQUIRED CONFINED-SPACE PROGRAM

Identification and characterization of confined spaces
Characterization of confined space hazards
Development of a program for entry permits
Training of affected employees
Duties of entrants, attendants, and supervisors
Providing of rescue and emergency services

TABLE 11-10
CRITERIA FOR A PERMIT-REQUIRED CONFINED SPACE

Contains or has the potential for a hazardous atmosphere
Contains a material that could engulf an entrant
Has a configuration that could trap or asphyxiate an entrant
Contains some other recognized serious safety or health hazard

TABLE 11-11
EMPLOYER REQUIREMENTS IN A PERMIT-REQUIRED CONFINED-SPACE PROGRAM

Prevent unauthorized entry
Identify and evaluate the hazards
Develop and implement criteria for safe entry
Provide support equipment
Evaluate conditions during entry operations
Provide one outside attendant during entry operations
Adapt program for multiple spaces
Designate and train support personnel
Manage rescue and support services
Manage contractor operations
Manage permit completion
Annually review and critique program

permit is required for entry. [18] The priority emphasis on testing the confined atmosphere with a direct-reading instrument is because of oxygen content and the presence of flammable or explosive gas or vapor and toxic-air contaminants. Danger signs must be posted to warn employees. A sign reading Danger. Permit-Required Confined Space. Do Not Enter. Or similar language will satisfy the OSHA signage requirements.

The permit program for entry of a confined space includes considerable detail and is outlined in 29 CFR 1910.146 (see Table 11-11). This is obviously an extensive and serious OSHA program that has been developed to avoid catastrophic confined-space deaths.

The actual permit system requires the signature of the entry supervisor before the entry begins. Two examples of permits, whose elements are OSHA-approved, are shown in Figure 11-7. These entry permits should remain valid for only a finite time period, for example, 8 hours, and the entry supervisor should then cancel the permit or revalidate it when appropriate. The canceled entry permits should be retained for at least 1 year. Each permit is job- and site-specific. Permits may be required as often as each entry.

Employers must provide all affected employees with safety training for the safe performance of confined-space entry. This includes entrants, attendants, entry supervisors, and members of the rescue service. The training should include respiratory protection, hazard testing, lockout and tagout procedures, ventilation systems, the use of personal protective equipment, and rescue and evacuation procedures. All of these personnel should be cross-trained in all of the various duties and each function. The attendant is the entrant's only link to the outside and as such should never enter the confined space to assist or rescue the entrant.

LOCKOUT AND TAGOUT PROCEDURES

When maintenance activities are required, all energy sources must be isolated and lockout and tagout procedures should be implemented. When equipment is at total rest, with all energy sources neutralized, it is energy isolation. There are many equipment energy sources, including electrical, steam, hydraulic, pneumatic, chemical, mechanical, and thermal.

The OSHA requirements for lockout and tagout control of hazardous energy sources is found in 29 CFR 1910.147. [18] This OSHA standard covers the servicing and maintenance of equipment, where the sudden and unexpected energization or start-up of the equipment or energy release could cause injury or death to employees. Employers must establish a safety program for affixing appropriate lockout or tagout devices to energy-isolating devices, and to otherwise disable equipment to prevent unexpected energization, start-up, or release of stored energy to prevent employee injury.

The term *lockout* is defined as the placement of a lockout device on an energy-isolating device to ensure the energy-isolating device and the controlled equipment cannot be operated until the lockout device is removed. The term *tagout* is defined as the placement of a tagout device on an energy-isolating device to inform and warn others that the energy-isolating device and controlled equipment must not be operated until the tagout device is removed. Examples of lockout devices are shown in Figure 11-8 on page 340. An example of tagout device is shown in Figure 11-9 on page 341. When tagout is used, the employer should provide a more rigorous periodic inspection of the safety system because tagout devices can be violated much more easily than lockout devices. Comprehensive equipment documentation of lockout and tagout system procedures, as shown in the example of Figure 11-10 on page 342, can be useful in safety system management. [1]

In 29 CFR 1910.147, OSHA requires that all employees comply with a lockout-

Confined Space Entry Permit
Date & Time Issued:_____   Date and Time Expires: _____
Job site/Space I.D.: _____   Job Supervisor _____
Equipment to be worked on: _____   Work to be performed: _____

Stand-by personnel_____ _____ _____

1. Atmospheric Checks: Time   _____
                         Oxygen   _____ %
                         Explosive _____ % L.F.L.
                         Toxic   _____ PPM

2. Tester's signature_____

3. Source isolation (No Entry):   N/A   Yes   No
        Pumps or lines blinded,   ( )   ( )   ( )
        disconnected, or blocked   ( )   ( )   ( )

4. Ventilation Modification:
        Mechanical   ( )   ( )   ( )
        Natural Ventilation only   ( )   ( )   ( )

5. Atmospheric check after
        isolation and Ventilation:
        Oxygen _____ %   >   19.5 %
        Explosive _____ % L.F.L. <   10%
        Toxic _____ PPM   <   10 PPM H$_2$S
        Time _____
        Tester's signature_____

6. Communication procedures: _____
    _____

7. Rescue procedures: _____
    _____
    _____
    _____
    _____
    _____

8. Entry, standby, and back up persons: Yes   No
        Successfully completed required
            training?   ( )   ( )
        Is it current?

9. Equipment: N/A   Yes   No
        Direct reading gas monitor -
            tested   ( )   ( )   ( )
        Safety harnesses and lifelines
            for entry and standby persons ( ) ( )   ( )
        Hoisting equipment   ( )   ( )   ( )
        Powered communications   ( )   ( )   ( )
        SCBA's for entry and standby
            person   ( )   ( )   ( )
        Protective clothing   ( )   ( )   ( )
        All electric equipment listed
            Class I, Division I, Group D
            and non-sparking tools   ( )   ( )   ( )

10. Periodic atmospheric tests
        Oxygen   _____% Time_____   Oxygen   _____% Time_____
        Oxygen   _____% Time_____   Oxygen   _____% Time_____
        Explosive _____% Time_____   Explosive _____% Time_____
        Explosive _____% Time_____   Explosive _____% Time_____
        Toxic   _____% Time_____   Toxic   _____% Time_____
        Toxic   _____% Time_____   Toxic   _____% Time_____

(continued)

FIGURE 11-7
OSHA SAMPLE PERMIT FOR PERMIT-REQUIRED CONFINED SPACE ENTRY.

We have reviewed the work authorized by this permit and the information contained herein. Written instructions and safety procedures have been received and are understood. Entry cannot be approved if any squares are marked in the "No" column. This permit is not valid unless all appropriate items are completed.
Permit Prepared by: (Supervisor) _____ _____
Approved By: (Unit Supervisor) _____ _____
Reviewed By (Cs Operations Personnel): _____ _____
                                              (printed name)   (signature)
This permit to be kept at job site. Return job site copy to Safety Office following job completion.
Copies: White Original (Safety Office)   Yellow (Unit Supervisor)   Hard (Job site)

_____

ENTRY PERMIT
PERMIT VALID FOR 8 HOURS ONLY. ALL PERMIT COPIES REMAIN AT SITE
UNTIL JOB COMPLETED

DATE:_____   SITE LOCATION/DESCRIPTION _____
PURPOSE OF ENTRY _____
SUPERVISOR(S) in charge of crews   Type of Crew Phone #

_____

COMMUNICATION PROCEDURES _____
RESCUE PROCEDURES (PHONE NUMBERS AT BOTTOM)_____

BOLD DENOTES MINIMUM REQUIREMENTS TO BE COMPLETED AND REVIEWED PRIOR TO ENTRY

REQUIREMENTS COMPLETED	DATE	TIME
Lock Out/De-energize/Try-out	____	____
Line(s) Broken-Capped-Blank	____	____
Purge-Flush and Vent	____	____
Ventilation	____	____
Secure Area (Post and Flag)	____	____
Breathing Apparatus	____	____
Resuscitator - Inhalator	____	____
Standby Safety Personnel	____	____
Full Body Harness w/"D" ring	____	____
Emergency Escape Retrieval Eq	____	____
Lifelines	____	____
Fire Extinguishers	____	____
Lighting (Explosive Proof)	____	____
Protective Clothing	____	____
Respirator(s) (Air Purifying)	____	____
Burning and Welding Permit	____	____

Note: Items that do not apply enter N/A in the blank.

RECORD CONTINUOUS MONITORING RESULTS EVERY 2 HOURS

CONTINUOUS MONITORING** TEST(S) TO BE TAKEN	Permissible Entry Level		
PERCENT OF OXYGEN	**19.5% to 23.5%**		___ ___ ___ ___ ___ ___ ___ ___
LOWER FLAMMABLE LIMIT	**Under 10%**		___ ___ ___ ___ ___ ___ ___ ___
CARBON MONOXIDE	**†35 PPM**		___ ___ ___ ___ ___ ___ ___ ___
Aromatic Hydrocarbon	† 1 PPM	* 5PPM	___ ___ ___ ___ ___ ___ ___ ___
Hydrogen Cyanide	(Skin)	* 4PPM	___ ___ ___ ___ ___ ___ ___ ___
Hydrogen Sulfide	†10 PPM	* 15PPM	___ ___ ___ ___ ___ ___ ___ ___
Sulfur Dioxide	† 2 PPM	* 5PPM	___ ___ ___ ___ ___ ___ ___ ___
Ammonia		* 35PPM	___ ___ ___ ___ ___ ___ ___ ___

*Short-term exposure limit: Employee can work in the area up to 15 minutes.
†8 hr. Time Weighted Avg.: Employee can work in area 8 hrs (longer with appropriate respiratory protection).

FIGURE 11-7 (continued)

REMARKS: _____
GAS TESTER NAME & CHECK # INSTRUMENT(S) USED   MODEL &/OR TYPE   SERIAL &/OR UNIT #

_____   _____   _____   _____

_____   _____   _____   _____

SAFETY STANDBY PERSON IS REQUIRED FOR ALL CONFINED SPACE WORK
SAFETY STANDBY PERSON(S)   CHECK #   CONFINED SPACE ENTRANT(S)   CHECK #

_____   _____   _____   _____

_____   _____   _____   _____

SUPERVISOR AUTHORIZATION - ALL CONDITIONS SATISFIED_____
DEPARTMENT/PHONE_____
AMBULANCE 2800   FIRE 2900   SAFETY 4901   GAS COORDINATOR 4529/5387

FIGURE 11-7 *(continued)*

tagout program. Upon observing a machine or other equipment that is locked out to perform servicing or maintenance, no employee shall attempt to start, energize, or use that machine or equipment. If employees disobey this mandate, a false sense of security will be created in the workplace, with potentially catastrophic results.

There is an OSHA-suggested guideline sequence to lockout procedures (Table 11-12). Following the maintenance activities, the equipment is ready to return to normal operation. Restoring such equipment to service involves another OSHA-suggested guideline sequence (Table 11-13 on page 341).

All lockout and tagout devices must be durable, weather-resistant, corrosion-resistant, and standardized for the facility. Lockout devices should be substantial to prevent easy removal without the use of excessive force or bolt cutters. Tagout devices should be substantial enough to prevent accidental removal; they state warnings, like Do Not Open, Do Not Close, Do Not Start, Do Not Energize, and Do Not Operate. The name, address, and phone number of the employee who applied the lockout or tagout device should be on that device.

Employers must provide each involved employee with appropriate training and retraining, as outlined in 29 CFR 1910.147.

Whenever a group or a number of employees need to lockout/tagout equipment, each employee should have an individual lockout/tagout device. Continuity of

TABLE 11-12
OSHA-SUGGESTED GUIDELINE SEQUENCE FOR EMPLOYER LOCKOUT PROCEDURES

1 Notify employees that required equipment maintenance will shut down and lock out equipment.

2 Identify the type of magnitude of the equipment energy sources, the related hazards, and their control.

3 Shut down the equipment.

4 Deactivate the energy-isolating devices to isolate the equipment from the energy sources.

5 Lock out the energy-isolating devices with assigned individual locks.

6 Dissipate or restrain any stored or residual energy sources.

7 Ascertain that there are no exposed personnel and that the equipment has been disconnected from the energy sources. Return equipment controls to off or neutral position.

FIGURE 11-8
EXAMPLES OF LOCKOUT DEVICES.
Photo courtesy of Idesco Corporation. (*Source:* National Safety Council.)

FIGURE 11-9
AN EXAMPLE OF A TAGOUT DEVICE.
(*Source:* National Safety Council.)
Photo courtesy of Idesco Corporation.

lockout-tagout protection must be ensured for shift changes and personnel changes. Whenever outside personnel of contractors are engaged in maintenance-related activities on-site, the employer must ensure that the lockout-tagout program is not compromised in any way.

HOT WORK

Hot work is an operation that can produce a spark or flame which has the potential to cause ignition of flammable vapors. The work location, the wind direction, and

TABLE 11-13
OSHA-SUGGESTED GUIDELINE SEQUENCE FOR RESTORING EQUIPMENT TO SERVICE

1 Check the equipment to be sure the components are operationally intact.

2 Check the immediate equipment area for housekeeping removal of nonessential items.

3 Check the work area to ensure only essential employees are present and that they have been safely positioned.

4 Verify that equipment controls are in a neutral position.

5 Remove the lockout devices and reactivate the energy-isolating devices.

6 Reenergize any stored or residual energy sources.

7 Notify operations personnel that the equipment maintenance has been completed and the equipment is ready for start-up.

LOCKOUT: TAGOUT SYSTEM™

CAUTION: Servicing or maintenance is not permitted unless this equipment is isolated from all hazardous energy sources. This is the exclusive responsibility of designated "Authorized Employees" (**see listing at bottom) who must follow the **complete** "Lockout/Tagout Procedure" as published by the company. **This sheet is limited and abbreviated: it must not be considered a substitute for the company's complete Procedure.**

LOCKOUT: TAGOUT DATA SHEET

For Equipment No. ___172___

Equipment Description: ___Blivet Sintering Machine with___
___Taurus Automatic Feeder___

HAZARDOUS ENERGY		ISOLATING DEVICES			CONTROL DEVICES (Check √)			
Type	Magnitude	Type	Location	I.D. No.	Lock	Tag	Both	Add'l Measures (See below)*
Electric	440 V.	Switch	Right-Front Post	S819			√	
Steam	170 psi	Valve	At wall 30 feet west	V47		√		
Hydraulic	3000 psi	Valve	On pump 10 feet east	V16			√	①

*ADDITIONAL SAFETY MEASURES (Refer To Table Above):

___① After closing and locking valve, drain hydraulic cylinder___
___on machine 172.___

**AUTHORIZED EMPLOYEES. Only the following are authorized to undertake the Lockout/Tagout Procedure on this equipment:

Jerzi Brodofski	Andre Schoenzer		
Paul Simpson	Nancy Mahabro		
Stanley Cheng	Joe Hershko		
Walter Paley			
John Wilson			

Prepared By: ___George Murdock___   **Date:** _8/8/90_   QUESTIONS?   PHONE _538_

© 1990 Idesco Corp., N.Y.C.   Q-SIGN® V9-11

FIGURE 11-10
EXAMPLE OF LOCKOUT/TAGOUT SYSTEM PROCEDURES.
(*Source:* National Safety Council.)

TABLE 11-14
RECOMMENDED REQUIREMENTS FOR HOT-WORK PERMITS

Equipment to receive the work

Type of work to be performed

Required protective equipment

Fire protection personnel and equipment on standby

Date and time of issue

Date and time of expiration

Name, address, and phone number of person who issued the permit

the presence of flammable vapors and liquids should be considered prior to performing hot work. [13]

Hot-work permits should be obtained before starting any work that could involve a source of ignition, like welding, grinding, or cuttng. These written permits should specify the criteria shown in Table 11-14. The authorizing person should make a careful analysis and area inspection before issuing a hot-work permit. There must be no unconfined flammable material present in the work area and no reasonable probability that flammables will enter the work area during the hot work. Hot-work permits should be reissued or reconfirmed for each work shift, if needed.

Before hot work is done on a vessel or in a confined space like a tank, all connections must be blinded and the vessel thoroughly cleaned and purged. Only a trace of flammable gas or vapor is permitted in the area, up to 10 percent of the LFL. Gas may be eliminated from a tank by water fill, steam injection, or air ventilation. [16] Continuous monitoring for flammable vapors should be done while hot work in a tank is being performed.

Whenever hot work is being performed on the exterior surfaces of vessels and piping, they should be gas-freed and isolated. Piping and vessels should be disconnected or blinded prior to hot work. A closed valve is not a positive block or blind.

There are certain specific instances where welding may be permitted on hydrocarbon-containing vessels and piping if the metal walls are sound and sufficiently thick and qualified personnel take strict precautions to avoid burning through the metallic wall. [13]

There are many personnel hazards to guard against during welding or cutting operations, as shown in Table 11-15.

TABLE 11-15
PERSONNEL HAZARDS DURING WELDING AND CUTTING OPERATIONS

Ultraviolet light rays from electric welding

Toxic fumes

Oil-contaminated clothing

Flammable gas or vapor

Oxygen from welding equipment leaking into confined spaces

Oxygen content increased in inerted confined spaces

FIGURE 11-11
PIPER ALPHA PLATFORM IN THE NORTH SEA.
(*Source:* Richard Folwell/Photo Researchers.)

Nonferrous or nonsparking hand tools are often specified in flammable environments as a fire prevention measure. In petroleum operations the use of nonsparking tools is not warranted as a fire prevention measure, according to the experience of the American Petroleum Institute. [17]

Maintenance activities can create a great many hazardous situations in the workplace. Failure to follow or control safe maintenance procedures, rules, and guidelines has led to many serious disasters at industrial facilities. One such incident, the Piper Alpha disaster, involved an offshore oil and gas production facility, in the U.K. sector of the North Sea. [7, 8] Many of the personnel were killed, considerable property was destroyed, and the environment was polluted.

THE PIPER ALPHA DISASTER

Occidental Petroleum was the operator of a large oil and gas production platform, Piper Alpha, in the U.K. sector of the North Sea (Figs. 11-11 and 11-12). Piper Alpha was connected by pipeline to the Claymore and Tartan platforms (Fig. 11-13). Oil and condensate were sent onshore to Flotta via pipeline. Piper Alpha gas was piped to the offshore gas manifold compression platform (MCP-01) and onto

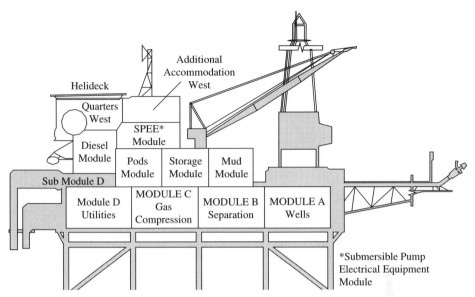

FIGURE 11-12
ELEVATION DRAWING OF THE PIPER ALPHA PLATFORM.

St. Fergus via pipeline. The Frigg gas production in Norway also flowed by another pipeline to MCP-01. The Piper Alpha complex was an integral part of the oil and gas production of the North Sea (Fig. 11-14 on page 346).

The Piper Alpha disaster occurred about 2200 hours on the evening of July 6, 1988, while the complex was producing 129,490 barrels per day of oil and condensate, and 53.5 million ft³/d of natural gas. It killed 165 of the 226 people on the platform at the time of the explosion. An organization chart is shown in Figure 11-15 on page 347. Two members of the Sandhaven, a fast-rescue craft, were also killed trying to rescue persons from the installation. The production platform was

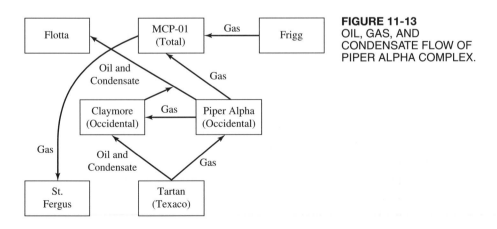

FIGURE 11-13
OIL, GAS, AND CONDENSATE FLOW OF PIPER ALPHA COMPLEX.

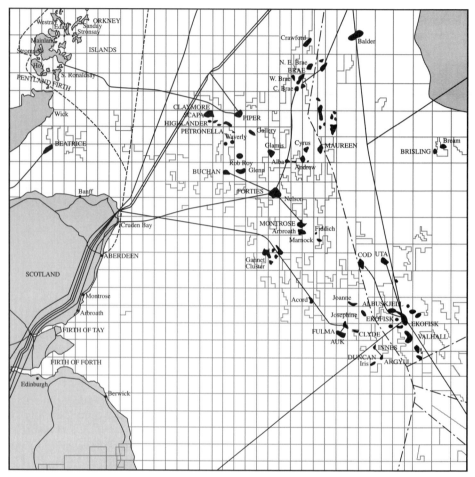

FIGURE 11-14
OIL AND GAS FIELDS IN THE CENTRAL SECTOR OF THE NORTH SEA.

totally destroyed. Wellhead fires were out of control for several days, causing pollution before they were extinguished. The death toll was the highest of any accident in the history of offshore oilfield operations.

The first event in this disaster was an initial explosion (Figs. 11-16 and 11-17 on pages 348 and 349). It happened in the gas compression module (C module) when a low-lying cloud of condensate ignited. Just 10 to 15 seconds later the platform was in flames.

Key witnesses died and most of the equipment was lost, making an exact explanation for the disaster difficult to obtain. This causal uncertainty often occurs in major explosions at complex facilities. As a result, a number of possible explanations were studied by a British government public inquiry under the direction of Lord W. Douglas Cullen. The most probable cause of the disaster was determined after careful consideration of all the evidence.

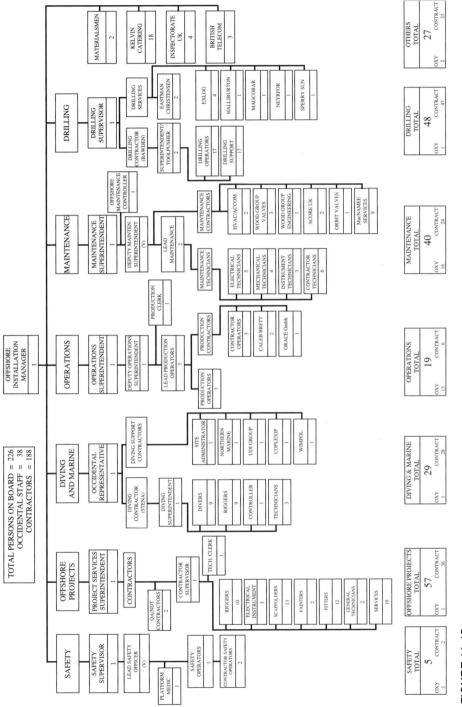

FIGURE 11-15
PIPER ALPHA PERSONNEL AT THE DISASTER.

347

FIGURE 11-16
INITIAL PIPER ALPHA EXPLOSION ON JULY 6, 1988.
(*Source:* Dave Caulkin/AP/Wide World Photos.)

The southeast quadrant of C module was where the ignition of the condensate vapor took place (Fig. 11-18 on page 350). This vapor cloud probably developed as shown in Figure 11-19 on page 351, based upon simulation tests made for the inquiry.

One of the two condensate injection pumps tripped at 2145 and shift personnel tried to restart the other pump, which had been shut down for maintenance. Unknown to them, a pressure safety valve on the other pump had been removed for recertification and had not been replaced. A blind flange that had been installed at the valve site was not leak-proof. Workers were unaware of the valve removal because of communication failures at shift change earlier in the evening and the safety failure of the work permit system related to the valve removal.

The initial explosion caused extensive damage in C module and immediately started a large crude-oil fire in B module, which was the oil separation module. This in turn engulfed the north end of the platform in dense black smoke. The leaking main oil line to Flotta quickly spread the fire to the area where the pipelines from the Claymore and Tartan platforms were connected by risers.

At 2220, a second major explosion was caused by the thermal rupture of the

FIGURE 11-17
PIPER ALPHA PLATFORM AFTER THE PHOTOGRAPH IN FIGURE 11-16.
(*Source:* Dave Caulkin/AP/Wide World Photos.)

riser on the gas pipeline from Tartan. At 2250, the riser on the gas pipeline to MCP-01 ruptured, and at 2320, the riser on the gas pipeline to Claymore also ruptured.

The offshore installation managers on Claymore and Tartan were not prepared to deal with an emergency on a platform that was connected to their own. The oil production of Claymore and Tartan was not shut down in a timely manner, helping further fuel the Piper Alpha fire. There had never been a pipeline safety exercise where one of these interconnected platforms was knocked out.

The initial explosion shut down the Piper Alpha main power supplies and control room. The fire-water system was rendered inoperable by the explosion. Helicopter or lifeboat evacuation was impossible because of smoke and flames. The emergency control systems were almost inoperable.

Most of the deaths were caused by smoke inhalation; only a few deaths resulted from burns. To remain in the accommodation meant certain death. Of the 62 people on the night shift, 39 survived. There was no systematic attempt to lead anyone to escape from the accommodation. Most of those who escaped either jumped into the sea or lowered themselves by rope. Thirty bodies were never recovered.

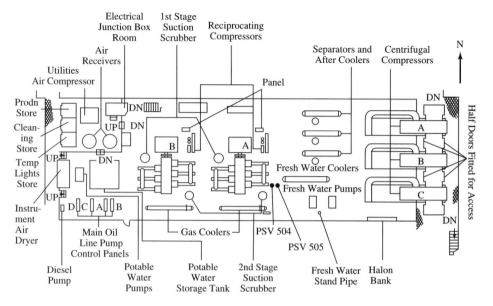

FIGURE 11-18
LAYOUT OF THE C MODULE AT PIPER ALPHA.

The inquiry identified numerous unsafe practices and procedures that had been present on Piper Alpha for some time. There had been inadvertent or unauthorized start-ups and operation of other equipment that was still under maintenance service. There had been incidents in the past of a lack of communication when there was a shift change.

The diesel fire pumps were kept on manual mode during certain periods; this hindered their operability in a dangerous way. An audit had recommended keeping them as automatic backups. The deluge heads in C module were blocked with scale anyway.

Management did not see that safety drills and personnel training were practiced regularly. In general, management's assessment of the risk of major hazards was superficial. All of the correct safety procedures and policies were in place. The management failed to follow or practice them.

The British Department of Energy was also to blame. Despite numerous visits to the Piper Alpha, safety inspections were superficial and of little use as a check of platform safety. The inspectors lacked experience in process safety and had been given inadequate guidance to be effective in assessing or monitoring the safety management of operators. The U.K. safety regulations for offshore platforms had become too restrictive, imposing solutions rather than objectives on the operators. Some regulations had become obsolete because of advances in technology.

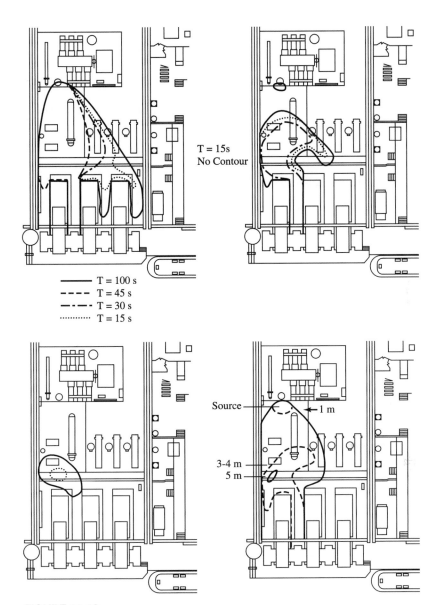

FIGURE 11-19
PIPER ALPHA SIMULATION TESTS FOR THE GROWTH OF THE
EXPLOSIVE VAPOR CLOUD.

BIBLIOGRAPHY

1 *Accident Prevention Manual for Business and Industry; Engineering and Technology,* 10th ed., National Safety Council, Itasca, Ill., 1992.
2 *Failure Analysis and Prevention,* vol. 11, ASM International, Materials Park, Ohio, 1992.
3 *Nondestructive Evaluation and Quality Control,* vol. 17, ASM International, Materials Park, Ohio, 1994.
4 Bloch, H. P., and F. K. Geitner, *Practical Machinery Management for Process Plants,* vol. 3, Gulf Publishing, Houston, 1990.
5 Bloch, H. P., and F. K. Geitner, *Practical Machinery Management for Process Plants,* vol. 4, Gulf Publishing, Houston, 1985.
6 *Cleaning Mobile Tanks in Flammable or Combustible Liquid Service,* 6th ed., No. 2013, API, Washington, D.C., 1991.
7 Cullen, W. D., *The Public Inquiry into the Piper Alpha Disaster,* vol. 1, British Department of Energy, London, England, 1990.
8 Cullen, W. D., *The Public Inquiry into the Piper Alpha Disaster,* vol. 2, British Department of Energy, London, England, 1990.
9 Dismuke, T. D., et al., *Handbook of Corrosion Protection for Steel Pile Structures in Marine Environment,* American Iron and Steel Institute, Washington, D.C., 1981.
10 Jones, L. W., *Corrosion and Water Technology for Petroleum Producers,* OGCI Publications, Tulsa, Okla., 1988.
11 Karassik, I. J., et al., *Pump Handbook,* 2d ed., McGraw-Hill Book, New York, 1986.
12 Muhlbauer, W. K., *Pipeline Risk Management Manual,* Gulf Publishing, Houston, 1992.
13 *Safe Welding and Cutting Practices in Refineries, Gas Plants, and Petrochemical Plants,* 5th ed., No. 2009, API, Washington, D.C., 1988.
14 Sawyer, J. W., *Sawyer's Turbomachinery Maintenance Handbook,* vol. 1, Turbomachinery International Publications, Norwalk, Conn., 1980.
15 Sawyer, J. W., *Sawyer's Turbomachinery Maintenance Handbook,* vol. 2, Turbomachinery International Publications, Norwalk, Conn., 1980.
16 Sawyer, J. W., *Sawyer's Turbomachinery Maintenance Handbook,* vol. 3, Turbomachinery International Publications, Norwalk, Conn. 1980.
17 *Spark Ignition Properties of Hand Tools,* 3d ed., No. 2214, API, Washington, 1989.
18 29 CFR 1910, OSHA, Washington, D.C., July 1, 1994.

PROBLEMS

1 Discuss and compare deterministic and probabilistic equipment problems.
2 What should be the basis for equipment inspection to determine the need for maintenance?
3 Differentiate between an equipment overhaul and a major repair.
4 What factors should be included in a maintenance program?
5 Compare ductile and brittle metal fracture failures.
6 A forged aluminum alloy fitting broke in service. The surface of the fitting had been anodized by a chromic acid process to protect it from corrosion. An analysis of the alloy showed it to be within the required composition of the specification. A brownish stain was observed on the fracture surface. Further analysis revealed the brown stain was residue chromic acid from the anodizing process. The strength and elongation

properties of metal samples of the fitting were comparable to the normal material specifications. What do you believe caused the fitting to fail? How would you reduce the risk of failure in the future?

7 A portable, galvanized-steel, oil-fired heater broke after a short time in service. The failure occurred in the portion of the heater shell that normally reached the highest temperature in service. The heater shell was made of shiny galvanized steel sheet that was 20 mil (0.5) mm thick. The gray, brittle area around the pipe was not shiny. What do you think might have caused the failure of the heater? How might the problem be corrected?

8 A valve in a dry automatic sprinkler system tripped accidentally, activating the system. The valve had a copper alloy clapper plate that was held closed by a pivoted iron latch in a drinking water environment. In an emergency the water pressure of 80 psi (11.7 kPa) on the bottom side of the clapper plate is normally released when a switch releases a weight that pivots the latch away from the clapper. This water pressure forced the clapper plate upward and the water flowed through the sprinkler system. What do you believed caused the system to trip accidentally? How could the problem be solved?

9 Zinc or magnesium anodes may be used to protect steel pipelines from galvanic corrosion. Why is this an effective method for corrosion protection?

10 From a materials viewpoint, why is H_2S in steel equipment a potential problem?

11 What is the major cause and failure mechanism of environmental-stress cracking in polymers?

12 Why is visual equipment inspection important to a maintenance program?

13 What are the more common methods for nondestructive inspection of equipment?

14 A 6-in^2 steel test coupon in a water exposure test experiences a 2-g weight loss over a 45-day exposure period. What is the corrosion rate in mils per year and in micrometers per year for this exposure?

15 Discuss and contrast preventive, predictive, and unscheduled maintenance activities.

16 Why is it important to preplan plant turnarounds well in advance?

17 What areas are addressed by OSHA for permit-required confined spaces?

18 In what hazard order should a confined space be tested with a calibrated instrument?

19 What type of confined-space training should be provided entrants?

20 Under what circumstances should an attendant enter a confined space to rescue an entrant?

21 Define energy isolation as applied to equipment maintenance.

22 Define lockout and tagout.

23 What is the OSHA-suggested guideline sequence for lockout procedures?

24 What is the OSHA-suggested guideline sequence for restoring equipment to service following lockout procedures?

25 Define hot work.

26 What are the recommended requirements for a hot-work permit?

27 What are the potential hazards to personnel during welding operations?

28 The lower flammability limit of a flammable hydrocarbon vapor is 1.5 vol % in air. What is the maximum concentration in air that can be safely used in a vessel to ensure a safe welding operation inside the tank?

29 The following questions relate to the Piper Alpha disaster.
   a How did this disaster occur?
   b Why did it occur?
   c Who was responsible?
   d If you were the safety supervisor, what would you have done differently?
   e If you were the offshore installation manager, what would you have done differently?

12

MANAGEMENT AND ENGINEERING CONTROLS

In the health and safety field, the concentration or intensity of a hazard and the exposure potential are usually measured and compared to a permissible level. In this risk assessment process, the occupational hazard and employee exposure must be controlled so that the permissible level is not exceeded. The types of controls available to accomplish occupational hazard reduction include management controls, engineering controls, and personal protective equipment. Prior to implementing any of these control systems, a thorough understanding of the problem situation should be obtained. Otherwise a reasoned approach to hazard control will be difficult and inaccurate. Usually management and engineering controls are preferable to personal protective equipment (PPE) because effective PPE requires continual human attention for proper implementation. [13, 21]

To make college students more aware of these options, engineering and business schools should devote more resources to teaching basic safety principles. The present curriculum should put more emphasis on the analysis of hazards, accidents, and risk in the workplace. College faculty usually have little, if any, background in this critical area. As the safety field is rapidly developing, adjustments need to be made in curriculum; otherwise, safety, health, and environmental protection will remain incidental to the academic goals of engineering and business. [11, 13]

Management can take at least three alternatives to correct workplace hazards. First, it may decide to take no action to correct the hazard because there is disagreement about the existence of the hazard; all of the alternatives have not been evaluated; or financial, production, or personnel constraints may be larger than the risk of taking no action.

Second, management can decide to modify the system. This generally improves

354

the safety of otherwise acceptable performance. Usually the modifications will correct the reported deficiencies through management and engineering controls or the use of personal protective equipment (PPE).

Third, management could consider the redesign of the workplace. However, this option usually involves a major capital expenditure and considerable inconvenience. Additional and different safety hazards may be introduced into the workplace by the redesign, unless the hazards are identified and corrected. It is imperative that management be able to identify the relative risk of the hazard, the location and type of hazard present, the potential consequences of the hazard, and the costs to control and correct the problem. Aging industrial plants need to be modernized for safety and efficiency. [15]

MANAGEMENT CONTROLS

The use of management controls is achieved through personnel management, monitoring and limiting worker exposure, training and continuing education, maintenance and housekeeping, and prudent purchasing. The best control is to eliminate or reduce the hazard at the source. Second best is to control the pathway of the hazard from the source to the recipient. The third alternative is to focus the controls on the recipient.

Management controls can be applied to any of these three areas: the hazard source, the pathway of travel, and the recipient work pattern and use of PPE. The types and forms of management controls are shown in Table 12-1 on page 356.

Management must exhibit a strong commitment to safety and lead by example. The organization must buy into the necessary changes. [27] Before any safety program can be initiated and be successful, it must have the full support and commitment of the top management of the organization. The executive officers and the board of directors should encourage and support safety by their leadership and provide the necessary resources to accomplish safety goals and objectives.

There are many ways that management can show commitment to safety. [2]

Set a good safety example.
Effectively manage safety and health programs.
Attend safety and health meetings.
Perform walk-through inspections.
Investigate accidents.
Investigate near-miss accidents.
Review safety performance at all levels.
Be open-minded and objective.

The front-line supervisor has the greatest influence on most employees' safety performance. Employees often see and interface with this level of management. How the front-line supervisor is perceived by employees will greatly affect their

TABLE 12-1
MANAGEMENT CONTROLS FOR OCCUPATIONAL HAZARDS

Exhibit management commitment and lead by example.

Motivate employees to recognize hazards.

Inspire employee ideas and suggestions.

Provide financial and nonmonetary resources.

Monitor workplace and measure performance.

Develop safe work procedures.

Train and educate workers.

Modify work procedures and exposure.

Practice good housekeeping.

Plan preventive maintenance.

Integrate purchasing.

Monitor regulatory situation.

attitude and commitment to safety. Strong safety leadership by a supervisor should ensure that desirable standards and attitudes are present in the workplace.

Supervisors have a responsibility to safeguard, educate, and train employees under their supervision. They are generally responsible for creating a safe workplace and implementing hazard recognition and control into work activities. Supervisors are aware daily about what is happening in the workplace, who is doing it, how tasks are performed, and what operating conditions exist. They must be certain their employees understand the hazards of the materials they handle, the equipment and tools they use, and the process operation. For example, to control the exposure of employees to chemicals and noise, management can reduce work times in highly contaminated areas of the workplace. Supervisors train and educate employees in safe work practices and other precautions. They provide employees with appropriate personal protective equipment, train them in recommended PPE use, and assure the proper use of the PPE in the workplace.

Employee safety performance is evaluated by the supervisor. Correcting an employee with poor safety behavior requires good judgment. Usually, the supervisor can convince employees of the need for safe job performance. Occasionally, corrective action must be administered by the supervisor, particularly when safety rules and work practices are violated. Supervisors should monitor work areas and employees daily for hazards that could adversely affect their safety and health. When a hazard is detected, it should be corrected as soon as is practical. Often hazards develop because of poor housekeeping practices that result from neglect or indifference on the part of employees and supervisors.

If a supervisor cannot convince an employee to follow safe work practices, then prompt and firm discipline should be administered. This is particularly true when an employee is deliberately disobeying rules or endangering lives. A supervisor who is lax, or is perceived to be lax, in the enforcement of safety rules undermines the entire safety and health program and encourages hazards and accidents.

Good housekeeping and preventive maintenance are important components of a successful safety program. Good housekeeping results in accident reduction, increased productivity, and improved employee morale. Employees perform better in a clean, orderly workplace. Good housekeeping should be incorporated into all operations and workplace tasks. This needs to be continually emphasized to all employees and not left to custodial personnel. If each employee practices good housekeeping as an integral part of the job, this attitude will assure an efficient work environment. A clean and orderly workplace greatly enhances employees' ability to accomplish work tasks without interruption and interference. In such a work environment, housekeeping becomes a normal, integral part of the operations. Less time and effort will be needed to keep the workplace clean and perform maintenance. Minimizing waste from scrap, off-spec material, and spillage results in savings and cost efficiency. Employees will be less likely to slip or fall, the risk of fires will be reduced, access to equipment will be simpler, and exits will be easier to reach in an emergency.

Preventive maintenance prolongs the useful life of equipment and facilities, decreases equipment downtime, and creates safer working conditions. Usually, preventive maintenance is performed when it will be least disruptive to operations. Management should control preventive maintenance activities by well-planned scheduling, keeping good records of the service and repairs, planning the repair and replacement of equipment components, and controlling the inventory of spare parts. Each piece of equipment should have records for the maintenance service schedule and the repair history for that equipment. During equipment maintenance, management should be especially attentive that proper safety procedures are followed by all involved parties. Periodic inventory of spare parts and materials should be conducted by management to assure that they are available when they are needed.

The greatest deterrent to workplace hazard is training and educating employees. Employee awareness can help ensure that regulations, standards, and desirable work practices are followed. The more employees are encouraged to participate in safety functions, the better for the employees and the organization. Management's attitude toward safety will greatly shape employees' safety attitudes. Management should introduce safety programs on the first day of employment.

Safety should be a primary criterion in the purchase of equipment and materials used on the job. Puchasing agents should be certain that the design, manufacture, and shipment of all items comply with regulatory and company standards. The features of all items should be compared with competitive products; some additional expense may be justifiable on the basis of added safety. In being cost-conscious, the purchasing agent should recognize that accidents, health problems, and incidents have direct and indirect costs that may far exceed any savings based solely on the purchase price.

All organizations must stay current on government and industry regulations, which can sometimes change dramatically. New regulations can greatly influence products, equipment, materials, and process and have the potential to affect the health and safety of employees and the long-term viability of a company and an industry.

Management should encourage employee involvement in developing safety plans. Participative, democratic, Theory Y management styles are greatly preferred over autocrative, dictatorial, Theory X management styles. All employees should be aware of the reasons and benefits behind an effective safety program. Otherwise, they may not make a positive contribution and could resist changes in work practices and methodologies.

Management has an obligation to the organization to provide information and guidelines to maintain the wellness of employees. The importance of diet, lifestyle, and physical activity to our wellness involves the food we eat, the things we do, and our level of physical activity. They affect a number of risk factors that are implicated in longevity—blood cholesterol, hypertension, and obesity. There has been accumulated considerable evidence that the main causes of life-threatening illnesses like cancer and heart disease are directly related to our environment. We have control over many of these factors during our lifetime and can do much to use this knowledge to enhance and prolong our health and life expectancy. [30] Medical and health surveillance systems for all employees can be used by management to maintain employee health, control disease, and minimize disabilities and lost times. [7, 25]

Management should be certain that safety, health, and environmental protection information is made an integral part of operations, purchasing, research, design, marketing, and training. Loss control is important in all phases of the organization, including plant layout, equipment selection, and process operation. Subcontractors should be informed about mandatory requirements to meet safety, health, and environmental standards.

The workplace should be monitored and the safety performance measured against goals, objectives, and standards. An analysis should be made of all operations to be performed to determine the appropriate safeguards. [24] An effective industrial-hygiene monitoring system to protect employee health, now and in the future, should be maintained by management.

ENGINEERING CONTROLS

Engineering controls eliminate or reduce the hazards by the initial design specification for the process, equipment, or facility; by subsequent modifications; or through substitution, isolation, ventilation, or dilution. Protection that can be built into the initial design is usually preferable to a method that relies upon continuous human implementation or intervention. Engineering controls should be provided to the extent feasible because they have to be supplemented by the use of PPE as the final alternative to achieve acceptable limits of hazard exposure. A thorough

TABLE 12-2
ENGINEERING CONTROLS FOR OCCUPATIONAL HAZARDS

Plan site selection.

Critique process design and plant layout.

Isolate hazard sources.

Provide external plant infrastructure.

Install monitoring and warning equipment.

Furnish accident prevention signage.

Critique operating conditions and procedures.

Substitute or eliminate hazardous materials.

Add mechanical guards.

Install ventilation systems.

Implement lockout-tagout procedures.

Use recommended tools and equipment.

knowledge of the production process is an important prerequisite for optimization of either management or engineering controls. [9]

There are many kinds of engineering controls (see Table 12-2). Site selection, process design, plant layout, and choice of raw materials and products are the primary factors that influence the alternatives for engineering control of hazards. It should be remembered, however, that human errors, rather than engineering defects, are responsible for most workplace accidents. [12]

Engineering controls should be introduced during the preliminary design of the plant processes and facility layout. Control measures can be more easily and cost-effectively integrated at this stage.

Siting factors include climate, terrain, space requirements, transportation, labor, and the local community. Prevailing winds and site topography may determine the optimum location for process operations and product and raw material storage in relation to administrative offices and the local community. Existing bodies of water, rainfall runon and runoff, and groundwater locations should be factored into site selection. [3]

Fire protection codes specify minimum distances between building, storage tanks, manufacturing areas, and property lines. A proper lightning protection system will reduce fires and explosions. [19]

Both present and future expansion needs should be considered in the site selection.

The site should be fenced to keep out trespassers, who may be injured or may interfere with the plant operations. Fencing within the facility protects employees and visitors from potentially dangerous areas like power stations and pits.

Transportation of people and materials on and off site should be done safely to accommodate the anticipated volume at the operating facility. Plant entrances should provide ample room for large trucks and railroad cars, and there should be good visibility at all entrances. Separate entrances should be provided for pedestrians and drivers. Fencing, traffic signals, subways, and pedestrian bridges are important when parking lots are remote from the plant or near railroad tracks.

Shipping and receiving should be compatible with the overall flow of the materials within the facility. Materials should flow efficiently onto the site, in and out of the production areas, and then off-site to markets and elsewhere. Tank cars and other bulk shipments of hazardous materials require additional planning for shipping and receiving. All roadways in the facility should be carefully laid out, constructed, and maintained. Heavy trucks require roadways up to 15 m (50 ft) wide for two-way traffic, with grades of 8 percent maximum. Ditches should be provided to carry away runoff water. Roadways should be at least 11 m (35 ft) from building entrances. Traffic signage and signals should be provided to limit speed and vehicle movement at hazardous locations. Walkways should be the shortest distance between buildings and preferably of concrete. Parking lots should be fenced and located so that employees will not have to cross a roadway to reach their building. The surface of the parking lot should be smooth and well-drained, with painted lines for designated stalls to reduce accidents and make full use of the parking space. Landscaping should be tasteful and not create blind spots. Outside lighting should enhance safety and security while aiding plant productivity.

The layout of buildings and the overall facility should result in the efficient use and flow of materials, processes, and personnel.

- Assure materials, products, and personnel flow efficiently throughout the plant.
- Locate offices of supervision, control rooms, and support facilities convenient for communication.
- Meet the appropriate codes and standards.
- Physically separate hazardous work areas.
- Provide appropriate waste disposal methods.
- Make employee services like cafeterias, lunchrooms, lockers, and parking accessible.
- Allow employees to move easily in and around the plant.
- Make all functions and operations in the plant as user-friendly as possible.

Flow sheets allow the materials and process for each stage of manufacture to be studied for the elimination and control of safety, health, and environmental hazards. Three-dimensional scale models are an effective way to control hazards. These 3-D models can easily be rearranged to arrive at an optimum layout for safety and efficiency in the final layout and design.

Hazardous operations that involve fire and toxic chemicals can be placed in separate areas. Congested areas can be avoided, sufficient headroom can be provided, and ample clearance for passageways and stairs can be maintained. Aisles should be at least 0.9 m (3 ft) wider than the widest vehicle to permit trucks and personnel to pass without colliding.

Machinery and other processing equipment may require mechanical guards to

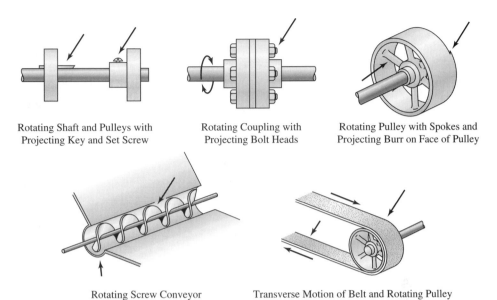

Rotating Shaft and Pulleys with
Projecting Key and Set Screw

Rotating Coupling with
Projecting Bolt Heads

Rotating Pulley with Spokes and
Projecting Burr on Face of Pulley

Rotating Screw Conveyor

Transverse Motion of Belt and Rotating Pulley

FIGURE 12-1
OPERATING HAZARDS FROM ROTATING, RECIPROCATING, AND TRANSVERSE
ACTIONS.

protect employees from potential operation hazards (see Figs. 12-1 to 12-3 on pages
361 and 362). [3] Effective protection devices at the point of operation cannot always
be designed and installed by the manufacturer because equipment is used in a variety
of ways and situations. Protective guards should be designed and installed by the
user company to safeguard employees in particular situations and to comply with
federal and local regulatory requirements. Safety protection devices should be
durable and substantial to prevent deterioration and shifting of the device. Guards include barriers, interlocking barriers, and automatic devices. Barrier guards bar access
to the dangerous parts of equipment and should remain fixed. Interlocking barrier
guards must be in a predetermined position before the equipment can be operated.
Automatic safeguard devices prevent operators from coming in contact with the hazard of the machine while it is operating or stop the machine in case of danger.

Power transmission guards should be fixed to protect against rotating members,
moving belts, reciprocating arms, gears that mesh, and cutting teeth. Punching,
shearing, and bending operations are also examples of power transmission action
or motion that requires security mounted guards.

Robots are now commonly used to perform certain operations, such as welding,
materials handling, and spray painting. Robotics introduces new hazards into the
workplace (see Fig. 12-4 on page 363).

• Being struck by moving parts within the operating envelope or movement
zone of the robot.

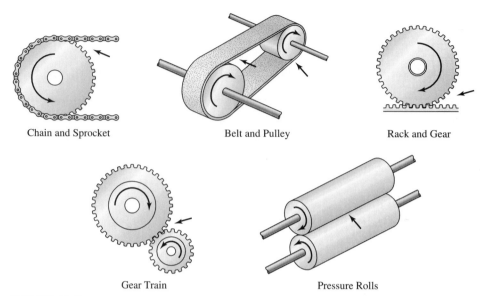

FIGURE 12-2
OPERATING HAZARDS FROM NIP POINTS. (*Source:* Used by permission of the National Safety Council, Itasca, Illinois.)

FIGURE 12-3
OPERATING HAZARDS FROM CUTTING, PUNCHING, SHEARING, AND BENDING. (*Source:* Used by permission of the National Safety Council, Itasca, Illinois.)

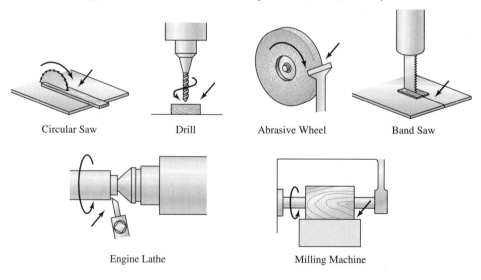

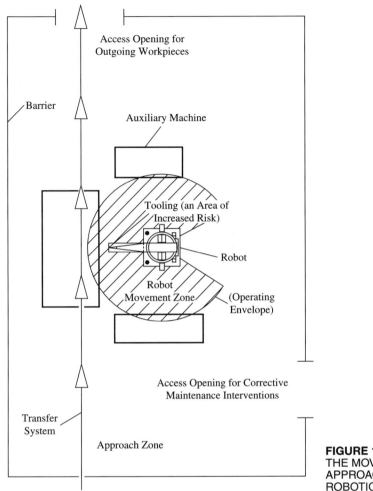

FIGURE 12-4
THE MOVEMENT AND
APPROACH ZONES FOR
ROBOTIC OPERATIONS.

- Being struck by dropped objects.
- Entrapment between moving robot parts and nearby objects in the approach zone. [3]

The movement zone is a hazard to anyone in it when the power is available to the robot. The approach zone allows employees to cross over into the movement zone. Robotics engineering controls should minimize the probability that an employee inside the movement zone could be injured by equipment operation. The robot should be physically restricted to the motion range needed to perform the particular operation. Mechanical stops to restrain robot travel at high speeds should be provided. An amber light on the robot should alert area personnel that the robot is energized. Fixed guards should be installed to define the perimeter of the robot movement zone, but should allow sufficient clearance for employees between the guard and the movement zone. Access gates should be interlocked to interrupt the robot drive energy when they are opened during the operating cycle of the robot. Warning signage should be placed around the guard perimeter to make all personnel aware of the hazards.

Electrical equipment is usually designed to be used in a specific type of service. Recommended codes and standards should be followed in the selection of electrical equipment, as discussed in Chapter 3. These include the National Electrical Code, the National Electrical Safety Code, and the codes of the National Fire Protection Association and the American National Standards Institute, as well as state and local codes. The National Electrical Code is probably the most important reference because it is usually required by insurance companies and local regulations in addition to federal law.

Engineering controls for electrical hazards should assure that electrical equipment is installed and used only for the intended purposes and conditions. It should be installed in less congested areas or, better yet, in special rooms. Interlocks, barriers, warning signage, guards, and other fail-safe devices should be included to protect employees and others nearby. Wiring must meet codes and standards for the current load and type of work.

All electrical switches should have voltage and amperage ratings compatible with the intended use of the equipment. Fuses and circuit breakers should be provided, when the safe carrying capacity of electrical conductors has been exceeded. Ground-fault interrupters should be provided to interrupt electrical circuits when employees are exposed to serious injury by coming in contact with the ground or ground equipment.

Electrical motors should be installed so not to interfere with the normal movement of employees or the flow of materials. They should be maintained and protected from dust, oil, moisture, overload, friction, misalignment, and vibration.

Electrical systems should be grounded to prevent excessive voltages from outside sources like line surges or lightning. This should be accomplished by bonding the conductor to a grounded electrode conductor. [10]

Electrical equipment should be grounded by attaching the metal enclosure to the main bonding jumper and to the service equipment with an equipment grounding conductor so that voltage differences will not occur. Double-insulated electrical power tools can be used instead of grounding, but not where moisture and a ground loop are present. Whenever water or other dampness and a ground loop are present, double-insulated tools msut be protected with a ground-fault interrupter.

Standard electrical equipment should not be installed where flammable gases, vapors, dust, or other ignitables are present. These hazardous locations are defined by the National Electrical Code and the National Fire Protection Association. Electrical equipment should be selected to conform with the requirements of the hazardous location.

TIMING OF HAZARD CONTROLS

The best time to introduce hazard controls is when the operating facility is in the conceptual stage. It is much easier to integrate management and engineering controls into the design stage than after the plant has been constructed and begun operation.

The layout of the facility, the process and equipment selection, raw materials selection, and product handling should be identified and characterized with respect to potential hazards. During the planning stage these situations should be reconciled with what is permissible and realistically achievable. There are usually two major guidelines that impact on plant design and hazard control: government regulations and economics. What is planned will need to be reconciled with what is permitted or feasible. [21]

Design, operations, and management personnel should be knowledgeable about government regulations. Hazard control measures are much simpler and less costly to include during design than after operations have begun.

CONTROL OF MATERIAL HAZARDS

Ideally the process should be an automated, closed system, with all hazards isolated from operating personnel. Not all industrial processes are adaptable to this approach. Multiple operations in the same work area can further complicate hazard control. An optimum arrangement, which involves a compromise, is usually the best long-term safe solution. Industrial hazards can often be minimized by process and equipment design that controls the hazard as much as is practical. Materials, products, and wastes need to be maintained within the closed process system. A process flow diagram should be developed, reviewed, and critiqued to assure desirable material balances.

The toxic materials that enter the work environment should be insignificant. Government regulations and guidelines usually define acceptable toxic levels with TLVs and PELs. Fire hazards are usually based upon the lower flammability limit and flash point for liquids and vapors. Equipment design techniques that can minimize the escape of hazardous materials into the environment include the following:

- Conduct the entire operation in an enclosed system.
- Carry out the entire operation automatically, with the operators in a remote location.
- Make the system relatively maintenance-free.
- Have the capability of removing hazardous residues without opening equipment to the environment.

In order to assure that workplace health hazards are controlled, sampling and analytical techniques should be developed during the design phase.

Easy and obvious solutions to problems during the design stage can become much more difficult and costly once the plant has been built. Equipment change and process compromise can be done readily with preliminary process flow diagrams, while the physical removal and replacement of equipment or its modification is expensive after installation and results in lost production.

Emissions and spills can result from many sources. Flanges, pumps, valves, joints, and fittings have seals, packing, and components that may leak. Fugitive emissions usually begin as low levels of contaminates that are not initially harmful to employees. Over time these fugitive emissions increase and accumulate in the workplace, resulting in harmful exposure. A well-planned, continuous maintenance program can mitigate leaks, but most of the leakage can be dealt with in the design stage.

Another source of air emissions of contaminates may occur when the process is upset or opened for maintenance or if safeguards fail to function properly. Collection of process samples and process parameter changes may also contribute to air pollutants. During design, if these situations can be anticipated, corrective measures can be built into the process.

During a plant shutdown or turnaround, process vessels and lines are often purged and cleaned prior to being opened for maintenance activities. While employee exposures may be high, they may be brief, making sampling and analysis imperative during these periods for employee protection.

Loading and unloading of barges, tank trucks, and tank railcars can create major toxicity and fire hazards unless adequate controls are provided. Vapor recovery or totally enclosed systems can be used to ensure that the vapor from the tank will be recovered. Proper grounding and bonding should be available to reduce the hazards of electrostatic electricity.

There are numerous workplace control methods for reducing or preventing exposure to hazardous materials. [21]

- Substitution of hazardous materials
- Modification of processes
- Isolation of processes or employees
- Local exhaust ventilation
- General dilution ventilation
- Airborne particulate removal
- Good housekeeping practices
- Medical controls
- Training in control methods
- Training in emergency response

Substitution of less toxic or nontoxic materials for highly toxic materials can be highly effective in hazard control. These changes can be in chemical composition, state of matter, or physical appearance. This can be a cost-effective and positive way of controlling material hazards. Manufacturing and marketing personnel may be reluctant to change the status quo in the workplace or with product specifications, but this can be overcome by improved safety, health, and environmental protection. Environmental, safety, and health concerns have forced a shift away from organic solvents to aqueous formulations. [16] Caution should be exercised in substitution of materials to ensure that new and unforeseen hazards are not introduced into the workplace.

A modification of a process can improve product quality, reduce operating costs, and protect employee safety and health. There are many ways to modify a process: changing equipment, adapting a new technology, using a different technique, automating the process. Often a recurring operations problem can lead to new operational practices and remedial action to achieve improved maintenance practices. [8]

Hazardous operations should be isolated as much as is practical to protect employees' safety and health. Process isolation may be a physical barrier like a wall, separate room, shield, enclosure, booth, or hood. Another form of materials isolation involves removal or recycling of hazardous material from the site prior to decommissioning and demolishing a unit or facility. [20]

If it is not practical to isolate operations from employees, it may be more feasible to isolate employees from the process. A particularly hazardous operation might be conducted when the least number of employees will be exposed. Control rooms that are enclosed and remote are used to monitor and control hazardous processes. Employees can be rotated to various locations during a shift to achieve a safe Time Weighted Average (TWA) over the entire workday.

Local exhaust ventilation captures airborne contaminates at their source before they escape into the general work area. A local exhaust system usually has a hood, ducting, an air cleaner, and an exhaust fan. Local exhaust systems require less airflow than general dilution ventilation systems. The process or equipment is confined and sufficient airflow is withdrawn to assure the airflow into the hood will entrain the contaminate into the airstream. The physical state of the contaminate, the process temperature and pressure, the ambient temperature, the method of contaminate generation, the toxicity and flammability of the contaminate, and the type and velocity of the release into the surroundings are the primary constraints to consider.

Hoods for local exhaust systems must be designed to capture the air contaminates. The fans and ducts must draw the proper amount of air through the hood. For circular or square duct openings, the air velocity for contaminate control is given by this equation. [17]

$$V = \frac{0.1Q}{X^2 + 0.1A} \qquad (12\text{-}1)$$

where: V = centerline air velocity at X distance from the hood, m/min
Q = airflow volume entering duct, m³/min
X = distance outward along hood axis, m
A = hood opening area, m²

Once a local exhaust ventilation system has begun operation, the performance of airflow rates and velocities should be verified and then periodically checked as a maintenance procedure.

General dilution ventilation changes air in the workplace to dilute air contaminates below acceptable levels. Exhaust fans are usually mounted in walls, roofs, or windows, with outside air entering through windows, open doors, or roof ventilators. It can be difficult to control employee exposures near the source of the contaminate because sufficient dilution may not have been achieved. The makeup air to replace exhaust air should be clean and comparable in temperature. The makeup air should be obtained from a location that does not have significant concentration of toxic gases and vapors that could be drawn back into the work area from the outside. Recirculation from exhaust vents or emissions stacks should be avoided. General ventilation should be avoided whenever major localized air contaminates are present because local exhaust ventilation should be used instead.

The removal of airborne particulates can be accomplished by a variety of wet methods, using water. Spraying, scrubbing, and wetting agents can be useful in reducing dust and other airborne particulates.

Spills should be cleaned up immediately. Dirt and dust should not be allowed to accumulate in the workplace. Repair or replace leaking containers or valves as soon as they are detected. All dirty rags or absorbents should be placed in an appropriate receptacle. Failure to practice good housekeeping can result in the creation of hazardous waste that requires costly disposal techniques and additional safety hazards. [31] Environmental protection should be an integral part of the development and operation of all operating facilities. [4]

Employees should learn to keep themselves and the workplace clean and orderly. Neatness, cleanliness, and order have a significant effect on the way the human mind works. If the employees and the workplace are clean and orderly, there is a good chance that workplace activities will be done well. [14] Personal hygiene is important to workers, who should have easy access to hand-washing facilities, emergency showers, and eyewash stations. Eating, drinking, or storage of food and beverages should be forbidden whenever toxic materials are present. Change rooms should be provided for employees to change into work clothes and special protective clothing. All work areas that contain hazards should be posted with appropriate warning signs.

Medical controls, which are a form of management control, serve a number of useful purposes in the protection of employees. A new-employee physical exam will establish a baseline of health and determine historical exposure to hazards. Later periodic physical exams should be compared to the baseline to detect any adverse effects that might be attributed to workplace hazards and their severity. Routine sampling of blood and urine are often used to identify potentially harmful exposures to chemical hazards.

A well-managed medical program will verify existing management and engineering controls or help identify hazard problems. Should an employee have symptoms of known hazard exposure, this will indicate a control or system failure that needs to be corrected.

Training and continuing education should be used to supplement and enhance management and engineering controls. Some of the training and education requirements are mandated by government regulations. Other training may be based on management policy or used to reinforce the government-mandated programs. In most process facilities the primary management safety responsibility belongs to the line management of operations, for example, the first-line supervisor, the shift supervisor, and the operations manager. These personnel are familiar with day-to-day plant operation. The main objective of training and education of supervisors is to teach them about safety and health hazards in their work areas. This includes hazardous materials, processes, and operating facilities. The cost-benefit relationship of hazard control should be included in supervisor training.

Nonsupervisory employees need education and training to effectively use safety, health, and environmental control measures. This includes both management and engineering controls. Detailed instruction manuals should outline proper procedures for all foreseeable situations. Safety meetings, booklets, postings on bulletin boards, and continuing education courses are useful training tools. All employees need to understand the reasons behind the implementation of management and engineering controls. Employees need to know proper operating practices if controls are to perform in accordance with their design and function.

Emergency response procedures also require appropriate training and education to make them effective. Fire control systems are intended to provide an early response to incipient fires. There is a need to train personnel in fire fighting, to have routine drills, and to establish procedures for safe evacuations. [23]

Some safety hazards are common throughout all industrial operations. Two of these, illumination and ergonomics, will be discussed in this chapter because of their importance to all aspects of safety, health, and environmental protection.

ILLUMINATION

Workplace hazards can be increased or reduced by the conditions of the work environment. Illumination is an example of a work environment factor that can have a tremendous influence on productivity, safety, health, and economics. Design engineers must provide adequate illumination in both office and operational settings to enhance the working conditions for personnel. Good lighting ensures safety, facilitates the performance of tasks, and creates an appropriate visual environment. It is more difficult to determine precise standards for lighting than for other safety areas because it is nearly impossible to be specific about the risks and the exact protection levels required.

Our sense of sight is extremely important to performance in and out of the workplace. Accidents can be caused by failure to see or failure to understand what is seen. Illumination levels must be adequate to provide a positive influence to job performance and employee morale. There are many factors to be considered in providing adequate lighting.

Visual needs of the task
Time duration of the work
Visual capabilities of the personnel
Potential visual hazards

Probability of errors
Consequences of the errors

Many factors can cause variations in illuminance at a workstation over time. Throughout the daylight hours the position of the sun can affect the natural lighting that is available. Most workplaces are used after normal daylight hours, making artificial lighting necessary to meet appropriate standards and management practices.

Other factors will cause variations in illuminance over time. These additional factors are all related to maintenance programs.

The light output from a bulb provides maximum lumens when the bulb is new. As the light is used, bulb performance deteriorates and the light output decreases. Tungsten lamps do not experience much light output reduction during their lifetime. Lumens from a fluorescent lamp may decrease to 90 percent of the initial output after 2000 hours of use. Scheduled bulb replacement is desirable.

Another factor that reduces light output from a luminary is the accumulation of dust and dirt on the reflecting or diffusion surfaces. Regular cleaning will restore this situation to like-new condition.

Reflection from other interior surfaces like ceilings, walls and floors reduces light output. As these surfaces become dirty, particularly if they were originally white, they lose their reflective output.

Sometimes accidents can be attributed to poor lighting. Many other accidents have resulted from glare, shadows, and excessive visual fatigue. Seemingly careless acts may in reality be the result of difficulty in seeing or understanding a hazardous situation as it really exists.

A well-conceived lighting design should create conditions that are comfortable to the eye so that all relevant information can be received through the sense of sight without introducing additional hazards to the workplace. Good lighting should enhance the accurate performance of tasks while nurturing a careful and responsible work attitude.

Illuminance meters have a photocell, a measuring circuit, and a display, which may be analogue or digital. The transfer of energy in the photosensitive material causes an electric current to flow in the measuring circuit so that it is displayed on the calibrated meter. The photocell sensitivity to light wavelengths is matched to the sensitivity of the eye. There are a wide variety of meters on the market to cover every conceivable measurement range. Care should be taken to protect meters from direct sunlight, which could cause damage.

The photocells used in illuminance meters should be cosine-corrected with a diffusing plate to ensure an accurate reading on the cell at all possible angles. Otherwise, the light reading may be understated. Photocells should be color-corrected with a special filter to match the color sensitivity of the meter with the capabilities of the human eye.

The meter should be calibrated at least annually at a photometric laboratory. Adjustments to the measuring circuits may be necessary to ensure accuracy.

To conduct a light survey, the meter should be zeroed while covered to exclude

TABLE 12-3
WORKPLACE DESIGN FACTORS TO USE DAYLIGHT FOR EFFECTIVE ILLUMINATION

Geographical location
Orientation of the workplace
Effect of topography, landscaping, and nearby structures
Sunlight intensity
Incidental daylight intensity
Lighting brightness based upon climate

incidental light. The surface of the photocell should be cleaned. For daylight readings, switch off artificial light. For artificial light readings, wait until after daylight hours. The survey data should be on a to-scale sketch of the area. To measure the illuminance over an area, divide the area into grids and measure the center of each grid square.

To determine the illuminance level at a workstation, place the meter at the point where the visual task occurs. Avoid your own shadow.

In the workplace, daylight should be used to take advantage of energy savings and to complement electrical lighting fixtures. The factors that should be considered for optimal use of daylight are shown in Table 12-3. [3] Minor adjustments in the layout of facilities may achieve significant energy savings during daylight hours through well-conceived and planned designs.

A well-conceived and planned office design can facilitate good lighting. Employees who need light for visual tasks should be placed near windows. Employees should not face windows because of glare. Walls and ceilings should conserve light by reflection to work areas. Lighting should be designed to reduce stress and eye fatigue.

The main objective of industrial lighting is to provide a safe working environment and comfortable vision to support operations. Workers need to implement, observe, and control equipment and processes. The electric lighting system must be designed and installed to assure good quality lighting over the entire work area. The types of industrial electric lighting are shown in Table 12-4. [3]

TABLE 12-4
TYPES OF INDUSTRIAL LIGHTING

General lighting	Relatively uniform illumination throughout the entire area
Localized lighting	Relatively uniform illumination throughout the work zones
Supplementary lighting	High illumination for small or restricted areas
Emergency lighting	Illumination for power outages to egress the building or area via exits and stairways

The Minimum to average illuminance ratio of general lighting is ≥0.8. The general area illuminance should be ≥0.33 of the average of the task areas. Adjacent interior areas should not vary in average illuminance by more than a 5-to-1 ratio.

Emergency lighting should have a power source that is independent of the normal power supply. The auxiliary power system should automatically activate when normal power fails. Emergency lighting systems are usually powered by diesel generators or batteries.

The quality of illumination is based upon the brightness distribution in the visual environment. The ability to see easily and accurately may be affected by brightness, glare, color, uniformity, diffusion, and direction. Poor-quality lighting can make the workplace uncomfortable and possibly hazardous. Even modest variations in quality, which may not be readily detected, can increase eye fatigue and reduce productivity.

The ease, speed, and accuracy of performing a task increase with an increase in the illumination of the task. The optimal illumination quantity depends upon the task.

Light flux, the flow of luminous energy, must be measured over the area of the affected surface. Illuminance is the light flux on a surface-per-unit area of that surface.

$$\text{Illuminance} = \frac{\text{light flux}}{\text{unit area}}$$

(12-2)

The quantity of illumination or illuminance is expressed in foot-candles (fc) or meter-candles (lux). A foot-candle is the illumination at a distance of 1 foot of 1 lumen per square foot. A meter-candle is the distance of 1 meter for 1 lumen per square meter. There are 10.76 lux per foot-candle. Typical illuminance levels for common indoor and outdoor lighting are shown in Table 12-5. Some typical industrial guideline levels of illumination are shown in Table 12-6. [3]

Visible light is a form of electromagnetic energy which is distinguished by its wavelength, as shown in Figure 12-5. The visible spectrum, 380 to 720 nm, is the small band of luminous energy that the eye responds to. The different wavelengths within this visible spectrum allow the eye to see the different colors (Fig. 12-6 on

TABLE 12-5
TYPICAL ILLUMINANCE LEVELS FOR INDOOR AND OUTDOOR LIGHTING

Lighting	Illumination level (fc)
Moonlight	0.1
Street lighting	0.5
Stairs	30
Office	50
Drafting room	75

TABLE 12-6
ILLUMINATION GUIDELINES FOR THE WORKPLACE

Workplace area	Foot-candles	Lux
General construction	10	108
Freight car interiors	10	108
Loading platforms	20	215
Freight elevators	20	215
Shipyards	30	323
Waste treatment	30	323
Rubber preparation	30	323
Rubber calendaring	50	538
Warehouses	10–50	108–538
Control rooms	75	807
Instrument shop	150	1614
Electrical shop	150	1614

page 374). As an example, wavelengths of 400, 540, 620, and 680 nm appear as violet, green, orange, and red, respectively. The sensitivity of the eye to the visible spectrum is greater in the middle range (yellow-green) than at the extremes (violet or red), as shown in Figure 12-6. For a given light flux, a yellow light would appear much brighter than violet or red.

The ability of a surface to reflect light is dependent upon the color of the surface and, to a lesser extent, on surface texture. The proportion of the light flux reaching the surface, that is reflected back into space, is called reflectance, which is expressed as a percentage. For example, the surface reflectance of white, gray, and black is about 90, 65, and 5 percent, respectively.

Surface texture has only a minor effect on the percentage of light reflected. On the other hand, surface texture determines how the reflected light is distributed. Polished surfaces change in appearance when viewed from different angles. Matte surfaces have a more uniform appearance from all viewing directions.

FIGURE 12-5
THE ELECTROMAGNETIC SPECTRUM.

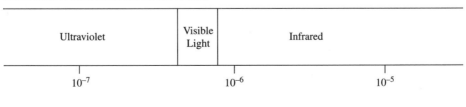

Wavelength (Meters)

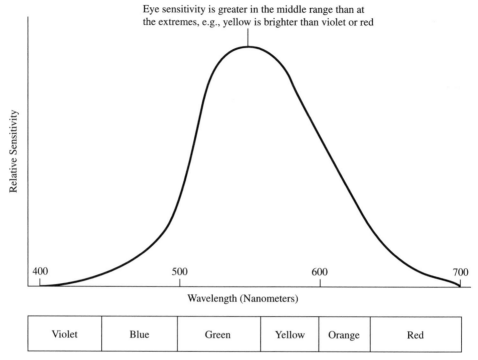

FIGURE 12-6
RELATIVE SENSITIVITY TO THE VISIBLE SPECTRUM.

Most accidents that are caused by poor illumination are avoidable with proper planning and design of electrical lighting and the use of daylight illumination. Other factors, such as glare and shadows, reduce vision, increase visual fatigue, and may lead to accidents. Regulatory safety colors may appear different in sodium or mercury lighting than in normal electrical sources.

The security system at a facility should include protective lighting. This allows night policing of outdoor areas by discouraging potential intruders. It involves lighting border areas of buildings to eliminate concealing shadows. Streetlights and floodlights are examples of protective lighting.

In office settings, the Illuminating Engineering Society recommends the lighting intensities shown in Table 12-7. [2] Adequate lighting is important to the health, safety, and morale of office personnel because this is their normal work environment. OSHA-recommended minimum-lighting intensities for hazardous waste operations are shown in Table 12-8. [28] These government regulations have been deemed to be adequate for these operating facilities. The contrast between Tables 12-7 and 12-8 is significant. The lower minimum-lighting intensities in waste operations may generally be considered for operating facilities. Control rooms at these operating facilities are similar to office settings and should be illuminated based upon the more demanding requirements of Table 12-7.

TABLE 12-7
RECOMMENDED ILLUMINATION LEVELS FOR OFFICES

Office area	Illumination Intensity (fc)
Stairways, corridors	20
Washrooms	30
Normal office work	100
Accounting	150
Detailed drafting	200

OSHA has other illumination requirements for certain workplace areas (Table 12-9 on page 376). [29] OSHA requirements vary in complexity. Some are quantitative, while others are much more qualitative and subject to interpretation. This demonstrates an important difference between regulations and adapting to the situation. Regulations of a government body must be adhered to precisely. The safety policy of an industry may be more flexible, depending upon the workplace situation and management decisions.

ERGONOMICS

Ergonomics studies the physical and behavioral interaction between people and their environments both inside and outside the workplace. It is also called human factors engineering. When there is an ergonomics mismatch, injury levels increase, production becomes inefficient, and everyone suffers. [3] The failure to apply ergonomics effectively in the workplace may result in the following losses. [22]

Lower production quantity and quality
Increased absenteeism and labor turnover
Increased accidents and injuries
Increased medical and material costs
Reduced ability to deal with emergencies

TABLE 12-8
OSHA-RECOMMENDED MINIMUM ILLUMINATION LEVELS FOR HAZARDOUS WASTE OPERATIONS

Operation area	Illumination level (fc)
General site	5
Warehouses, corridors	5
General shops	10
Medical facilities	30

TABLE 12-9
OSHA ILLUMINATION REQUIREMENTS FOR CERTAIN WORKPLACE SITUATIONS

Workplace requirement	OSHA illumination requirement
Exit sign	$\geq$5 fc on the surface
Labor camps	$\geq$20 fc at 30 in from the floor for toilet and storage rooms
	$\geq$30 fc at 30 in from the floor for kitchen and living quarters
Electrical equipment	Provide adequate illumination for all work areas around service equipment, switchboards, panel boards, and motor control centers.
Trucks	$\leq$2 lm/ft^2 in an area; requires directional truck lighting

The U.S. Department of Labor compiles injury and illness incidence rates for the private sector of the economy based upon the following relationship:

$$\text{Incidence rate} = \frac{\text{(number of cases) (200,000)}}{\text{total employee hours during period}}$$

also

$$\text{Incidence rate} = \frac{\text{(number of lost work days) (200,000)}}{\text{total employee hours during period}} \qquad (12\text{-}3)$$

The 200,000 constant is based on 100 full-time workers at 40 hours per week and 50 weeks per year.

The 1991 private-sector incidence rates were 8.4 for total cases and 3.9 for lost workdays. Lost workdays are defined as days the employees are away from work or limited to restricted work activity because of occupational injury or illness. Total cases are all work-related deaths and illnesses resulting in loss of consciousness, restriction of work, job transfer, or medical treatment. [1]

The objective of a sound ergonomics program is to develop a system where the workplace, work methods, equipment, and surroundings are compatible with the physical and behavioral limitations of the employees. The better the fit of the program with the workplace, the better the safety and efficiency in that workplace.

Each employee has human attributes that relate to his or her ability to perform any given task. Physical characteristics include height, weight, and other anthropometric data. Workers also have previously developed work habits and methods that influence their approach to performing a task.

The other half of the ergonomics program is the workstation, equipment, and work objectives that the employee must deal with. By matching what the employee brings to the task with what the task brings to the employee, a sound ergonomics program should result.

A survey to analyze the work environment should precede the development of an ergonomics program. This analysis should include the physical and nonphysical aspects of the working tasks. Physical work (handling loads, manipulating equipment) usually strains muscles and the human skeleton. Nonphysical work involves the human senses and mental capabilities in obtaining and using information to deal with work situations.

Whenever a new process, piece of equipment, or modification is to occur in the workplace, ergonomics should be an integral part of the planning. Existing operations should also be studied periodically to determine the presence of adverse trends (Table 12-10).

Analysis of ergonomics for new processes, equipment, or production modifications needs to be done during the design phase. Often cumulative trauma disorder of the hand and wrist only is apparent from the long-term exposure to repetitive tasks. The following factors should be used to evaluate these situations:

Production flow and rates
Tools and equipment required
Movement and handling by personnel
Personnel attributes and physical limitations
Work environment dimensions

By performing an analysis with these factors the workplace and surroundings should be designed to achieve a safe and efficient operation. The capabilities of the employee must adapt to the work environment and allow correct body posture at the workstation. The work surface chair height, location of materials, supplies, and tools must all be compatible with the needs of the employee.

Periodic review of accident and injury records is desirable to identify problems related to manual materials handling or cumulative trauma disorders (CTDs). CTDs are the leading cause of occupational illness in the United States. [3] Carpal tunnel syndrome, tendinitis, and tennis elbow are among the more common forms of CTDs. The major risk factors for CTDs include high levels of repetitive movement, vibration, or hand force; awkward work positions; and exposure to temperature extremes.

TABLE 12-10
INDICATORS OF POTENTIAL ERGONOMIC PROBLEMS IN THE WORKPLACE

Increase in accidents or injuries

Absenteeism

Turnover of employees

Employee complaints

Poor product quality

Customer complaints

Increase in overtime

Presence of cumulative trauma disorder

Absenteeism and high turnover of employees often are the result of excessive physical or mental stress. When employees or customers complain about the workplace or product quality, management can gain valuable insight into possible ergonomic problems. Overtime and greatly increased production rates may cause employee fatigue and result in errors or mistakes in judgment.

NIOSH has noted a significant increase in musculoskeletal injuries when workers are lifting heavy or bulky objects, when workers are lifting objects from the floor, or when lifting is a frequent part of the job. NIOSH has developed an equation to determine the acceptability of a lift. It applies to a two-hand symmetrical lift in front of the body with no trunk twisting.

$$\text{AL} = 40 \left(\frac{15}{H}\right) \left(1 - 0.004|V - 75|\right) \left(0.7 + \frac{7.5}{D}\right) \left(1 - \frac{F}{F_{max}}\right) \tag{12-4}$$

$$\text{MPL} = 3\,(AL) \tag{12-5}$$

where: AL = action limit, maximum acceptable nominal lifting risk, kg
   MPL = maximum permissible limit, minimum for hazardous lifting conditions, kg
   H = horizontal distance at the start of the lift between the center of the hand grasp and the midpoint between the ankles, cm
   V = vertical distance at the start of the lift between the hand grasp and the floor, cm
   D = vertical travel distance between the origin and destination of the lift, ≥ 25 cm
   F = Lifts per minute
   F_{max} = maximum lift frequency, based upon V and lifts per hour

The values for the maximum lift frequency for stooped lifts ($V = \leq 75$ cm) for 1-h and 8-h time periods are 15 and 12, respectively. For standing lifts ($V = >75$ cm) of 1-h and 8-h time periods, the F_{max} values are 18 and 15, respectively.

Lower-back injuries from overexertion are a major occupational safety problem. Many workers' compensation claims are based on lower-back injuries, which can be difficult to verify for insurance purposes. It is important to reduce or eliminate the physical stresses of lifting in the workplace. This can be done by calculating the action limit and the maximum permissible limit, using equations (12-4) and 12-5). The AL and MPL can then be compared to NIOSH guidelines to determine the need for management and engineering controls (Fig. 12-7).

At the action limit, at least 99 percent of the male workers and 75 percent of all female workers will be able to perform the lifting task safety. Below the AL, the lifting task is considered acceptable. Above the AL, management controls like employee selection and training should be applied in addition to engineering controls. At the maximum permissible limit, about one-fourth of men and less than 1 percent of women can safely perform the lift. Above the MPL, engineering controls are required to safely perform the lift.

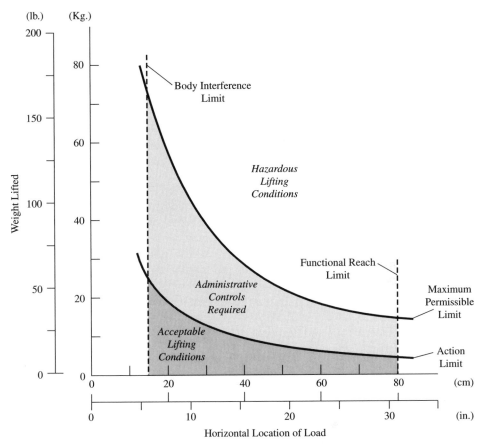

FIGURE 12-7
ACCEPTABLE LIFTING LIMITS.
(*Source:* Used by permission of the National Safety Council, Itasca, Illinois.)

When body movements are used to exert pressure on objects, muscles contract or extend to perform the work. In dynamic muscular effort like walking, the muscles help circulate blood, which furnishes nutrients and oxygen. As a result, walking is an exercise that can be done for extended periods of time with ease.

When static work positions are maintained for a prolonged period of time, the muscles are in a state of contraction, which restricts the flow of blood. This deprives the muscle cells of vital nutrients and oxygen, as well as allowing waste products to accumulate in the muscles. The resulting muscle fatigue is painful, and static muscular efforts cannot be maintained for extended periods of time. Long-term, these static positions can cause deterioration in joints, ligaments, and tendons.

The use of vibrating tools can cause numbness, white fingers, and loss of dexterity in fingers and hands. Hand-arm vibration syndrome is a chronic progressive disorder that may go undetected for months or even years. NIOSH recommends medical monitoring of all workers exposed to vibration to identify symptoms as they occur.

Carpal tunnel syndrome impairs the median wrist nerve, causing numbness, tingling pain, and loss of control in the hands. The carpal tunnel is inside the wrist. When the wrist is bent, nerve and finger flexor tendons in the carpal tunnel rub on the tunnel surfaces. This rubbing thickens the membranes surrounding and lubricating the finger flexor tendons and compresses the median nerve, resulting in the symptoms. These types of CTDs of the upper extremity can be anticipated by a job performance study of the risk factors in Table 12-11. If any of these risk factors are present during the performance of tasks in the workplace, a potential safety and health problem may exist.

The biomechanics of the musculoskeletal system during workplace activities involves the physical, biological, engineering, and behavioral sciences. This is an interdisciplinary field that should match employee capacities to the task requirements. Such an occupational biomechanical study should evaluate job stress to quantify the degree of risk. This study should then be the basis for the design of a new task or the redesign of an existing task. The biomechanical modeling of lower back and upper extremities is important to understanding workplace situations that may create CTDs, as shown in Table 12-12.

The workplace layout and equipment design should be based upon anthropometric information. Employees consist of men and women in all sizes and shapes. There is no such thing as an "average employee" in ergonomics. The workplace should be designed and operated so as to accommodate most of the user population. This means accommodating characteristic differences like body weight, dimensions, and mobility. The workplace must be safe, comfortable, and efficient for the employees.

Body dimensions and weights for adults vary considerably, as shown in Table 12-13 on pages 382–383 and Figure 12-8 on pages 384–385. [3] This anthropometric information will also vary with age and clothing. A well-conceived and designed work environment should cover up to 90 percent or more of the adult work-

TABLE 12-11
RISK FACTORS FOR UPPER-EXTREMITY CUMULATIVE TRAUMA DISORDERS

Stressful hand and arm postures during the work cycle, such as wrist or fingers bending, pinched fingers, wrist or arm twisting, and overhead reaching

Hand tools and vibrating tools that cause stressful conditions

Forces of exertion to hold or operate hand and vibrating tools

Frequency and duration of stressful conditions

Repetitive arm, hand, and leg motions

Extended hand and arm reaches

Static arm, hand, and leg work

Restriction of movement because of clothing and personal protective equipment

---

TABLE 12-12
THE APPLICATION OF WORKPLACE BIOMECHANICAL STUDIES TO CTDS

Evaluation of present workplace conditions

Improvement of workplace conditions

Guidelines for lifting, static, and vibration work

Development of workplace layout

Criteria for employee selection

Requirements for employee training

Ongoing evaluation and feedback of the program effectiveness

ers selected for the tasks to be performed. This usually means making the workplace reasonably flexible and adjustable.

Many employees will work daily in a relatively small work area, which should be well planned, designed, and operated. Productivity, efficiency, health, and safety will be directly related to this work area.

Seated workers should have workstations with the following components:

- A chair that is comfortable and compatible with the job performance
- Easy access to tools, supplies, and equipment in the area
- All work surface items easily handled on or just above the work surface
- No heavy lifting required from a seated position

Standing workers should have workstations that take into account the following:

- The standing surface and shoe soles should be cushioned.
- Work height should be 5 to 10 cm below the hanging elbow of the employee.
- Heavy work should have low work surfaces; light work should have higher work surfaces.
- Tools, supplies, and equipment should be stored slightly below elbow height.
- Heavy objects that need to be pushed should weigh no more than 8 to 16 kg, depending on the sex of the employee.
- Some head movement is desirable.

Workers who alternate between sitting and standing positions will experience less fatigue.

Many employees depend on indicators or instrument readings to control operations in the workplace. The information from these instruments is important to worker safety and operation efficiency. The proper location of controls and displays, ease of maintenance, and correct working heights should be considered in equipment and instrumentation design, purchase, and installation.

Instrumentation displays can be visual or auditory: a dial gauge, a digital readout, or an audio alarm. Whatever the method, the message to the employee must be clear and understandable. Confusing information or format can lead to employee errors, which may be catastrophic in the workplace.

Visual displays usually provide quantitative or qualitative readings or identify

TABLE 12-13
ANTHROPOMETRIC DIMENSIONS FOR ADULTS

Measurement (in)	Males		Females	
	Median	± σ	Median	± σ
Standing				
1 Forward functional reach				
a Includes body depth at shoulder	32.5	1.9	29.2	1.5
b Acromial process to functional pinch	26.9	1.7	24.6	1.3
c Abdominal extension to functional pinch	24.4	3.5	23.8	2.6
2 Abdominal extension depth	9.1	0.8	8.2	0.8
3 Waist height	41.9	2.1	40.0	2.0
4 Tibial height	17.9	1.1	16.5	0.9
5 Knuckle height	29.7	1.6	28.0	1.6
6 Elbow height	43.5	1.8	40.4	1.4
7 Shoulder height	56.6	2.4	51.9	2.7
8 Eye height	64.7	2.4	59.6	2.2
9 Stature	68.7	2.6	63.8	2.4
10 Functional overhead reach	82.5	3.3	78.4	3.4
Seated				
11 Thigh clearance height	5.8	0.6	4.9	0.5
12 Elbow rest height	9.5	1.3	9.1	1.2
13 Midshoulder height	24.5	1.2	22.8	1.0
14 Eye height	31.0	1.4	29.0	1.2
15 Sitting height, normal	34.1	1.5	32.2	1.6
16 Functional overhead reach	50.6	3.3	47.2	2.6
17 Knee height	21.3	1.1	20.1	1.0
18 Popliteal height	17.2	1.0	16.2	0.7
19 Leg length	41.4	1.9	39.6	1.7
20 Upper-leg length	23.4	1.1	22.6	1.0
21 Buttocks-to-popliteal length	19.2	1.0	18.9	1.2
22 Elbow-to-fist length	14.2	0.9	12.7	1.1
23 Upper-arm length	14.5	0.7	13.4	0.4
24 Shoulder breadth	17.9	0.8	15.4	0.8
25 Hip breadth	14.0	0.9	15.0	1.0
Foot				
26 Foot length	10.5	0.5	9.5	0.4
27 Foot breadth	3.9	0.2	3.5	0.2
Hand				
28 Hand thickness, metacarpal III	1.3	0.1	1.1	0.1
29 Hand length	7.5	0.4	7.2	0.4

TABLE 12-13
ANTHROPOMETRIC DIMENSIONS FOR ADULTS *(CONTINUED)*

Measurement (in)	Males Median	± σ	Females Median	± σ
Hand (cont.)				
30 Digit two length	3.0	0.3	2.7	0.3
31 Hand breadth	3.4	0.2	3.0	0.2
32 Digit one length	5.0	0.4	4.4	0.4
33 Breadth of digit one Interphalangeal joint	0.9	0.05	0.8	0.05
34 Breadth of digit three Interphalangeal joint	0.7	0.05	0.6	0.04
35 Grip breadth, inside diameter	1.9	0.2	1.7	0.1
36 Hand spread, digit one to two, 1st phalangeal joint	4.9	0.9	3.9	0.7
37 Hand spread, digit one to two, 2d phalangeal joint	4.1	0.7	3.2	0.7
Head				
38 Head breadth	6.0	0.2	5.7	0.2
39 Interpupillary breadth	2.4	0.2	2.3	0.2
40 Biocular breadth	3.6	0.2	3.6	0.2
Other Measurements				
41 Flexion-extension, range of motion of wrist, degrees	134	19	141	15
42 Ulnar-radial range of motion of wrist, degrees	60	13	67	14
43 Weight, in pounds	183.4	33.2	146.3	30.7

Source: National Safety Council.

on-off positions. Simple designs are usually better because there is less likelihood of confusion. The display should increase with the same motion as the control mechanism. Displays should be grouped by function or sequence in the operation. They must be clearly labeled to avoid confusion, particularly during an emergency. Quantitative readings are better done with digital displays for accuracy. Qualitative readings are better suited to a moving graph or printer to more easily detect trends.

Audio displays should be used for immediate action. They should have a simple, short message or alarm to get the attention of employees. The audio intensity should be well above the background level to be easily recognized in the workplace.

All displays are used by employees to control the operations of the affected system. The controls must be compatible with the capabilities of the operator. Con-

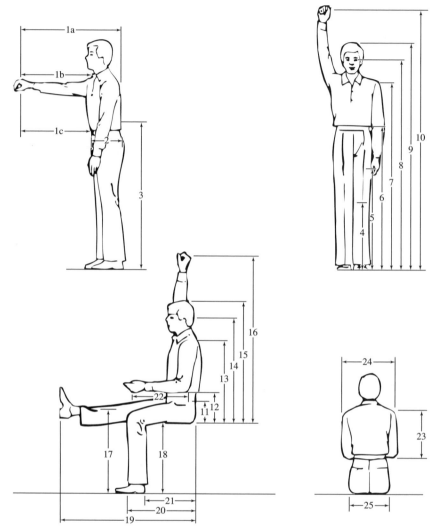

FIGURE 12-8
ANTHROPOMETRIC ILLUSTRATIONS FOR THE DIMENSIONS IN TABLE 12-13.
(*Source:* Used by permission of the National Safety Council, Itasca, Illinois.)

trols should operate and function as expected. For example, lights should turn on when the switch is up, valves should open counterclockwise, and indicators should increase to the right or with a clockwise movement. When these movements or situations are reversed, the opposite is expected. When an employee misreads a poorly designed gauge or misuses a control, the safety of the workplace is placed in jeopardy. The probability of this situation developing can be greatly reduced by coding the controls by color, shape, or location. Controls should be convenient and accurate and have a rapid response. The use frequency and sequence of controls is

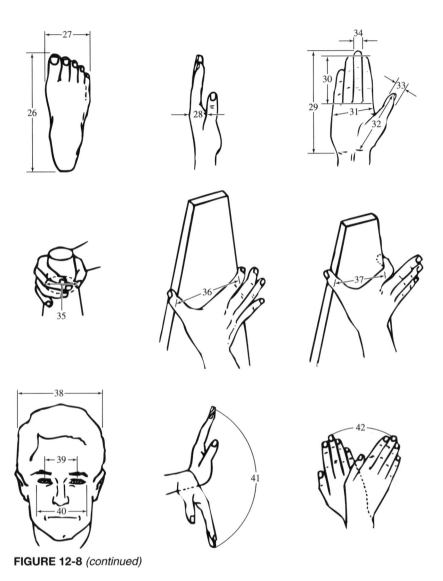

FIGURE 12-8 *(continued)*

usually the first consideration. Control sequences should be grouped together to avoid the need for significant physical movement of operators during the manipulation of the controls.

The top management of the organization must be committed to the ergonomics program for it to be successful. This management commitment is needed to provide the appropriate resources and motivation for the program to succeed. Participative management styles are highly desirable to identify and mitigate potential hazards. Feedback is particularly helpful in achieving the desired results.

FIRST AID

First aid is the immediate medical care given to a person who has unexpectedly fallen ill or been injured. First aid is often administered in emergencies by non-healthcare providers. Knowledge and implementation of appropriate first aid may make the difference between life and death, temporary or permanent disability, and short- or long-term hospitalization. Establishing a plan to manage emergencies is crucial for the health and well-being of employees. In addition, employee first aid knowledge benefits family members and the community at large. [32]

According to the American Red Cross, more than 2 million people are hospitalized annually because of accidental injuries, many of which are work-related. The most common causes of work-related injuries are listed in Table 12-14. This section explains first aid for each of these injuries, with the exception of carpal tunnel syndrome, an injury resulting from chronic stress of the wrist that takes weeks or even years to develop. Multiple injuries are discussed in relation to cardiopulmonary resuscitation and the bystander role. [5]

According to the United States Department of Labor, Bureau of Labor Statistics, work-days lost because of injuries and illnesses are greater for certain occupations than others. The most hazardous male-dominated occupations are in construction and transportation; for women, nursing and housekeeping services predominate. Injuries are also high on assembly-lines, where men and women are equally employed.

OSHA Requirements

First aid for employees who become sick or are injured on the job comes under the OSHA regulations in 29 CFR 1910.151. Such first aid may consist of attention to

TABLE 12-14
NUMBER OF NONFATAL OCCUPATIONAL
INJURIES AND ILLNESSES

Nature of Illness/Injury	Private Industry (thousands)
Sprains, strains	1022.7
Bruises, contusions	222.7
Cuts, lacerations	173.6
Fractures	143.6
Heat burns	41.0
Carpal tunnel syndrome	33.0
Chemical burns	15.7
Amputations	12.4
Multiple injuries	67.5
All other	598.8

Source: Bureau of Labor Statistics, U.S. Department of Labor, 1992.

minor injuries that require minimal treatment up to and including emergency aid for major injuries until professional medical personnel arrive. After initial evaluation of an accidental injury, appropriate provisions for on-site care can be made with the advice of a physician. [18]

OSHA requires that medical personnel be readily available for consultation regarding matters of workplace safety and health. In situations where the workplace is not within a reasonable distance of a medical facility, OSHA also requires that an appropriate number of personnel be trained in first aid procedures and that approved first aid supplies be available on-site. In locations where the potential exists for eyes or skin to contact corrosive materials, OSHA requires that eyewash stations and safety showers be available for immediate emergency care.

Sprains and Strains

Sprains and strains are the most frequent job-related injury. Back injuries account for about one-fifth of all sprains and strains. Typically the symptoms of back injuries (i.e., back pain, leg numbness, and muscle weakness) occur 2 to 3 days after the initial injury. At this time, employees usually are forced to stay home and not report for work; therefore, on-the-job first aid is not an issue. However, all employees should be taught methods to *prevent* back injuries, of which proper body mechanics is essential. The Red Cross believes that 90 percent of all back injuries are preventable.

A *sprain* is the partial or complete overstretching or tearing of relatively long bands of tissue, called ligaments, which connect bones to bones at a joint. Sprains of the ankle, knee, wrist, and finger joints are common. A *strain* is the overstretching or tearing of muscles or the ends of muscles, known as tendons, which connect muscles to bones. Strains typically occur at soft tissues between joints. Strains are often caused by lifting heavy objects and usually involve the muscles of the neck, back, thigh, or lower leg. [26]

First aid for sprains or strains can easily be remembered with the acronym RICE: *R*est, *I*mmobilize, *C*ool, and *E*levate. (This acronym is also helpful in providing first aid to persons with fractures, described later in this chapter.)

Rest the affected area (usually an extremity). Have the person sit or lie down or stop performing the task at hand. Suggest that no further pressure be placed on the affected area.

Immobilization of the affected area can be accomplished in a variety of ways. One way is to apply a rigid or soft apparatus to the extremity. Tape a board, magazines, folded newspapers, bulky blanket, or pillow around the extremity so that the sprained joint can move only slightly. An anatomical splint may also be used: if a leg or ankle is involved, bind the person's legs together; if an upper extremity is involved, bind the arm to the chest. Splinting should be done only if the injured person must be moved and only if it does not cause more pain and discomfort to the person. Splint the injured part *in the position in which you found it.* Splint the injury well above and below the affected area. Remember that the ground may also be used as a splint if it successfully keeps the injured part from moving.

Many materials can serve as splints, and any number of materials can be used as binders: tape, belts, torn pieces of cloth, clothesline, etc. Ace bandages can also be applied. If a person sprains an ankle while wearing a shoe, it is best to leave the shoe on. This helps prevent swelling by compressing on the torn small blood vessels (capillaries) and minimizes bleeding into the soft tissue around the ankle.

Two major steps to remember when applying any bandage that encircles an extremity are to begin wrapping at the furthest end of the extremity and to exert less pressure with the wrap as one goes up the extremity toward the body. Leave a small portion of the extremity showing at the distal end. After wrapping the extremity, it is important to look at and feel the exposed fingers or toes for color, warmth, sensitivity, and movement. Do they look the same as the unaffected fingers and toes? Is the temperature the same? Can the person feel you touch their fingers or toes? Can the person move them when you ask them to? If possible, and you feel capable of doing so, check for a pulse in the foot or wrist. If pulses were present before you applied the wrap and are not detectable after wrap application, then the bandage is too tight. Remove and rewrap the bandage.

Cool the affected area immediately after a sprain or strain by applying ice. This causes the damaged blood vessels to narrow and slows the leakage of blood into the soft tissue. A towel or cloth should be placed between the cold source and the person's skin to prevent damage to the skin. Cool devices should be applied for 20 to 30 min every 4 h for 24 to 48 h.

Elevate the extremity, if possible. Elevation decreases blood flow to the area, thus minimizing blood leakage and swelling.

Bruises and Contusions

Contusion is the medical term for a bruise. Contusions fall under the category of soft tissue injuries. Soft tissue comprises skin, fat, and muscle beneath the skin. Contusions are often painful and cause swelling and discoloration at the site of injury; however, the skin is not broken. The swelling and reddish color are caused by bleeding of small blood vessels under the skin. Over time the area turns different colors, usually dark red, then purple, and finally yellow as the blood is cleared by the body.

First aid for contusions includes applying an Ace wrap compression bandage, if possible, and immediate application of ice or a cold pack for 24 to 48 h. A towel or cloth should be placed between the cold source and skin. Following cold application, heat should be applied for approximately 24 hours (20–30" q4° x 24°). The affected area may be gently massaged after heat application.

Many contusions need no more than first aid treatment. However, a bruise that is very large or very painful may indicate more and severe, deeper tissue damage that requires medical examination.

Cuts and Lacerations

An injury to soft tissue is known as a *wound*. Wounds may be closed (no break in the skin, as in bruises) or open (with an obvious break in the skin). There are several kinds of open wounds.

A *scrape* or *abrasion* is an area of the skin or mucous membrane that has been rubbed away. These are usually painful because scraping away of the upper layer of skin exposes pain nerve endings. There usually is minimal bleeding.

A *cut* or *laceration* is a smooth or irregular tear in the skin. Cuts are usually caused by sharp-edged objects. However, they can also occur as a result of a hard blow from a blunt object that splits the skin. Lacerations that extend deep into the body can damage nerves, muscles, large blood vessels, or major organs. Cuts may or may not be painful, but they generally bleed profusely.

An *avulsion* is the partial, forcible tearing away of a body part or structure, usually fingers, toes, arms, or legs. An avulsion often causes a piece of skin to hang over the area. Avulsions are usually deep and therefore bleeding is significant.

A *puncture* is a hole in the skin that is made by a sharp, pointed object. A gunshot wound is considered a puncture. Puncture wounds usually do not bleed very much unless large blood vessels or internal organs are penetrated. If the object that made the puncture remains embedded in the person, it is called *impaled*. Occasionally an object can pass completely through a person creating an entry and exit wound.

All open wounds damage the body's primary defense system, the skin. As a result, all open wounds are at high risk for developing infections. The goal of first aid treatment of open wounds is to prevent infection and control bleeding. To accomplish this dressings and bandages must be applied.

A *dressing* is an absorbent pad placed directly over the wound. Ideally it is sterile. However, in an emergency, sterile dressings may not be available; therefore, clean material is adequate. A bandage is any material used to wrap or cover any part of the body. Bandages may be used to hold dressings in place when tape is not available.

If a minor wound is bleeding, allow some blood to flow. This flushes dirt and debris from the wound. Then, use soap and water to clean the wound, but do not forcibly scrub the wound. After cleaning, apply an antibiotic ointment and a clean dressing with a bandage.

Do not take time to wash major wounds; the person could bleed too much. Control severe bleeding by placing a sterile covering over the wound. If no sterile products are available, any clean cloth will do. (Do not use cotton batten because it fluffs off and is difficult to remove from a wound.) Apply direct pressure at the site. The injured person may assist in this if able. Use your own bare hand to apply pressure *as a last resort*. Next apply a bandage snugly over the dressing. If possible, elevate the wound above the level of the heart, but do not elevate an extremity if you think it might be broken.

If the wound continues to bleed profusely, apply pressure on a nearby artery (pressure point). This maneuver constricts the artery and thus slows bleeding. If the original dressing becomes saturated with blood, do not remove it. Instead, apply another dressing and bandage over it. Removing the original dressing and/or bandage could result in more severe bleeding. Do not apply a tourniquet to stop bleeding unless the risk of hemorrhaging to death is imminent. Improperly applied tourniquets can result in extensive soft tissue damage. All persons with major open wounds need to be transported to a medical facility for further treatment.

If a body part has been severed, try to locate it. Once found, wrap it in a moist, clean cloth and place it in a plastic bag. Submerge the plastic bag in ice water. Do not place the tissue directly against the ice because this could cause freezing of the body part. Be sure the severed body part is transported to the hospital with the victim. Physicians may be able to reattach it.

If an object is impaled in a victim, **do not remove it.** The impaled object helps to minimize bleeding by applying direct pressure at the site. The object needs to be immobilized in place so that additional nerve and soft tissue damage is minimized. Secure the object with soft, bulky materials taped or wrapped to the victim's body.

Because most wounds bleed to some extent, it is important to be aware of transmission of infectious diseases associated with exposure to blood. Serum hepatitis (hepatitis B) and acquired immune deficiency syndrome (AIDS) are the two most common diseases associated with blood. Transmission of both these diseases can be minimized.

A wound provides a route of exit for the infectious organism. An open wound or cut (visible or not) provides a convenient mode of entry for an infectious organism to enter your body. When trying to control bleeding from another person's wound, it is important to reduce your risk of infection.

To minimize the chance of disease transmission during first aid treatment follow these guidelines:

1 Wear disposable gloves.

2 Wash your hands before and after administering care, even if you have worn gloves.

3 Avoid being splashed by blood.

4 Cover the wound with a dressing and/or bandage.

5 Do not allow your skin to touch objects that have blood on them.

6 Do not wash bloody hands in a sink used to prepare food.

7 Avoid eating, drinking, touching any part of your body, or handling personal items such as pens and combs immediately after giving wound care, unless you have thoroughly washed your hands with soap and water.

Fractures

A *fracture* is the sudden breaking of a bone. The break may be complete, a chip, or a crack. It may be caused by a fall, a blow, or a twisting motion. Like wounds, fractures may be closed (no break in the skin) or open (obvious break in the skin). Also like wounds, infection and bleeding are common with open fractures. Fractures can occur in any bone of the body, but fractures of the extremities are the most common. [17]

First aid treatment of all fractures is similar to the first aid treatment for sprains and strains: *RICE.* The goal is similar to all injuries to the body: minimize further damage (i.e., bleeding, soft tissue damage, and nerve involvement).

Rest the extremity. Be sure the person is no longer applying stress to the injured area.

Immobilize the extremity. Use the ground as a splint to immobilize the affected part if the person is already lying down. Splint an injured leg to the uninjured leg by placing a rolled blanket or rolled clothes between the legs and tie the legs together at the ankle, below and above the knee, and below the groin. Splint an injured arm to the chest by securing the arm to the chest in the position in which the person was found. If the situation does not permit these measures, use any rigid material (rolled newspapers, magazines, hightop sneaker, etc.) to splint the part. Regardless of the type of splint used, be sure to check the distal (furthest) part of the extremity for circulation (pulses, color, warmth) and nerve involvement (sensitivity and movement) before and after splint application. Be sure to splint all fractures *in the position found!* Do not attempt to place the broken bone in a normal position. Simply prevent the part from moving to minimize further damage.

Finally, apply ice or a cold pack to the fracture site. Continue to keep the person lying down and calm. Cover the person with a blanket. Do not apply so many blankets that the person perspires. Continue to check the exposed toes or fingers for color, warmth, sensitivity, and movement. If any of these signs should worsen, remove and reapply the splint. Continue to check the distal extremity.

Fractures of the spine (broken back) are treated similarly to other fractures with some exceptions. Suspect a spinal injury when a person:

- falls from a height greater than his or her own height
- is in a motor vehicle accident, particularly if seat belts were not worn
- is found unconscious for no known reason
- may have suffered a severe blunt force injury
- has a head or trunk puncture
- was thrown from a motor vehicle
- has a broken helmet
- has been struck by lightning
- was injured during contact sports

The main goal in first aid care of a person suspected of having a spinal injury is to prevent spinal cord damage. To do this the spine must be immobilized (prevented from moving). If the person *must* be moved, provide in-line stabilization (support the victim's head with your hands), align the victim's head and body if this can be done without resistance or pain, or log-roll the victim (turn as one unit) onto a rigid surface. This requires at least two people, although three or four are preferable.

Heat and Chemical Burns

Burns are also a special type of soft tissue injury. Burns may be caused by heat, chemicals, electricity, or radiation. Thermal (heat) burns are the most common. The effects of burns vary according to the type, duration, and intensity of the agent and the part of the body involved. Burns may affect only the part of the body exposed (local injury) or affect the entire body (systemic injury), depending on the severity and location of the burn.

A burn will initially injure or destroy the first layer of skin. If exposure to the causative agent persists, destruction of deeper layers of the body continues. Burns are often described by their depths: superficial, partial, and deep. Deep burns are the most serious.

Because burns damage the skin, infection, loss of body fluids, and loss of body heat may occur. In addition, if superheated air or airborne chemicals were inhaled, breathing may be impaired.

Before discussing the care of a burn, it is important for the first aid provider to assess the environment for possible dangers. First aid should be performed only when the environment is considered safe. For example, if a person has been electrocuted, *do not touch the person if wires or other conducting sources are still connected to the body.* If possible, turn off the power source. If you *must,* move the wires away from the body with a nonconducting material such as wood or paper and then proceed with first aid care. Also, be sure not to enter a room where chemicals have been aerosolized. You could be overcome by the fumes. If chemicals have spilled on a person, do not use your bare hands to remove any clothing that has unknown chemicals on it or you may burn your own hands.

The first step in caring for thermal burns is to stop the burning. If the person's clothes are on fire, yell for him to *stop, drop, and roll.* If this is not possible, try to extinguish the flames with water. Finally, if water is not available, smother the flames with a blanket, coat, or some other suitable material. This last method may result in deeper burns because it smothers the fire against the victim's body.

After the fire is out, flush the area with large amounts of cool water. Do not apply ice because ice causes additional body heat loss. If the person cannot be immersed in or doused with water, apply water-soaked cloths to the burned areas. Be sure to re-wet the cloths periodically to keep them cool. Do not attempt to remove the person's clothing.

After cooling the burn for several minutes, cover the burned area loosely with a clean, dry cloth. This protects the wound from dirt and debris, and therefore possible infection, and keeps air out, which helps reduce the pain. Do not apply ointments of any kind, as they may seal in heat and are difficult to remove later. Do not use home remedies because they can cause infections or undesirable chemical reactions. Do not break blisters because unbroken skin is the best prevention against infection.

Chemical burns are treated similarly to thermal burns. Flush the burn with large amounts of cool running water until medical help arrives or for at least 15 minutes. If an eye has been chemically injured, hold the affected eyelid open and flush the eye so that the water runs from the nose out to the side of the face. To do this, turn the victim's head so that the injured eye is closer to the ground. This prevents any of the chemical from reaching the other eye. Remove any clothing on which the chemical has spilled. Use gloves or some other material, not your bare hands, to grasp and remove the contaminated clothing.

Although some chemicals react with water to produce heat, flushing the chemical away under pressure washes the chemical off the skin and dilutes it sufficiently to minimize burning. Do not submerge the part with the chemical burn in water unless there is a frequent exchange of water and you have no other choice. The chemical could increase in concentration in the water and cause additional soft tissue damage. In an emergency, flooding with lots of water is the best remedy. All chemical burns require medical evaluation.

Electrical burns cause burning both outside and inside the body because current is conducted through the body. These burns are often deep and have an entrance and exit wound. Depending on the force of the current, persons with electrical burns, including those struck by lightning, are also considered to have a spinal cord injury until proven otherwise. (See "fracture" section.) It is not necessary to cool the burn. Do cover the wounds with dry, clean cloth. Keep the person from getting chilled. If the person is awake, try to keep him or her calm and reassure them that help is on the way. All electrical burns are considered serious and require medical evaluation.

Regardless of the cause of the burn, all burned extremities should be elevated above the level of the heart if possible to help control swelling. Severe burns of the face, neck, hands, feet, or genitals are considered critical and need medical evaluation. If a person has been exposed to a burn agent that might have affected the lungs, help the person sit up and watch their breathing closely until the emergency medical team arrives. Do not give any person with a severe burn anything to drink because the gastrointestinal tract often stops functioning properly. Drinking fluids might cause the person to vomit and aspirate (inhale) the vomitus. In addition, this person may require surgery and an empty stomach is important for the same reason.

Radiation burns are usually caused by overexposure to the sun. The best treatment is prevention. Do not sunbathe between 10 A.M. and 2 P.M. Use sunscreens with a sun protection factor (SPF) of up to 15. Lotions with SPFs over 15 have not been proven to offer additional protection. Apply sunscreens 15 to 30 minutes before exposure and every 60 to 90 minutes thereafter. Those who are frequently exposed to the sun have an increased risk of skin cancer.

Amputation

Traumatic amputation of a body part is treated similarly to an avulsion wound. Bleeding from an amputation is not difficult to control because the severed blood vessels begin to close in the area of the amputation. A tourniquet may be placed just above the amputation site to minimize, but not completely eliminate, bleeding. Cover the wound with a water-soaked cloth and then a dry cloth. The severed part should be wrapped in a clean cloth, placed in a plastic bag, and submerged in ice water before being transported with the person to the acute medical care facility. Cover the person with a blanket and keep him or her calm by talking in a reassuring manner.

Multiple Injuries

If you encounter an *adult* who is unresponsive or who you suspect as having multiple injuries, follow your "ABCs": airway, breathing, circulation.

First establish unresponsiveness by gently shaking the person and shouting, "Are you OK?" If there is no response, have someone call in the emergency. If no one else is available, make the call yourself. Do not hang up until the dispatcher hangs up.

When returning to the victim, kneel close beside the body, tilt the head back with one hand, and lift the chin with the other hand. Bend down and place your ear next to the person's nose and mouth and listen and feel for any air movement from the nose or mouth. Look for chest movement. If none is detected, open your mouth wide, take a deep breath, and place your mouth directly over the victim's mouth. While pinching the victim's nose, exhale into the victim's mouth. Watch for the chest to rise. Do this two times.

Next check to see if the person has a pulse. Place the tips of your fingers against either side of the neck in the groove near the Adam's apple. If a pulse is present but the person is still not breathing, give one slow breath into the victim's mouth about every 5 seconds. Check for breathing and pulse approximately every minute. This is called *rescue breathing* and should be continued as long as a pulse is present but the person is not breathing spontaneously.

After giving the two slow breaths, if you do not feel a pulse, place the lower part of your hand on the breastbone, approximately between the nipples. Position your shoulders directly over your hands. Compress the breastbone in about 2 inches. Do this about 15 times and then give the person two more slow breaths. Repeat this sequence for about one minute, and then check for a pulse again. This is *adult* cardiopulmonary resuscitation (CPR). Once initiated, CPR must be continued until emergency medical help or someone who knows how to perform CPR relieves you, the person revives, a medical doctor pronounces the person dead, or you become exhausted.

Concerns regarding liability in providing first aid can be answered via the Good Samaritan laws. This law gives legal protection to those who, in good faith, provide emergency care to ill or injured persons. If you provide care that any reasonable and prudent person in the same situation would, the Good Samaritan law protects you. This is a legal protection from being successfully sued for any injury to the victim. Each state's Good Samaritan law varies slightly. If you are interested in more information, consult your local library, a lawyer, or the Red Cross about your state's law.

Airway Blockage by Food

Because choking is a relatively common occurrence, one should know the steps of the Heimlich maneuver. The universal signal for an obstructed airway is two hands clutching the throat. When a person grasps his throat, approach and ask, "Can you speak?" If the person shakes the head "no," explain that you will be helping. Then stand behind the person, place the thumb side of your fist (with thumb buried in hand) against the middle of the abdomen, just above the bellybutton. Grasp your fist with your other hand. Give quick upward thrusts into the abdomen until the object is coughed up or the person becomes unconscious. If the person becomes unconscious, gently lower the body to the floor with the face up. Position the head using the head tilt and chin lift maneuver described in the "multiple injuries" section. Examine the mouth and throat for obstructions and remove them if found. Attempt to exhale air into the person. If no air goes in, reposition the person's head. If air still does not go in, straddle the person's thighs and place the heel of one hand just above the bellybutton. Place your other hand on top of the first. Briskly push up and in five times. Return to the person's head. Lift the jaw and tongue and sweep the mouth with your finger, from the furthest cheek, across the tongue, to the nearest cheek. Perform the head tilt and chin lift maneuver and attempt to give the person a breath. Continue this sequence of breaths, thrusts, and sweeps until breaths go in. Once the object is dislodged and breaths go in, begin the CPR sequence, as previously described. If the person begins to breathe and the pulse returns, place him on his side and continue to check him frequently until emergency medical help arrives.

First Aid Kit

A variety of first aid kits can be purchased from any drug store. Your local Red Cross chapter may also sell them, or you can make your own. Personalize the kit as your needs dictate. Check the kit regularly. Check expiration dates on items and check batteries periodically. Be sure to include emergency phone numbers as needed. [6]

The American Red Cross kit contains the following items:

activated charcoal
antiseptic ointment
syrup of ipecac
triangular bandages
blanket
scissors and tweezers
small flashlight and extra batteries
hand cleaner
adhesive bandages (assorted sizes)
disposable gloves
plastic bags
cold pack
adhesive tape
gauze pads and roller gauze (assorted sizes)

The author wishes to acknowledge and thank Joan D. Wentz for the preparation of the first aid segment of this chapter.

BIBLIOGRAPHY

1 *Accident Facts,* National Safety Council, Itasca, Ill., 1993.
2 *Accident Prevention Manual for Business and Industry: Administration and Programs,* 10th ed., National Safety Council, Itasca, Ill., 1992.
3 *Accident Prevention Manual for Business and Industry: Engineering and Technology,* 10th ed., National Safety Council, Itasca, Ill., 1992.
4 *Accident Prevention Manual for Business and Industry: Environmental Management,* National Safety Council, Itasca, Ill., 1995.

5 *American Red Cross Community First Aid and Safety,* Mosby Lifeline, St. Louis, 1993.

6 *American Red Cross Emergency Response,* Mosby Lifeline, St. Louis, 1993.

7 Anton, T. J., *Occupational Safety and Health Management,* 2d ed., McGraw-Hill, New York, 1989.

8 Bailey, J. M., "Well Control Problem Leads to Better Workover Procedures," *Oil & Gas Journal,* January 16, 1995.

9 Cralley, L. V., and L. J. Cralley, *In-Plant Practices for Job-Related Health Hazards Control,* Wiley, New York, 1989.

10 Hajian, H. G., and R. L. Pecsok, *Working Safely in the Chemistry Laboratory,* American Chemical Society, Washington, D.C., 1994.

11 Hammer, W., *Occupational Safety Management and Engineering,* 4th ed., Prentice-Hall, Englewood Cliffs, N.J., 1989.

12 Kamp, J., "Worker Psychology," *Professional Safety,* May 1994.

13 Kavianian, H. R., and C. A. Wentz, *Occupational and Environmental Safety Engineering and Management,* Van Nostrand Reinhold, New York, 1990.

14 Kelley, F. R., *Safety Is Attitude,* Management Safety Systems Seminar, St. Louis, 1995.

15 Kirschner, E. M., "Chemical Industry Modernizes Aging Plants for Safety and Efficiency," *C&E News,* July 10, 1995.

16 Kirschner, E. M., "Environment, Health Concerns Force Shift in Use of Organic Solvents," *C&E News,* June 20, 1994.

17 Mourad, Leona, *Orthopedic Disorders,* Mosby-Year Book, St. Louis, 1991.

18 Office of the Federal Register, 1985, *Code of Federal Regulations,* 29 CFR 1910.151, Office of the Federal Register, Washington, D.C.

19 Perkowski, M., "Protect Plants from Lightning," *Chemical Engineering Progress,* July 1995.

20 Pina, J. J., "Preventing the IFSs," *Professional Safety,* September 1995.

21 Plog, B. A., *Fundamentals of Industrial Hygiene,* 3d ed., National Safety Council, Itasca, Ill., 1988.

22 Pulat, B. M., and D. C. Alexander, *Industrial Ergonomics: Industrial Engineering and Management Press,* Norcross, Ga., 1991.

23 *Recommended Practice for Fire Prevention and Control on Open Type Offshore Production Platforms,* 3d ed., No. RP-14G, API, Washington, D.C., 1993.

24 *Safe Welding and Cutting Practices in Refineries, Gas Plants, and Petrochemical Plants,* 5th ed., No. 2009, API, Washington, D.C., 1988.

25 Slote, L., *Handbook of Occupational Safety and Health,* Wiley, New York, 1987.

26 *Taber's Cyclopedic Medical Dictionary, 14th ed.,* F. A. Davis, Philadelphia, 1981.

27 Thayer, A. M., "Chemical Companies Extend Total Quality Management Boundaries," *C&E News,* February 27, 1995.

28 29 CFR 1910.120, July 1, 1994.

29 29 CFR 1910, OSHA, Washington, D.C., July 1, 1994.

30 *The Wellness Encyclopedia,* Houghton Mifflin, Boston, 1991.

31 Wentz, C. A., *Hazardous Waste Management,* 2d ed., McGraw-Hill, New York, 1995.

32 Wentz, Joan D., Personal Communication, September, 1994.

PROBLEMS

1 What is the airflow volume into a local exhaust ventilation duct for dust removal for a system with the following parameters: the circular duct is 12 cm in diameter and the air velocity at the duct opening is 1100 m/min.

2 What is the air velocity in problem 1 at a point 3 cm from the duct end? At 12 cm from the duct end?

3 Discuss the three major alternatives that may be used to correct workplace hazards.

4 Why are older manufacturing plants more likely to have serious safety hazards than newer, more modern facilities?

5 What are the three areas where management controls could be applied?

6 What are the types of management controls for occupational hazards?

7 Why is it important for management to lead by example in the control of workplace hazards?

8 Why must the organization accept and buy into the changes necessary for hazard control?

9 Discuss the organizational management person who has the greatest influence on employee safety performance.

10 Why is good housekeeping important to a safe workplace?

11 How can management control preventive maintenance activities?

12 How can purchasing agents help assure safety in the workplace?

13 A large electrical transformer at a demolition site contains salvageable copper. During the salvaging of the copper the PCB reservoir is punctured, causing a large PCB spill and a costly site cleanup. How could the spill have been prevented?

14 Ten 55-gal contaminated open-head drums without lids were stored overnight in a fenced hazardous waste storage area. A heavy rainstorm that night left 3 in of rainwater in each drum. This contaminated rainwater required costly disposal as a hazardous waste. How could the rainwater disposal have been avoided?

15 What important longevity risk factors are affected by diet, lifestyle, and physical activity?

16 Why should workplace safety and health be monitored by management and measured against goals and objectives?

17 What is the greatest cause of accidents?

18 What are the more common engineering controls for workplace hazards?

19 Why is the plant site selection critical to safety, health, and environmental protection?

20 Why is it best to introduce engineering controls into process operations during the design stage?

21 What factors should be considered when planning the layout of an industrial facility?

22 Why are process flow sheets and 3-D models important to safe operations?

23 Describe typical mechanical operating hazards that may require machinery guards.

24 Discuss the workplace hazards that are introduced by robotics.

25 Why should special precautions be taken in the selection of electrical equipment where flammable and explosive materials may be present?

26 Why is it desirable to have an automated closed-process system?

27 Discuss equipment design techniques that can reduce the escape of hazardous materials into the environment.

28 Discuss workplace control methods to reduce exposure to hazardous materials.

29 Why is good housekeeping important to occupational safety?

30 Discuss the importance of management of implementing medical workplace controls.

31 Why is training and continuing education important in planning and implementing management and engineering controls?

32 In 1995, Shell Chemical Company published a brochure entitled *Health, Safety and Environmental Performance Report for the Six Codes of Management Practices of Responsible Care.* What type of management control was exhibited by this action?

33 An operating plant is to be shut down and demolished. It is known to contain asbestos insulation, PCB capacitors, and PCB fluorescent light ballasts. How could hazardous spills and contamination be avoided in this situation?

34 Employee psychology plays a key factor in workers' compensation, particularly in cumulative trauma claims, such as lower-back pain, carpal tunnel syndrome, and other injuries with aches and pains. Explain how management control job attitudes and personal traits can affect the employee's decision to either keep working despite the pain or to file a workers' compensation claim.

35 Name workplace situations requiring management and engineering controls that could generate static electrical sparks and create problems if flammable vapors and gases are present.

36 Discuss reasons for establishing a safety and health program in the workplace.

37 The dry-cleaning industry in the early 1900s switched from petroleum solvents to perchloroethylene. Recently, there have been attempts to eliminate perchloroethylene by using water-based wet-cleaning methods. Discuss the reasoning behind these developments in materials substitution.

38 Why did trichloroethane (TCA) replace trichloroethylene (TCE) in many industrial solvent applications?

39 The chlorofluorocarbon CFC-113 and 1,1,1-trichloroethane (TCA) solvents are being banned from production and import because they are ozone depleters. Citrus water–based solvents are replacing these solvents in some applications. What would be some of the advantages of this substitution?

40 Why did the use of hydrocarbon solvents in the United States decline from 8.3 million pounds in 1987 to 4.7 million pounds in 1992?

41 What are the advantages and disadvantages of vapor degreasing of metal and electronics parts?

42 The air in a workplace contains 100 ppm of acetone (TLV = 750 ppm), 150 ppm of methyl ethyl ketone (TLV = 200 ppm), and 50 ppm of butane (TLV = 800 ppm). Is this atmosphere safe? How would you reduce the contaminate levels to satisfy the mixture TLV?

43 An employee must be present at three workstations during an 8-h shift. Workstation A contains 700 ppm of pentane (TLV = 600 ppm). Workstation B contains 650 ppm of pentane. Workstation C contains 0 ppm pentane. The employee must be present at each of these workstations for at least 2 h during an 8-h shift. How would you apply management and engineering controls to ensure employee safety?

44 You are the safety manager in a large manufacturing plant in Houston. What is the relative order of importance of the following items as you perceive them? What would be the order of importance if you were the company president and were located at the headquarters in New York? Discuss the logic of your response.

a Reduce rate of injuries.	**g** Reduce potential liability.
b Improve labor relations.	**h** Enhance employee morale.
c Ensure regulatory compliance.	**i** Increase rate of production.
d Improve operating costs.	**j** Improve job pride.
e Provide personal satisfaction.	**k** Enhance customer relations.
f Aid public relations.	**l** Improve product quality.

45 You are the production manager in a department that is experiencing higher-than-expected safety problems in a new process operation. Which of the following learning techniques would you use to apply management controls to this situation?

a Visual instructions, with pictures, movies, exhibits

b Presentations and doing the real thing

c Verbal instructions and manuals

d Participating in a discussion or giving a talk

46 A worker is injured on the job, and the medical bill is $500. Which of the following, based on the industry, best approximates the real total cost of that accident to the company?

a $500 to $1,000   **c** $4,000 to $10,000
b $1,000 to $4,000   **d** $4,000 to $50,000

47 The first law of good work is to maintain a clean and orderly workplace. How can management achieve this important concept?

48 A noisy workplace has been found to exceed the OSHA 90 dBA limit for 8 h of exposure. The 8-h TWA was determined to be 117 dBA, and the company has only 30 days to achieve compliance. Noise reductions are possible with enclosure, vibration isolation, and absorption engineering controls. Which from the following sets of these options will achieve compliance ($\leq$89 dBA, 8-h TWA management objective) in the most cost-effective way?

Option	Noise reduction (dBA)	Cost
Enclosure		
1	15	$4000
2	12	$3000
3	9	$1500
Vibration isolation		
4	10	$2000
5	8	$1000
6	6	$500
Absorption		
7	13	$3000
8	11	$1000

49 Determine the total light flux that must be furnished a 5 m^2 surface area to obtain an average illuminance of 400 lux.

50 A worker measures the illuminance at a workstation by pointing an illuminance meter at the workstation. Discuss the relevance of the light reading from the meter.

51 White, gray and yellow are used to paint various surfaces in a plant location. What would be the visual appearance of these colors in daylight and under a yellow sodium light? How could this situation become a safety hazard?

52 Are glossy or matte surfaces preferred around workstations to reduce glare?

53 The daylight illuminance on a surface is measured at noon in December and at noon in June. How should these measurements compare?

54 How could the glare that is associated with a television or VDT be reduced for the viewer?

55 Develop a checklist of safety-related questions to help identify departures from desired lighting conditions in the workplace.

56 Name several industrial lighting interiors that require frequent luminary cleaning intervals.

57 What would be an effective lamp replacement strategy and schedule?

58 An employee is lifting 30-kg parts from a tall container and placing them on a table for processing. The worker does this task for 8 hours per day at a rate of 1.5 parts per minute. The horizontal distance is 50 cm, the vertical distance is 25 cm, and the vertical travel distance is 55 cm. According to the NIOSH guideline, is this an acceptable lift?

59 A company has 35 recordable accidents in 1.4 million employee hours. What is the company's incidence rate?

60 A company has 10 lost work days recorded for 1.25 million employee hours. What is the company's incidence rate?

61 What are the major risk factors for cumulative trauma disorder?

62 What manual material lifting and handling factors can contribute to an increase in musculoskeletal injuries?

63 Describe how carpal tunnel syndrome occurs.

64 Why does a static work position for a prolonged period of time usually result in painful muscle fatigue?

65 What is usually the basic or root cause for both the direct and indirect causes of accidents?

66 What are the OSHA requirements for first aid?

67 How are strains and sprains different?

68 What is RICE and when is it implemented?

69 When and how are splints applied?

70 Describe how to apply a bandage to an extremity.

71 Identify two methods to control severe bleeding.

72 What steps should be taken when a body part is severed?

73 Describe first aid care for a person with an impaled object.

74 What guidelines should be followed to protect yourself from bloodborne infectious diseases?

75 What is an anatomical splint?

76 What is the most important consideration when providing first aid to a person suspected of having a spinal cord injury?

77 What are the first aid steps provided for persons with thermal or chemical burns?

78 What actions should be used to prevent radiation burns from the sun?

79 What are the steps of CPR and the ABCs?

80 Describe the Heimlich maneuver.

81 What should be included in a first aid kit?

13

PERSONAL PROTECTIVE EQUIPMENT

Engineering and management controls can reduce or even eliminate many occupational safety hazards. Whenever these techniques can solve safety hazards, they are more desirable than the use of personal protective equipment (PPE). If the hazard is still present in the workplace after the engineering and management controls have been implemented, then the only viable alternative is to use PPE as the primary control measure. [7] The protective device provides a protective barrier between the hazard and the employee. If the device fails or is improperly used, the employee will be directly exposed to the hazard and may even have a false sense of security during the exposure. [1,12] PPE is the last line of defense. [4]

SELECTION OF PPE

OSHA requires a hazard assessment of the workplace to determine the presence of hazards requiring PPE (29 CFR 1910.132). This assessment should provide written verification, which identifies the actual workplace being evaluated by the person doing the evaluation and the date the evaluation occurred. This written verification then becomes the certification of hazard assessment. [8,25]

Many forms of PPE have become standards in the workplace and are required in the manufacturing and construction industries. [2,39] Hard hats and safety shoes are PPE that are widely used to protect workers from physical harm. In many industries, safety glasses are required. These are examples of PPE that normally supplement the use of engineering and management controls in dealing with physical hazards like flying or falling objects. The required PPE helps employees to avoid injuries to the body and safeguards employees in the event of accidents. [10]

A hard hat provides impact protection against blows to the head. Goggles can protect eyes from dust and other airborne particulate matter. Gloves can prevent hand cuts, abrasions, and punctures. Protective footwear can reduce the impact of falling objects and puncture wounds to the feet. Fluorescent vests and flags increase the visibility of workers to others.

When chemical hazards are present, PPE should only be selected when more desirable engineering and management controls are not feasible or to provide additional protection. PPE for chemical hazard control may also be used as an interim control measure until better engineering and management controls have been implemented.

The attitude of employees and visitors toward the proper use of PPE is affected greatly by the attitude of management. Supervisors must lead by example. Once the proper PPE has been selected for the workplace, there should be a training program to educate employees in the correct use and maintenance of the equipment. There should be written policies that enforce the use of the PPE.

Several criteria enter into the selection process for PPE in the workplace. The degree of protection that a particular type of PPE affords under the range of working conditions plays an important role. Initial cost and long-term economics, the practical application of the PPE, and government or industry regulations are also factors to consider. Usually when the PPE is required for the job, all or part of the cost is paid by the employer. The sharing of these costs is determined by the employer, the industry, and the locality. Cost sharing is more likely for items that may be worn off the job site or workplace.

TRAINING FOR PPE USE

A number of factors influence worker acceptance of the requirements to use PPE properly (Table 13-1). In an industry where employees are accustomed to wearing PPE, compliance may be automatic. When a new PPE is introduced, employee acceptance is more delicate. Management must provide the employees with a reasonable explanation for the PPE. Employees may resist any change in traditional work practices. The best approach for a PPE trial in the workplace may be to let the supervisors try it first. If the supervisors accept the PPE change, other employees will also.

It is usually helpful to meet with employees or their representatives to review a contemplated PPE program and reach an understanding about potential problems in implementing the program. An explanation of the purpose and benefits of the program is essential to its ultimate success. [17]

Employee macho or vanity may enter into PPE acceptance. Much resistance to change can be overcome if employees are allowed to select PPE from several available styles. A variety of safety shoes and glasses are now available for just this reason. [21]

The actual PPE training program should contain all of the elements in Table 13-2. It is important that management inform and interact with employees prior to the actual introduction of PPE into the workplace. Management must enforce the PPE

TABLE 13-1
FACTORS THAT INFLUENCE EMPLOYEE ACCEPTANCE OF PPE

Perceived example of management

Ease, comfort, and convenience of use

Understanding the need for its use

Economic and disciplinary losses for the lack of use

Perceived acceptance by other employees

policy, preferably through a progressive disciplinary action plan, ranging from unpaid leave of absence to employee termination. [1,19,39]

PROTECTION OF THE HEAD

Protective gear should protect the head from the impact and penetration of falling or flying objects in the workplace. Electrical shocks and burns should also be a consideration. OSHA requires that helmets meet ANSI requirements and specifications for industrial head protection, Z89.1-1986. [28] Headgear must withstand the impact of an 8-lb iron ball dropped from a 5-ft height without cracking or denting. Helmets must have a suitable suspension system to help maintain the position of the helmet on the wearer. ANSI requirements contain three helmet classes (Table 13-3 on page 404). The commonly used terms *hard hats, safety hats,* and *safety*

TABLE 13-2
ELEMENTS OF A PPE TRAINING PROGRAM

Standard and regulatory requirements

Hazard characterization in the workplace

Implementation of engineering and management controls

Description of the need for PPE

Explanation of the PPE selection

Discussion of the PPE capabilities and limitations

Demonstration of the use and proper fit of the PPE

How to decontaminate PPE

How to care, maintain, and repair PPE

When and how to dispose of PPE

Hands-on PPE practice

Explanation of PPE written policy, regulations, and enforcement

Discussion of PPE cost and purchase

Discussion of record-keeping requirements

TABLE 13-3
ANSI CLASSES FOR HELMETS

Class A: Head protection from the impact of falling objects and from low-voltage electrical shock

Class B: Head protection from the impact of falling objects and from high-voltage electrical shock

Class C: Head protection from the impact of falling objects

helmets are technically incorrect under the ANSI standard. An example of several helmets and suspension systems are shown in Figure 13-1 and 13-2. [38]

Prior to use, helmets should be inspected for cracks, stress cracking, abrasion, chalking, and wear, which could reduce their safety. Ultraviolet radiation from sunlight or welding operations can degrade plastic helmets with prolonged exposure. Chemical exposure can reduce helmet life.

Helmets should be washed in warm soapy water at least monthly and rinsed thoroughly. When removing hydrocarbons or other chemicals from helmets, use only solvents recommended by the manufacturer because some solvents can damage the helmet. The sweatband and other parts of the suspension system should be replaced periodically.

FIGURE 13-1
PLASTIC HELMETS FOR OCCUPATIONAL HEAD PROTECTION. (Courtesy of Willson Safety Products.)

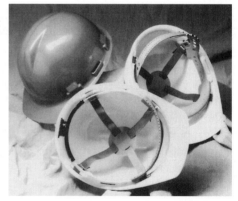

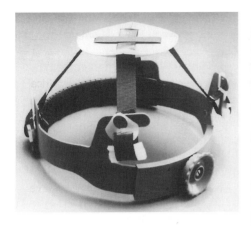
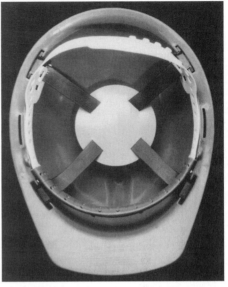

FIGURE 13-2
HELMET SUSPENSION SYSTEM FOR OCCUPATIONAL HEAD PROTECTION. (Courtesy
of Willson Safety Products.)

PROTECTION OF THE FACE AND EYES

Eyes are irreplaceable. To protect our eyes, the eyes close automatically or blink
when exposed to irritation, impact, bright light, and heat. Blinking will not protect
us from workplace hazards like airborne particles, chemical splashes, or radiation.

The eyes and face require occupational protection from injury by physical
agents, chemical agents, or radiation. OSHA requires that eye and face protection
meet ANSI requirements. [26] PPE is required whenever there is a reasonable
probability of preventable injury. [1,9]

In addition to protecting against workplace hazards, eye and face PPE should fit
snugly and be comfortable, durable, capable of being kept clean, disinfected, and
in good repair. If a person requires corrective lenses, the spectacles or goggles
must satisfy that need.

Selection of eyewear should be based on the level of protection needed, the
comfort in wearing the PPE, and the ease of repair. Many styles of safety eyewear
are now available that are capable of satisfying individual tastes. [17]

There are four general types of glasses used for protection.

Style A safety glasses, which are common safety glasses used for moderate
frontal impact from particles or chips, but offer no protection from above or from
the side

Style B safety glasses, which have close-fitting wire-mesh side shields to provide above and side protection

Style C glasses, which have semi–side shields to provide frontal impact protection, but are vulnerable to particles that may enter between the side shields and the face

Chipping goggles, which protect from severe impact hazards from chipping and heavy grinding and usually have vents to prevent fogging

Industrial safety glasses can be made of either glass or plastic. Glass has greater resistance to abrasion and scratching and more resistance to most chemicals. Plastic lenses have greater resistance to high-speed projectiles and breakage by sharp objects and will be more shatter-resistant at temperature extremes. Some examples of safety glasses are shown in Figure 13-3. [6,38] Examples of safety goggles from Encon Safety Products are shown in Figure 13-4 on page 408. [6]

Contact lenses represent a special case in this safety field. As long as the eyes of wearers of contact lenses are protected, there do not appear to be any special safety hazards. When the workplace has the potential of chemical vapors or splashes, radiant heat, metal fumes, or high levels of particulates in the air, the use of contact lenses should be restricted.

Generally, face shields should be worn over proper basic eye protection. They protect the face and neck from flying particles, chemical sprays and splashes, and, in some instances, antiglare protection. Face shields are available with or without crown and chin protection, with replaceable window styles ranging from clear transparent to tinted transparent to combinations of wire and filter screens. The face shield materials should be noncorrosive metals or engineering plastics with good mechanical strength, nonirritating to the skin, and able to be easily disinfected. An example of a face shield is shown in Figure 13-5 on page 409. [16]

Around acids, bases, or other sight-threatening chemicals, acid hoods, face shields, or nonventilated chemical goggles should be used. The hoods can be ventilated to reduce the heat generated by the wearer. Face shields with acid environments should provide additional protection over goggles and not be used as primary protection to eyes. An example of a face shield with goggles is shown in Figure 13-6 on page 409. [16]

Welding operations present safety problems in the control of infrared and visible radiation. UV radiation may also be present. Under OSHA welders may choose protective filter lenses based upon the proper filter shade number for the welding operation. [32] Employees exposed to welding hazards must wear appropriate protective clothing, as required by OSHA. [25,26] Fumes and gases from welding may present severe health hazards like exposure to cadmium or fluorides. [32]

PROTECTION OF HEARING

The workplace has long been a source of noise-induced hearing loss. If employees are exposed to a steady noise level of more than 85 dBA for a full 8-h shift on a daily basis, they are at risk of hearing loss. This is the basis for the OSHA hearing conservation standard. [23]

Hearing protection devices attenuate noise levels. Noise reduction rating (NRR) levels developed by the Environmental Protection Agency must be shown on the

FIGURE 13-3
SAFETY GLASSES FOR OCCUPATIONAL EYE PROTECTION. (Courtesy of Encon Safety Products and Willson Safety Products.)

hearing protector package. The NRR number is subtracted from the noise level in the workplace to indicate the theoretical noise level received by the ear of the protected employee. The NRR system has several practical limitations. It was developed under ideal laboratory conditions, the fit of the PPE on the employee will not be perfect throughout the shift, and noise levels and frequencies will not be uni-

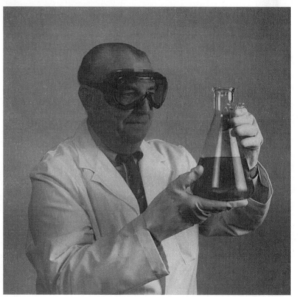

FIGURE 13-4
SAFETY GOGGLES FOR
OCCUPATIONAL EYE PRO-
TECTION. (Courtesy of Encon
Safety Products.)

form in the workplace. The resulting NRR may be as much as 50 percent less than the NRR value on the label.

There are several types of hearing PPE for use in the workplace: inserts or earplugs, hearing bands, and earmuffs. Aural inserts, commonly called earplugs, are inexpensive and either formable or molded. The formable earplugs will fit all ears because they are made of plastic foam that expands to fit the contour of the ear canal. The noise reduction of formable earplugs is fairly high, with typical NRR values of 26 to 33 dB. They are usually comfortable.

Molded earplugs have an economic advantage because they can be worn repeatedly. They are made of soft rubber or plastic to seal the ear canal; this makes them more uncomfortable than formable earplugs. Molded earplugs have typical NRR

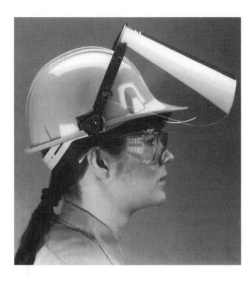

FIGURE 13-5
FACE SHIELD FOR FACE PROTECTION.
(Courtesy of North Safety Products.)

values in the 21 to 29 dB range. Custom-molded earplugs are available to fit the contour of an individual employee's ear canal, thereby reducing the level of discomfort and possibly raising the NRR value.

Hearing bands allow on-off use with reusable soft seals on the ear canal to block out noise. The band attaches to the pods or pads and can be worn behind the neck or under the chin. The employee retains easy access to the ear seals throughout the day for protection as needed. The typical NRR values for hearing bands range from 17 to 25 dB. Examples of earplugs and hearing bands are shown in Figure 13-7 on page 410. [38]

Earmuffs cover the entire external ear to form an acoustical barrier between the noise and the ear canal. They are comfortable and provide a wide range of protec-

FIGURE 13-6
FACE SHIELD AND GOGGLES FOR FACE
AND EYE PROTECTION. (Courtesy of North
Safety Products.)

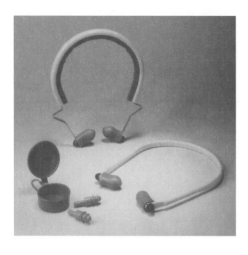

FIGURE 13-7
EARPLUGS AND HEARING BANDS FOR
HEARING PROTECTION. (Courtesy of Willson Safety Products.)

tion based upon earmuff size, shape, seal, and type of suspension. Typical NRR values for earmuffs range from 19 to 29 dB. Examples of ear muffs are shown in Figure 13-8. [38] If the noise exposure is frequent over an extended time period, earmuffs may be the best choice because of their comfort and convenience. Other necessary PPE like safety helmets and eye glasses or goggles may cause the efficiency of the hearing PPE to be compromised, making earplugs or ear bands better choices.

PROTECTION FROM FALLS

When employees work in elevated situations, they need fall protection while the job is being done. At heights of 4 feet or more, OSHA regulations apply to the

FIGURE 13-8
EAR MUFFS FOR HEARING
PROTECTION. (Courtesy of Willson Safety Products.)

workplace under 29 CFR 1910.23, 1926.105, and 1926.500. An analysis of the job site and the task to be performed, the number of workers at the site, the length of time on the site, the environmental conditions, the potential backup safety systems, and the possible rescue methods should all enter into the determination of the fall protection system needed. [1]

The mobility of the worker must match the capabilities of the fall protection system. Equipment must be used in accordance with the recommendations of the manufacturer. Employees who are working at the elevated conditions must be qualified and trained to properly use the equipment. The equipment should be installed, inspected, and maintained as recommended. A plan should be developed to rescue an employee who has fallen, while using the protection system.

There are passive and active fall protection systems, both of which can protect employees at elevated work conditions. Passive components and systems such as nets do not require any action by the employee and can be installed to protect employees all of the time. Active components and systems such as harnesses, lanyards, and lifelines require action or manipulation by employees to activate the fall protection. The equipment must be worn or connected by the employee to be effective. Once a fall hazard has been identified and cannot be eliminated, a fall protection system must be adopted to control the hazard.

Nets provide fall protection for personnel and catch falling debris from above. Personnel nets must be manufactured and tested under ANSI A10.11-1989 and OSHA 29 CFR 1926.105. The mesh openings must not be greater than 6 in by 6 in, and the nets should be no lower than 25 ft below the work surface. Personnel nets are normally used in bridge or tall-building construction.

Debris nets are designed to protect employees or other people from falling tools, foreign objects, and construction debris. Usually debris nets are 1/4 in by 1/3 in mesh. They also afford some protection for falling employees as long as they are kept free of fallen debris.

Active fall protection systems have an anchor point and connecting components to the employee. The anchor point is critical because it is the position on the independent structure where the fall arrest device is attached. OSHA requirements for an anchor point is at least a 5000-lb minimum static load strength for a 6-ft fall. Active fall systems must still allow employees the mobility to perform their job.

Harnesses encompass the torso to distribute the fall arrest force over the enclosed body parts. A full-body harness combined with a lanyard or lifeline evenly distributes the fall arrest forces among the shoulders, legs, and buttocks of the employee, reducing the probability of further injury. A full-body harness keeps the worker upright and comfortable, while awaiting rescue. Waist safety belts are not recommended for fall protection because all of the fall arresting forces are centered on the abdomen, greatly increasing the probability of damage to internal organs. Workers may also slip out of waist belts while awaiting rescue, resulting in more serious injury or even death.

Lanyards are short, flexible ropes, straps, or webs that connect the employee to the anchor. The lanyard length determines the amount of worker free fall before the protective device stops the fall. Some lanyards have shock-absorbing devices; others are retractable as the employee moves around the workstation.

There are both horizontal and vertical lifelines. Horizontal lifelines provide an overhead fixture point as the employee moves horizontally. Vertical lifelines provide a drop line to which a rope-grabbing device is attached. Retractable lifelines are self-contained devices that act as an automatic taut lanyard with the rope extending out as the distance increases and retracting as the worker moves closer.

Fall arresters and shock absorbers use friction in the rope to disperse fall energy and slow or brake an employee's fall to prevent injury. This gradual delay action keeps the body from experiencing a severe jolt or shock, when the fall is arrested. Fall-arresting systems are usually composed of an anchor point, a vertical lifeline, a fall arrester, a harness, and a lanyard with a shock absorber. A major injury can result from the forces that act on the body at the instant the fall is stopped or from collision with obstacles. [1]

Most fall protection systems contain nylon or polyester harnesses or ropes, which are flexible, strong, and abrasion-resistant fibers. Nylon is probably the better choice for most applications because it is lighter and more weather-resistant than polyester. The major advantage of polyester over nylon is resistance to most mineral acids, chemicals, and oxidizing agents. All systems with synthetic materials should have limited exposure to bright light and ultraviolet light and should be stored in a cool dry place. These systems are best cleaned by washing in soapy water to remove loose debris, rinsing with fresh water, and drying in a cool, ultralight light-free area. These systems should have regular, periodic inspection and prevention maintenance programs under trained personnel.

PROTECTION OF RESPIRATORY SYSTEM

Employers are responsible for protecting employees from harmful air contamination. Engineering and management controls like enclosure, segregation, ventilation, and material substitution should be used first, when feasible. If such controls are not feasible or are only in the process of being implemented, then appropriate respirators should be used to protect the health of the employees. OSHA 1910.134 requires a written program with operating procedures in the selection and use of respirators to match the hazard. Employees must be trained to use the respirators, which should be inspected, cleaned, maintained, and stored in convenient locations. [27]

The proper selection of respiratory PPE is a three-step process, involving hazard identification and evaluation, and the selection of appropriate PPE based upon the hazard characterization. The identification and evaluation of airborne hazards should be based upon the comparison with established limits like OSHA permissible exposure limits or ACGIH threshold limit values. More stringent internal workplace limits may also be set. The workplace atmosphere may be hazardous for a variety of reasons: because of toxic gases, vapors or fumes, particulate matter, oxygen deficiency, or some combination of these contaminates.

If the workplace has an oxygen-deficient atmosphere, the hazard is the absence of essential oxygen in confined or poorly ventilated spaces. When oxygen concentrations are less than 19.5 percent in air during emergencies, workers must wear respirators. [22]

There are two main types of respirators to protect against hazards: air-supplying and air-purifying respirators.

Air-Supplying Respirators

Air-supplying respirators provide air to the workers. One type of air-supplying respirator is a self-contained breathing apparatus (SCBA) with a transportable supply of breathing air or oxygen; it provides protection from toxic chemicals and oxygen deficiency. With SCBA the wearer of this PPE can carry up to a 4-h supply of air or oxygen.

Closed-circuit SCBAs remove carbon dioxide from exhaled air and restore the original oxygen content to the inhaled gas. This type of PPE is most applicable to mine rescues or for use in confined spaces, where the atmospheres are oxygen-deficient or immediately dangerous to life or health (IDLH).

Open-circuit SCBAs exhaust air to the atmosphere instead of recirculating it. A cylinder of compressed air supplies air to a regulator that reduces the air pressure prior to delivery to the face piece. Open-circuit SCBAs have a shorter service life, typically 30 to 60 min, and are used in fire fighting, most industrial emergencies, and at hazardous waste sites. Examples of SCBAs are shown in Figures 13-9 to 13-11 on pages 414–415. [6,16,36] Emergency situations, particularly in confined spaces may call for escape self-contained breathing apparatus (ESCBA), as shown in Figure 13-12 on page 415. [6]

There are two types of open-circuit SCBA: demand and pressure demand. When the wearer of a demand SCBA inhales, a negative pressure is created in the face piece, allowing air to flow into the face piece "on demand" by the wearer. A demand open-circuit SCBA should not be used in IDLH atmospheres because a leak around the face piece could allow the contaminated atmosphere to flow directly into the face piece.

A pressure demand open-circuit SCBA is designed to maintain positive pressure in the face piece at all times. Any leakage at the face piece will be outward into the contaminated atmosphere.

Another type of air-supplying respirator has an airline to provide air to a face piece, hood, or complete body suit. Airline respirators are available as demand, pressure demand, or continuous-flow models. The demand and pressure demand models are similar in operation to demand or pressure demand SCBA. The continuous-flow airline model maintains airflow at all times. These airline systems usually include a constant flow, full-face respirator, a rotary vane pump, quick-disconnect coupler, and at least 50 ft of breathing hose. A filtration unit, waist belt, and hair net or hood are optional additions to airline systems. It is important that the system be low maintenance. No carbon dioxide, oil vapor, or oil mist should be introduced into the system. Examples of line systems are shown in Figures 13-13 to 13-16 on pages 416–417. [16,38]

Airline respirators should not be used in IDLH atmospheres or from atmospheres that the wearer can escape without the use of a respirator because the wearer may not escape if the airline supply fails. The length of the air hose also limits the distance between the wearer and the air source. The air supply to the air-

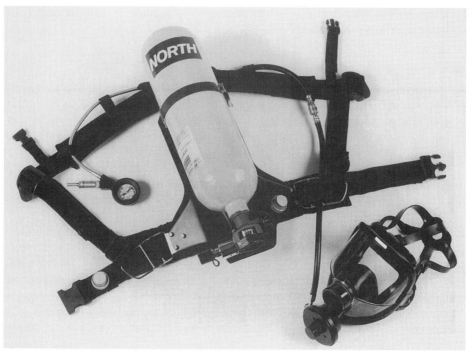

FIGURE 13-9
SCBA SYSTEM FOR RESPIRATORY PROTECTION. (Courtesy of North Safety Products.)

FIGURE 13-10
DONNING OF SCBA FOR RESPIRATORY PROTEC-TION. (Courtesy of Encon Safety Products.)

FIGURE 13-11
WEARING OF SCBA FOR RESPIRA-
TORY PROTECTION. (Courtesy of
Willson Safety Products.)

FIGURE 13-12
ESCAPE SCBA FOR RESPIRATORY PROTECTION. (Courtesy of Encon Safety Products.)

FIGURE 13-13
AIRLINE SYSTEM FOR RESPIRA-
TORY PROTECTION. (Courtesy of
Willson Safety Products.)

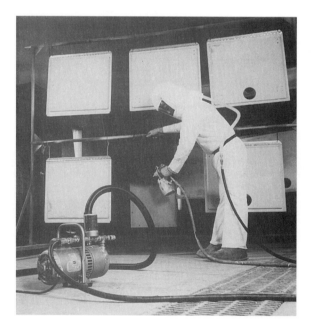

FIGURE 13-14
OPERATING AIRLINE SYSTEM
FOR RESPIRATORY PROTEC-
TION. (Courtesy of Willson
Safety Products.)

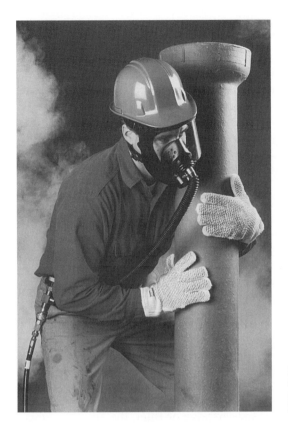

FIGURE 13-15
OPERATING AIRLINE FOR RESPI-
RATORY PROTECTION. (Courtesy of
North Safety Products.)

FIGURE 13-16
HOODS FOR AIRLINE SYSTEMS.
(Courtesy of Willson Safety Products.)

line must be continually monitored for purity, particularly carbon monoxide contamination.

Some air-supplied airline respirators also have a small SCBA auxiliary air supply tank to guard against potential air compressor failure. These combination systems are used when the airline fails and the wearer must escape from an IDLH atmosphere.

Air-supplied respirators, with breathable air supplied from a clean source, are used for most confined-space entries. SCBA provides the highest level of respiratory protection in confined-space entry. In an emergency, many rescue personnel do not want to wait until it is safe to perform a body recovery before attempting entry into a hazardous location.

The cylinder size and weight of SCBA can be problems in confined-space entry, because they limit the mobility of the wearer. With airline SCBA, the air hose may become vulnerable to damage from heat, chemical exposure, and physical stress. Escape SCBAs with a 5- to 15-min air or oxygen cylinder offer an alternative to this difficult emergency dilemma facing rescuers because they are more compact and less restrictive for confined-space entry. [14]

Air-Purifying Respirators

The air-purifying respirator can purify the air by removing gas, vapor, or particulate contaminates. Since there is no fresh air supply it should never be used in an oxygen-deficient atmosphere. It is classified as either a gas-and-vapor respirator or a particulate respirator. It consists of a face piece and a cartridge or filter that is tailored for the specific air contaminate. The useful life of this type of respirator is limited by the removal capacity of the air cartridge or filter, the air contaminate concentration, and the breathing demand of the wearer.

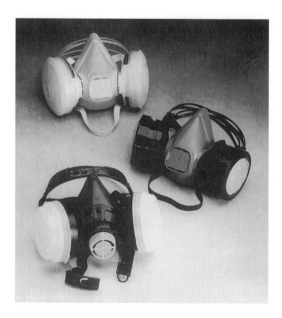

FIGURE 13-17
CHEMICAL CARTRIDGE RESPIRATORS FOR RESPIRATORY PROTECTION. (Courtesy of Willson Safety Products.)

FIGURE 13-18
DONNED CHEMICAL CARTRIDGE RESPIRATOR.
(Courtesy of Willson Safety Products.)

   Chemical cartridge respirators (Figs. 13-17, 13-18, and 13-19) remove gas and vapor contaminates by passing the contaminated air through cartridges with activated charcoal or material to sorb or react with the contaminate. [16,38] The selection of the proper cartridge to match the contaminate is critical to the success of this type of PPE (see Fig. 13-20 on page 420 and color codes in Table 13-4 on page 421). Chemical cartridge respirators are available as half-mask or full-face models. Donning these respirators is simple (Fig. 13-21 on page 421). [38] The chemical cartridges should not be used to protect against very low concentrations of extremely toxic gases or to protect against eye irritants without full-face mask respirators. The detection of irritations or odor by the user is the basis for changing the cartridge.

FIGURE 13-19
DONNED CHEMICAL CARTRIDGE RESPIRA-
TOR WITH HEAD AND EYE PROTECTION.
(Courtesy of North Safety Products.)

FIGURE 13-20
VARIOUS TYPES OF CHEMICAL CARTRIDGES. (Courtesy of Willson Safety Products.)

TABLE 13-4
SELECTED CARTRIDGE COLOR CODES FOR AIR-PURIFYING RESPIRATORS

Contaminate hazard	Color code
Acid gases	White
Organic vapors	Black
Ammonia gas	Green
Acid gases and organic vapors	Yellow
Radioactive materials, except tritium and noble gases	Purple

Source: OSHA 29 CFR 1910.134.

The particulate respirator is also an air-purifying respirator with a mechanical filter to screen out dust, fog, fumes, or smoke particles by passing the contaminated air through a pad or filter. When the user finds difficulty in breathing, the filter has become clogged and should be changed. Particulate respirators are available for many applications: particle dusts like coal, flour, and free silica; paint spray; mercury and chlorine vapor; welding fumes; and radionuclides (Figs. 13-22 and 13-23 on page 422). [16] Appropriate combinations of gas, vapor, and particulate cartridges can be used together as conditions warrant.

Gas masks are air-purifying respirators to remove specific air contaminants from air that is not oxygen-deficient. They are capable of removing gases, vapors, and particulates. Gas masks have canisters for contaminate removal, allowing a larger capacity than with cartridges.

Powered air-purifying respirators (PAPR) are also available; they supply positive-pressure filtered airflow for natural breathing to help reduce fatigue. These PAPR systems are adaptable to helmets, hoods, full-face masks and half masks (Figs. 13-24 and 13-25 on page 423). [38] The battery pack can be recharged overnight.

Stored respirators must be protected from dust, sunlight, moisture, and chemicals; otherwise, they may become ineffective. They should be wiped carefully with a damp cloth and dried thoroughly prior to storage. After cleaning and inspection, each respirator should be placed in an individual, sealed plastic bag and then in a steel storage cabinet.

All employers with respirators must have a written respiratory protection plan

FIGURE 13-21
DONNING A CHEMICAL CARTRIDGE RESPIRATOR. (Courtesy of Willson Safety Products.)

FIGURE 13-22
DISPOSABLE DUST AND MIST RESPIRA-
TOR FOR RESPIRATORY PROTECTION.
(Courtesy of North Safety Products.)

on the selection and use of respirators. The fit of the face piece should be tested prior to its use. This test involves two people—the wearer of the PPE and the person administering the test. The comfort of the respirator face piece on the wearer should be checked. Each time the respirator is worn, the fit should be checked with pressure tests.

A qualitative test with an irritant smoke or odorous vapor allows the wearer to detect any noticeable penetration of the smoke or vapor. This is an easy test to perform; it is quick and convenient and fairly reliable.

A quantitative test is performed with instrumentation to detect the leakage of the nontoxic test atmosphere. This test, which does not rely on subjective human response, is recommended prior to work in IDLH or other toxic atmospheres.

After a respirator has been used it should be cleaned and sanitized by washing with detergent with bactericidal disinfectant in warm water, followed by a water rinse and air drying. The washing can be done by hand or in a clothes washer, if a wash rack is provided to hold the face piece in a fixed position. Following cleaning and sanitizing, the respirator should be reassembled and inspected for proper working condition and repair or replacement of faulty parts. The employer must

FIGURE 13-23
DISPOSABLE DUST, MIST, AND FUME RESPI-
RATOR FOR RESPIRATORY PROTECTION.
(Courtesy of North Safety Products.)

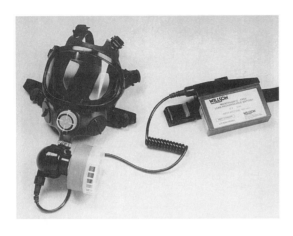

FIGURE 13-24
POWERED AIR-PURIFYING
RESPIRATOR SYSTEM FOR RES-
PIRATORY PROTECTION. (Cour-
tesy of Willson Safety Products.)

have a complete training and medical surveillance program so that employees
know how to properly use and maintain their respirators. Only physically qualified
personnel should use respirators.

PROTECTION OF THE FOOT

About 5 percent of all occupational disabling injuries involve the foot or toe. The
two critical functions of the foot are mobility and total body structural support. The
foot is a complex body part, with 26 bones and almost no cushion or muscle layers, making it vulnerable to many job-related injuries.

When the workplace requires the handling of heavy materials or work near
falling, rolling, or sharp objects, safety-toe shoes should be worn by the employee.

FIGURE 13-25
DONNED POWERED AIR-PURI-
FYING RESPIRATOR SYSTEM.
(Courtesy of Willson Safety
Products.)

FIGURE 13-26
SAFETY FOOTWEAR FOR MEN FOR FOOT PROTECTION. (Courtesy of Red Wing Shoe Company.)

OSHA requires safety-toe footwear to meet the ANSI standard Z41-1991, Personal Protection-Protective Footwear. [5,29] Three levels of safety-toe footwear comply with the ANSI standard, based on the minimum impact strength of the toe: 75, 50 and 30 pounds, respectively. The most common classification is the 75, which is based upon a 75-lb weight dropped 18 in. Metatarsal or overfoot guards are usually worn with safety shoes to protect the foot from heavier impacts.

Larger companies make safety shoes available through an internal shoe department to ensure a comfortable fit and satisfactory protection from hazards. Some of the many styles of safety footwear are shown in Figures 13-26 and 13-27. [18] Chemical and electrical resistance are also optional safety features.

Electrical-hazard shoes minimize the hazards from contacts with electrical current when the path of the current runs from the point of contact to the ground. These shoes will reduce the potential of electrical shock when worn in dry conditions with circuits not exceeding 600 volts. Electrical-hazard footwear has a rubber sole and heel with no exposed metal to conduct electricity. The protective metal toe box is insulated to prevent electrical shock and burns.

Safety footwear is also available that will help dissipate static charges. Static dissipative footwear conducts built-up static electrical charges to a grounded floor.

FIGURE 13-27
SAFETY FOOTWEAR FOR WOMEN FOR FOOT PROTECTION. (Courtesy of Red Wing
Shoe Company.)

Chemical-protection footwear provides protection ranging from dirt, mud, and
water to hazardous contaminates, corrosive splashes, and a variety of organic
chemicals. This type of footwear includes disposable and nondisposable overboots,
rubbers, boots, and shoes. Urethane, PVC, neoprene, rubber, and polyethylene are
common construction materials. Some examples of chemical-protection footwear
are shown in Figure 13-28. [3]

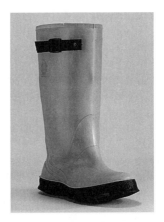

FIGURE 13-28
CHEMICAL-RESISTANT SAFETY FOOTWEAR. (Courtesy of
Bata Shoe Company.)

PROTECTION OF THE HANDS, ARMS, AND BODY

There is no substitute for your hands and fingers. They cannot be replaced and must be treated with care. Hands and fingers are complex and highly versatile, having strength, flexibility, and sensitivity, and can be coordinated to perform complicated jobs and tasks in the workplace. Loss of any of the vital parts of the hand—nerves, skin, nails, bones, muscles, ligaments, or tendons—affects its functioning.

Gloves, coveralls, aprons, and coats are commonly used in the workplace to protect the hands, arms, and body of workers from cuts, abrasion, chemicals, electrical shock, and temperature extremes. [31] The material for gloves depends largely upon the working conditions and the degree of protection desired. Inexpensive cotton or canvas gloves are satisfactory for most light work. The handling of abrasive or sharp-edged materials usually requires leather- or metal-reinforced gloves. Many plastic or elastomeric materials are used to provide chemical resistance to gloves. The proper glove size is important because tight-fitting gloves can cause fatigue, while loose-fitting gloves can be hazards. Some examples of gloves and their various applications are shown in Figure 13-29. [16]

The electrical-insulating qualities of PPE requires strict testing under OSHA 29 CFR 1910.137. [30] This is particularly important in damp or other environments that could contribute to electrical shock.

When assessing chemical resistance of glove materials, a chemical compatibility chart should be used to compare the chemical breakthrough time for each glove material under consideration (see Table 13-5 on page 428). The breakthrough time is based upon the elapsed time between initial chemical contact with the glove outside surface and the measured detection of the chemical on the glove insider surface (see Fig. 13-30 on page 429).

The term *permeation* is a process by which the chemical molecules move through the protective clothing material, including absorption of chemical molecules into the outside material surface, diffusion of the molecules through the material, and desorption of the chemical molecules from the inside surface of the material into the interior environment. Steady-state permeation, meaning a constant rate of permeation after breakthrough and equilibrium have occurred, is usually assumed.

Published breakthrough test data for materials is based upon standard laboratory test conditions. Actual breakthrough tests in the workplace, using the gloves or other PPE, are encouraged to increase the certainty of matching the PPE to the hazard. Chemical gloves should extend well past the wrists, leaving no gap between the glove and the shirtsleeve or coat. See Table 13-6 on page 429 for general guidelines to match glove materials to chemical hazards.

Employees need protection from on-the-job exposure to hot and cold extremes. Wool, cotton, and leather gloves are commonly used to handle hot and cold objects and surfaces. Other types of materials have been developed for use in PPE that must deal with more extreme hot and cold environments. For example, polyester and polyolefin combination PPE can be used with dry ice, low temperature freezers, or cryogenic atmospheres. Protection from high-temperature extremes can be obtained with materials like Zetex, Kevlar, and Nomex. Asbestos is strictly con-

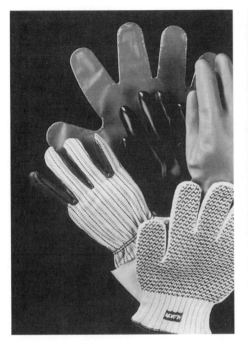

FIGURE 13-29
GLOVES FOR OCCUPATIONAL HAND PROTECTION. (Courtesy of North Safety Products.)

TABLE 13-5
CHEMICAL COMPATIBILITY CHART FOR SELECTED GLOVE MATERIALS

Chemical	Viton (10 mil)	Butyl (17 mil)	Neoprene (22 mil)	Nitrile (22 mil)
Acetaldehyde	NR	9.6 h	21 min	NR
Acetic acid (50%)	NT	NT	>480 min	NT
Acetone	NR	>17 h	12 min	NR
Ammonium hydroxide (29%)	NT	>480 min	ND	
Aniline	10 min	>8 h	>480 min	NR
Benzene	6 h	31 min	16 min	NR
Butyl acetate	NR	1.9 h	52 min	75 min
Carbon disulfide	>8 h	7 min	NT	30 min
Carbon tetrachloride	>13 h	NR	31 min	150 min
Chloroform	9.5 h	NR	12 min	NR
Cyclohexane	>7 h	1.1 h	159 min	NT
1,2-Dichloroethane	6.9 h	2 h	33 min	NT
Ethyl alcohol	NT	>8 h	>480 min	4 h
Ethyl ether	12 min	8 min	18 min	2 h
Formaldehyde (37% in water)	>16 h	16 h	>480 min	ND
n-Hexane	>11 h	NR	39 min	ND
Hydrochloric acid (37%)	NT	NT	>480 min	ND
Hydrofluoric acid	NT	NT	>480 min	2 h
Methylene chloride	1 h	24 min	6 min	NR
MEK	NR	>8 h	22 min	NR
n-Pentane	>8 h	NR	38 min	ND
Phenol (85% in water)	>15 h	>20 h	>480 min	NR
Sodium hydroxide (50%)	NT	NT	>480 min	ND
Sulfuric acid (25%)	NT	NT	>480 min	ND
Tetrachloroethylene	>17 h	NR	28 min	>5 h
Toluene	>16 h	21 min	14 min	10 min
1,1,1-Trichloroethane	>15 h	NR	27 min	90 min
Vinyl chloride	4.4 h	NR	NT	NT
Xylene	>8 h	NR	23 min	75 min

Key: ND = none detected; NR = not recommended; NT = not tested

trolled by OSHA in the workplace and has been eliminated from high-temperature PPE. [36] Tyvek from spun-bonded polyolefin fibers is now widely used in general protection coveralls and other protection garments to deliver high tear resistance and excellent particulate barriers. Combinations of Tyvek with Saran PVC, Teflon, and polyester are available for additional chemical protection, as are polypropylene, PVC, and Viton combinations.

Rain and corrosive and hydrocarbon splash protection clothing is usually based on PVC, rubber, neoprene, polyurethane, and chlorinated polyethylene–coated fabrics like nylon and polyester. This type of PPE has broad application, from aprons and bibs to fully encapsulated suits.

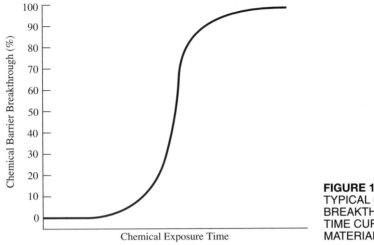

FIGURE 13-30
TYPICAL CHEMICAL
BREAKTHROUGH
TIME CURVE FOR PPE
MATERIALS.

Aluminized fabrics are used to reflect heat transfer from radiation. Depending on the supporting fabric, hot-metal splashes may also be repelled by certain aluminized fabrics.

WASTE DISPOSAL

There should be procedures for the disposal of PPE and wastes derived from cleaning PPE. PPE should be thoroughly washed after each use. Employees should be provided shower and changing facilities. PPE that has already been permeated by contaminates should not be reused because the ability to act as a predictable breakthrough barrier has been lost.

HAZARDOUS WASTE OPERATIONS

The hazardous waste business has more complex hazards than the normal chemical-processing operation because of the wide potential variation of waste hazards

TABLE 13-6
GENERAL GUIDELINES FOR SELECTION OF GLOVE
MATERIALS FOR CHEMICAL HAZARDS

Glove material	Recommended chemical usage
Natural rubber, PVC	Acids, bases, alcohols
Neoprene	Oxidized acids, phenol, aniline
Butyl	Ketones, esters, ethers
Nitrile	Aliphatics, oils
Viton	Aromatics, chlorinated solvents

TABLE 13-7
OSHA AND EPA GUIDELINES FOR WORKPLACES UNDER 29 CFR 1910.120 FOR
HAZARDOUS WASTE AND EMERGENCY RESPONSE SITUATIONS

Level A	Level B	Level C	Level D
Positive-pressure full-face SCBA or positive-pressure air respirator with escape SCBA	Same as level A	Full-face or half-mask air-purifying respirator	Safety glasses or chemical splash goggles
Fully encapsulating chemical-resistant suit	Hooded chemical-resistant coveralls, 1- or 2-piece suit, jacket and overalls	Same as level B	NA
Coveralls*	Same as level A*	Same as level A*	Coveralls
Long underwear*	NA	NA	NA
Chemical-resistant inner and outer gloves	Same as level A	Same as level A	Gloves*
Chemical-resistant safety-toe boots	Same as level A	Same as level A	Same as level A
Hard hat*	Same as level A*	Same as level A*	Same as level A*
Disposable protective suit, gloves, and boots	Disposable boot covers*	Same as level B*	Same as level B*
NA	Face shield*	Same as level B*	Same as level B*
NA	NA	Escape mask	Same as level C*

*Optional

encountered. OSHA has regulations for hazardous waste operations and emergency responses under 29 CFR 1910.120. [13,20,24] These regulations cover cleanup of hazardous waste sites; operations at treatment, storage, and disposal facilities; and emergency response to hazardous releases. There is a complex written PPE program required under 29 CFR 1910.120.

OSHA and the EPA worked together to develop guidelines for PPE selection for different levels of employee protection, when working under 29 CFR 1910.120. These guidelines, shown in Table 13-7, apply to hazardous waste sites and operations and emergency response situations, but they have general application to hazardous material operations. [15,24]

As shown in Figure 13-31, level A PPE should be used with a buddy system. The highest level of skin, respiratory, and eye protection is required. [11] This requires the characterization of a hazardous substance that is harmful to the skin or can enter the body by absorption through the skin and is harmful to the eye and

FIGURE 13-31
LEVEL A PERSONAL PRO-
TECTIVE EQUIPMENT. (Cour-
tesy of Kappler Safety Group.)

respiratory system. It is designed to provide maximum skin and respiratory protection against chemical gas, vapors, and splashes. Level A PPE should also be used in confined, poorly ventilated areas where the presence or nonpresence of hazardous substances has yet to be determined.

Level B PPE should be used with a buddy system. It provides the highest level of respiratory and eye protection, but less skin protection. Gas or vapors could contact the skin. This requires the characterization of an appropriate hazardous substance or an atmosphere with less than 19.5 percent oxygen.

Level C PPE should be used when an air-purifying respirator can remove the hazardous level of air contaminates that have been characterized (Fig. 13-32 on page 432). [11] Respiratory protection is downgraded to an air-purifying respirator. Skin exposure to these contaminants does not represent a health and safety problem.

Level D PPE should be used when the atmosphere contains no known hazard and the work to be performed is not expected to result in inhalation or other contact with hazardous chemical levels. This is standard work clothing.

These levels are OSHA-recommended guidelines and may be adapted to the conditions of the workplace. If necessary, a reasoned approach, with some combination of levels A, B, C, and D protection may be used, depending upon the workplace conditions and the current personnel situation.

ELECTRICAL AND TELECOMMUNICATIONS OPERATIONS

There are numerous potential electrical and physical hazards at telecommunications centers and field installations. The PPE requirements for employees in telecommunications are regulated under OSHA 29 CFR 1910.268 and in some instances are unique to that industry. [33]

OSHA requires employees to be protected with proper PPE from electrical circuit hazards under 29 CFR 1910.333 and 29 CFR 1910.335. For example, conductive apparel like jewelry should not be worn if employees may contact exposed en-

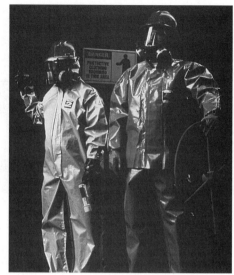

FIGURE 13-32
LEVEL C PERSONAL PROTECTIVE EQUIPMENT. (Courtesy of Kappler Safety Group.)

ergized parts unless the jewelry is rendered nonconductive by covering, wrapping, or otherwise providing insulation. [34] Employees must have appropriate PPE to protect their bodies from on-the-job electrical hazards. [35]

BLOODBORNE PATHOGENS

Occupational exposure to blood or other potentially infectious materials is regulated by OSHA under 29 CFR 1910.1030. The PPE used must not permit blood or other potentially infectious materials to reach the clothing, skin, eyes, mouth, or nose of the employee. This applies to gloves, gowns, coats, face masks, eye protection, mouthpieces, respirators, and other ventilation devices. [37]

SAFETY SHOWERS AND EYEWASH FOUNTAINS

The most important equipment for personal protection from fire and corrosive chemicals are safety showers and eyewash fountains (Figs. 13-33 and 13-34). [6] Employees should know the exact location of every safety shower and eyewash station in the work area and how to operate them. If a worker's clothing ignites or if a worker is splashed with a corrosive chemical, every moment counts in avoiding a serious injury or death. [9]

If clothing or skin is afire or wetted by a corrosive, get under the shower head and pull the chain or ring. If the clothing is contaminated, remove it while continuing to shower, as shown in Figure 13-35 on page 434. [6] If the eyes are affected, look upward at the shower head to flush them.

FIGURE 13-33
SAFETY SHOWER AND EYEWASH
FOUNTAIN. (Courtesy of Encon
Safety Products.)

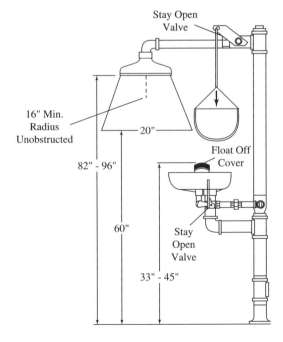

FIGURE 13-34
DIMENSIONAL REQUIREMENTS
FOR SAFETY SHOWERS AND EYE-
WASH FOUNTAINS BASED ON
ANSI Z358.1, 1990. (Courtesy of
Encon Safety Products.)

FIGURE 13-35
OPERATION OF SAFETY SHOWER AT 20
GPM AND 40 PSIG. (Courtesy of Encon
Safety Products.)

There are several types of eyewash fountains, commonly called eyewash stations, depending on the workplace location and availability of utilities (see Figs. 13-36 to 13-39). [6]

If a corrosive or other chemical irritant splashes your eye or face, go to the eyewash station, remove your glasses, initiate the water flow, bend over until the eyes are directly in the streams of water (Fig. 13-40 on page 436). [6] Hold the eyelids open with your hands and flush for at least 15 minutes to remove all of the foreign substance.

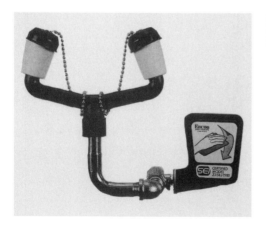

FIGURE 13-36
PIPE-MOUNTED EYEWASH STATION.
(Courtesy of Encon Safety Products.)

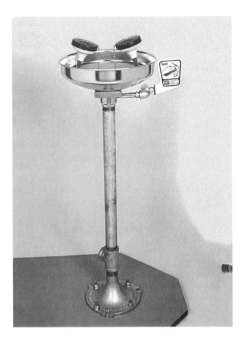

FIGURE 13-37
FLOOR-MOUNTED EYEWASH STATION.
(Courtesy of Encon Safety Products.)

FIGURE 13-38
AIR-PRESSURIZED, SELF-CONTAINED EYEWASH STATION.
(Courtesy of Encon Safety Products.)

FIGURE 13-39
GRAVITY-FED SELF-CONTAINED EYEWASH
STATION. (Courtesy of Encon Safety Products.)

FIGURE 13-40
OPERATION OF EYEWASH STATION. (Courtesy
of Encon Safety Products.)

BIBLIOGRAPHY

1 *Accident Prevention Manual for Business and Industry,* 10th ed., National Safety Council, Itasca, Ill., 1992.
2 Anton, T. J., *Occupational Safety and Health Management,* 2d ed., McGraw-Hill, New York, 1989.

3 Bata Shoe Company, personal communication, Belcamp, Md, 1995.
4 Crowl, D. A., and J. F. Louvar, *Chemical Process Safety: Fundamentals with Applications,* Prentice-Hall, Englewood Cliffs, N.J., 1990.
5 Dessoff, A. L., "Get in Step with the Right Footwear," *Safety and Health,* December 1995.
6 Encon Safety Products, personal communication, Houston, 1995.
7 Fawcett, H. H., *Hazardous and Toxic Materials,* Wiley, New York, 1988.
8 Foster, R. W., "Hazard Assessment for Personal Protective Equipment: An Investigative Process," *Professional Safety,* April 1996.
9 Hajian, H. G., and R. L. Pecsok, *Working Safely in the Chemical Laboratory,* American Chemical Society, Washington, D.C., 1994.
10 Hammer, W., *Occupational Safety Management and Engineering,* 4th ed., Prentice-Hall, Englewood Cliffs, N.J., 1989.
11 Kappler Safety Group, personal communication, Guntersville, Ala., 1995.
12 Kavianian, H. R., and C. A. Wentz, *Occupational and Environmental Safety Engineering and Management,* Van Nostrand Reinhold, New York, 1990.
13 Krieger, G. R., *Accident Prevention Manual for Business and Industry: Environmental Management,* National Safety Council, Itasca, Ill., 1995.
14 Lewis, D., "Confined Space Rescue: Should You Ever Remove Respiratory Protection?" *Safety and Health,* January 1996.
15 Mitchell, J. H., "Work Constraints Associated with Advanced Levels of Personal Protection," *Professional Safety,* April 1994.
16 North Safety Products, personal communication, Cranston, R.I., 1995.
17 Plog, B. A., et al: *Fundamentals of Industrial Hygiene,* 3d ed., National Safety Council, Itasca, Ill., 1988.
18 Red Wing Shoe Company, personal communication, Red Wing, Minn., 1995.
19 Roughton, J. E., "Complying with the Standard," *Professional Safety,* September 1995.
20 Roughton, J., "Protecting the Hazardous Waste Worker," *Pollution Engineering,* June 1995.
21 Swanson, S., "How to Motivate Workers to Wear PPE," *Safety and Health,* March 1996.
22 29 CFR 1910.94, OSHA, Washington, D.C., July 1, 1994
23 29 CFR 1910.95, OSHA, Washington, D.C., July 1, 1994
24 29 CFR 1910.120, OSHA, Washington, D.C., July 1, 1994
25 29 CFR 1910.132, OSHA, Washington, D.C., July 1, 1994
26 29 CFR 1910.133, OSHA, Washington, D.C., July 1, 1994
27 29 CFR 1910.134, OSHA, Washington, D.C., July 1, 1994
28 29 CFR 1910.135, OSHA, Washington, D.C., July 1, 1994
29 29 CFR 1910.136, OSHA, Washington, D.C., July 1, 1994
30 29 CFR 1910.137, OSHA, Washington, D.C., July 1, 1994
31 29 CFR 1910.138, OSHA, Washington, D.C., July 1, 1994
32 29 CFR 1910.252, OSHA, Washington, D.C., July 1, 1994
33 29 CFR 1910.268, OSHA, Washington, D.C., July 1, 1994
34 29 CFR 1910.333, OSHA, Washington, D.C., July 1, 1994
35 29 CFR 1910.335, OSHA, Washington, D.C., July 1, 1994
36 29 CFR 1910.1001, OSHA, Washington, D.C., July 1, 1994
37 29 CFR 1910.1030, OSHA, Washington, D.C., July 1, 1994
38 Willson Safety Products, personal communication, Reading, Pa., 1995.
39 Wortham, S. D., "Learn the ABCs of PPE," *Safety and Health,* August 1995.

PROBLEMS

1 Why are engineering and management controls usually more desirable than the use of personal protective equipment to solve occupational safety hazards?

2 Explain the primary function of PPE.

3 Why is it desirable to perform a hazard assessment of the workplace?

4 What has the greatest effect on the attitude of employees toward the proper use of PPE?

5 What factors influence worker acceptance of the proper use of PPE?

6 What elements should be included in an effective PPE program?

7 Discuss the ANSI classification for helmets for head protection.

8 When is PPE required for eye and face protection?

9 Discuss the advantages of glass versus plastic safety lenses.

10 When is the use of face shields appropriate?

11 What are the principal criteria used by OSHA in choosing eye PPE for welding operations?

12 Discuss the rating system for hearing PPE.

13 What are the types of hearing PPE that are generally available?

14 Discuss the two main types of earplugs.

15 Why are earmuffs usually more comfortable to wear than earplugs?

16 Describe the two general types of fall protection systems.

17 What is the advantage of a full-body harness with a lanyard for fall protection?

18 What is the process for the proper selection of respiratory PPE?

19 At what oxygen level are respirators required?

20 Discuss the operation of the two main types of respirators for workplace hazards.

21 What is the appropriate cartridge color code for an air-purifying respirator for the following contaminates?

   a HCl gas

   b NH_3 gas

   c TCA vapor

   d Uranium gas

22 During an emergency you put on an air-purifying respirator that has green cartridges. The workplace environment you now enter has been flooded with nitric acid fumes. What would probably happen to you?

23 What is the primary function of safety-toe shoes?

24 Why are there so many different materials of construction for gloves in the workplace?

25 Define chemical breakthrough time as it applies to gloves.

26 For hazardous waste operations, what is the main distinction in level A PPE protection?

27 What OSHA standard specifically applies to hazardous waste operations?

28 Why should employees become familiar with the location of safety showers and eye-wash stations in the workplace?

29 You are splashed in the face and eyes by muratic acid. How long should you flush your eyes at the eyewash station?

30 A rubber safety glove has a 12-h breakthrough time for a hazardous workplace chemical, based upon a steady-state permeation rate of 3 mg/m^2 per second. The glove has been in use for two consecutive 8-h shifts by the same worker, who has 280 cm^2 of protected skin per hand. How many grams of chemical has reached both hands of the worker during the two shifts?

31 Why should metal helmets not be worn around electrical hazards?

32 Why is electrically conductive footwear worn?

33 What is the OSHA standard for each of the following?
  a Hearing protection
  b Head protection
  c Foot protection
  d Eye protection
  e Respiratory protection
34 What hazard would be present if you wore an air-purifying respirator in an oxygen-deficient atmosphere?

14

AUDITS, INCIDENTS, AND EMERGENCY PLANNING

Auditing is a procedure for a periodic, systematic, documented, and objective evaluation of operations and practices in meeting safety, health, and environmental requirements. Nearly every type of situation can be audited by tailoring the audit to the specific needs of management and government.

BENEFITS OF AUDITS

There are many potential benefits to a well-conceived audit program, as shown in Table 14-1. [10,31] One of the most important benefits is the identification of hazards. [14,16]

An organization must be in regulatory compliance. An audit usually identifies areas of noncompliance that should be remedied before they can create penalties. Often noncompliance will become apparent in an audit because of ignorance or an inadvertent error of management.

Once the hazards have been identified by an audit, a course of action should be devised to allow the hazard to be corrected in a practical, cost-effective manner. In this way compliance is achieved, while allowing the organization to use a reasoned approach.

As workplace hazards are eliminated or reduced, there will be an accompanying reduction of risk and associated liabilities. The audit allows management to prioritize hazards from the audit in terms of their relative risk and potential liabilities. [15]

Hazards identified by audits adversely affect productivity and efficiency. This interrelationship makes it important for personnel to remediate hazards because of the potential benefits to traditional operating goals.

TABLE 14-1
POTENTIAL BENEFITS OF A WELL-CONCEIVED SAFETY, HEALTH, AND
ENVIRONMENTAL AUDIT PROGRAM

Recognition of existing workplace conditions
Better regulatory compliance
Correction of identified hazards
Reduction of risk and liabilities
Increased productivity and efficiency
Control of costs
Improved employee and community relations

Hazards from audits increase direct and indirect costs in the workplace. Reduction of these costs allows management objectives to be reached and profitability to be increased.

Employees will appreciate the reduction of workplace hazards because it protects their health and reduces accidents and injuries. As a result, employee morale improves and with it productivity. The resulting safer workplace creates added safety for local communities and the public. These situations will improve both employee and community relations, increasing the credibility of management.

PLANNING AN AUDIT

There should be careful consideration and planning before the actual performance of the audit itself. Considerable time, money, and personnel resources must be committed to a well-conceived audit. The needs should be defined by management well in advance of the actual audit because the results will be used to correct hazards and deficiencies in the workplace. If the audit objectives are unclear or improper, or if the audit team is inexperienced or otherwise inadequate, the results will be inaccurate or unreliable. In this case, management will not have a true picture of the situation in the workplace. The resources that should be committed because of the audit results are usually far greater than the resources to do the audit itself. Hence, the need to obtain an accurate snapshot of the workplace.

The primary purpose of an audit is to measure the degree of success in meeting safety, health, and environmental objectives, regulations, standards, guidelines, and other criteria. Audits may be used to determine the ability of the present management system to effectively comply with the goals and objectives of the organization.

Audits are a snapshot of the compliance situation for the location under study. If no compliance violations or other problems are found during the audit, the facility should be considered exemplary. If numerous violations and problems are discovered during the audit, the facility is being mismanaged. Usually the findings of audits are somewhere between these two extremes. [31]

Periodic audits are needed to track facility compliance over time. An advantage of periodic audits is to determine the status of violations since the prior audit. After any audit, the facility will prepare an action plan of issues to address to bring the facility into compliance and to rectify other problems. An audit should assess the progress the facility has made in solving the action plan items from the last audit.

The audit criteria usually include regulatory requirements, acceptable good-management practices, and internal organization policies and procedures. Insurance companies encourage audits to reduce risk. Audits can identify more efficient and cost-effective ways to operate and manage facilities. As such, they should be viewed as investments in the future.

Audits are a three-stage process involving preaudit, on-site, and postaudit activities (Table 14-2). This format greatly enhances the potential success of the audit. While there is no guarantee of a successful audit, most of the better programs share certain common characteristics (Table 14-3 on page 444).

An audit should be done in a methodical manner by trained, expert personnel who have the ability to ask probing, open-ended questions in a personable, objective manner. Such audits can assure continuous critique and improvement at all facilities in the organization.

Organizations benefit from audits by assuring that personnel systematically address problems. Future problems can be anticipated and prevented. The establishment of an effective compliance system reduces future liabilities. These audits aid in identifying, programming, and budgeting resources. Further, personnel gain experience and training and thus improve their job performance.

Planning is important for a successful audit. Numerous preaudit activities must be accomplished before the actual site visit. The audit team leader has the responsibility during preaudit activities to tailor the entire audit program to the needs of the facility and the overall organization. This requires considerable interaction between the team leader, who works with the facility and overall organization management, and the audit team.

Preaudit activities begin with the team leader assuring commitment, defining goals and objectives, and obtaining the necessary resources to perform the audit. The facility to be audited must be selected and scheduled for the visit. Contact personnel must be identified for each protocol. Sufficient background information about the facility should be obtained to adapt the organization audit plan to the specifics of the facility. The preaudit activities should determine how to handle privileged information that will arise during the actual visit. If this issue is not dealt with prior to the site visit, the integrity of the audit could be jeopardized. The team leader needs to make the facility personnel aware of audit team procedures and what the team expects from the facility.

THE ACTUAL AUDIT

On-site activities begin with an opening meeting at the facility. The plant manager or most senior facility manager should open the meeting with a reaffirmation of management commitment to the audit and the assembled audit team. Background information and a review of the audit plan should be presented at the opening

TABLE 14-2
ACTIVITIES OF AN AUDIT

Preaudit activities	On-site activities	Postaudit activities
Obtain management commitment.	Opening meeting at facility.	Resolve outstanding compliance issues.
Obtain management commitment.	Confirm background information.	Issue the draft report of findings and recommendations.
Develop goals, policies, objectives, and procedures.	Review audit plan.	Review and incorporate draft comments into report.
Obtain resources to perform audits.	Complete orientation tour of facility.	Prepare final report.
Develop audit capabilities and techniques.	Examine permits, records, and reports.	Distribute final report.
Train audit personnel.	Interview facility personnel.	Prepare action plan to address deficiencies.
Prioritize and select facility.	Inspect facility for compliance.	Implement recommendations in action plan.
Schedule facility for audit.	Complete protocols and checklists.	Follow up on close-out action plan items.
Select competent audit team members.	Identify noncompliance and deficiencies.	Retain all documentation and records.
Assign team responsibilities.	Formulate preliminary findings.	Identify time frame for future audits.
Make travel arrangements.	Verify all findings.	Plan audit with facility.
Have periodic discussions with key facility personnel.	Obtain background information.	Complete list of findings.
Adapt audit plan to facility.	Prepare draft report for closing meeting.	Develop protocols and checklists.
Present findings at closing meeting.	Determine the handling of privileged information.	Identify contacts for recommendations and actions.
Inform facility of audit procedures and expectations.		

meeting, which is followed by a tour of the facility by the audit team. The team members then disperse within the facility to begin records review, prearranged interviews with selected facility personnel, and compliance inspections. During these activities the team leaders should remain in contact with team members and facility management to resolve questions and conflicts that may occur. The length

TABLE 14-3
COMMON CHARACTERISTICS OF SUCCESSFUL AUDITS

Top management commitment

Reflects organizational goals

Well-defined scope and objectives

Written criteria, protocols, and checklists

Knowledgeable, personable team members

Cooperative facility personnel

Planned appointments with facility personnel

Preaudit questionnaire to facility

Open communication throughout audit

Focus on corrective action from findings

---

of the team visit is predetermined, usually 5 days unless the site and issues are very large or complex.

At the end of each day the team leader reviews the activities and preliminary findings with the team members. It is necessary for representatives of the facility to be present at these daily team meetings to remain fully informed. All findings should be verified, listed by protocol, and presented at the closing meeting or exit interview at the end of the week. Typical audit protocols are shown in Table 14-4. These protocols are not meant to be all-inclusive, but should be tailored by management to the needs of the organization.

TABLE 14-4
SAFETY AND ENVIRONMENTAL PROTOCOLS FOR AUDITING REGULATORY COMPLIANCE AND ORGANIZATIONAL MANAGEMENT PRACTICES

Hazardous materials

Fire hazards

Electrical hazards

Operating process hazards

Machinery hazards

Noise hazards

Illumination hazards

Ergonomic hazards

Solid waste

Air emissions and wastewater

Groundwater and drinking water

Hazardous waste

Natural and historic resources

An audit protocol lists the procedures followed during the audit to gather information about the operations and management practices at a facility. The protocol provides consistency because it is a checklist, with some room for subjective interpretation. Once established, a protocol can be used to audit other similar facilities and perform later audits at the same facility on a consistent basis. Within each protocol issues regarding regulations, guidelines, policies, and procedures must be determined. Audit protocols must be kept up to date to reflect changes in these areas. If the facility is provided with an advance copy of the protocols that will be used, it can be much better prepared to deal with the actual audit.

Management commitment is an important first step for successful audits. Management must provide adequate audit resources and be willing to finance corrective actions. A typical audit will find deficiencies and possibly major violations at facilities. Management must be prepared to deal with these findings or face serious consequences. Resources are needed to perform the audits, correct deficiencies, and modify programs for future compliance.

An audit generates several types of findings. A *significant* finding is a compliance finding that poses a direct and immediate threat to human health, human safety, the environment, or a business activity. A *major* finding is a compliance finding that poses a direct threat and requires action, but not necessarily immediate action. A *minor* finding is usually an administrative problem, not a threat to health, safety, or the environment. A management practice represents poor behavior or an unwise action not directly covered by law, regulation, or procedure. Any major findings should be immediately brought to the attention of the management for a solution.

Action needs to be taken immediately on significant findings, and fairly quickly on major findings. The timing of action on minor findings and management practices is usually determined on a case-by-case basis.

After the on-site visit, postaudit activities commence. The findings and recommendations are finalized and transformed into a facility action plan to correct deficiencies. Then the action plan is completed.

It is important to retain all documentation and records from the audit at the facility to aid in preparation for future audits and future regulatory compliance discussions. The time frame for the next periodic audit should be identified to establish continuity for future compliance efforts.

Both the EPA and OSHA are supportive of audit programs by corporations. It may be in the overall best interests of the organization to combine audits into a comprehensive environmental, health, and safety program. Often it is difficult to separate the issues and potential impacts of concern for the environment, health, and safety.

The EPA and OSHA have their own audit programs for compliance with their regulations. Internal audits by organizations can use similar criteria by obtaining these agencies' guideline documents. For chemical process safety audits, the EPA's Chemical Safety Audit Program (1989) and OSHA documents related to 29 CFR 1910.119 provide useful information. [4,5]

Risk assessment should play an important role in the focus of an audit at a facility or other operation. [17,22,26] For example, natural gas pipelines are usually aging and may require significant additional investment to maintain their safety in the future. Whenever there is an incident with a gas pipeline, the results are often disastrous and well-documented by the news media. [11]

Since gas pipelines pass through many remote areas, which cannot be constantly monitored with precision, companies have enlisted the public in identifying pipelines and leaks and educated communities about what to do if a leak occurs. [20,25]

INCIDENT INVESTIGATION

The term *incident* refers to all accidents and near-miss events that did or could cause injury or death, property damage or loss, or environmental harm. By determining what went wrong and why it went wrong, we can greatly reduce or eliminate their recurrence. [3,4,13] Some typical examples of near-miss incidents include the following:

- Process parameter excursions beyond established critical limits
- Reportable material releases
- Releases of flammables that do not ignite
- Equipment failures that could have been disastrous
- Emergency shutdowns that occur safely
- Activation of relief valves, deluge systems, or other such safety protection systems
- Operator errors that are revised or changed to safe situations without incidents

All employees, including management, need to remember that what is not reported cannot be investigated. It follows that what is not investigated cannot be changed and what is not changed cannot be improved. [3]

The main objective of incident investigation is to prevent the recurrence of an incident. This can usually be achieved by establishing a management information system (MIS) that does the activities listed in Table 14-5. Using this systematic approach, the MIS selects the optimum approach to solve the problem and maintains control over dynamic workplace situations. [6]

TABLE 14-5
PREVENTING RECURRENCE OF AN INCIDENT

Identify the root and contributing causes.

Evaluate and prioritize the causes.

Identify the preventive measures to reduce the risk of recurrence.

Evaluate and prioritize the preventive measures.

Select the recommendations and corrective action plans.

Implement the recommendations and corrective action plans.

Critique the results and appropriate feedback throughout the MIS.

The root cause is the primary reason for an unsafe act, event, or condition that results in the incident. [24] Root causes are usually related to a management judgment, decision, or environment that allowed or permitted the act, event, situation, or condition to occur. Determining the root cause, or multiple root causes, in an objective manner can be a sensitive issue unless management truly views the incident investigation as an opportunity to improve safety in the workplace.

Incidents are opportunities to improve management systems, not opportunities to assign blame. Incident investigation links the lessons learned from prior operations and historic incidents to safe designs and operations in the future. This is an important concept because a common finding of root causes of major incidents has been the failure to communicate the lessons learned at one facility to another facility within the same organization. [3]

BENEFITS FROM INCIDENT INVESTIGATION

Whenever an incident occurs, it should be traced back to its root causes, for example, unsafe conditions, faulty equipment, and human error. A well-conceived incident investigation is a systematic process that results in the benefits shown in Table 14-6. The realization of these benefits depends a great deal upon the commitment of management because they must accept responsibility and provide the resources for the preventive measures needed to correct the causes of the problems. [30]

THE AUDIT TEAM

Selecting the proper personnel for an audit is important to its success. Audit personnel may need different qualifications from an incident investigation team.

An audit team requires the type of functional background shown in Table 14-7 on page 448. The actual requirements will depend on the scope and depth desired

TABLE 14-6
POTENTIAL BENEFITS OF INCIDENT INVESTIGATIONS

All incidents will be reported and investigated.
Root causes of incidents will be determined.
Appropriate preventive measures and corrective action will be identified.
Operations knowledge, techniques, and facilities will be improved.
Process and equipment safety will be enhanced.
Safety awareness of employees will be increased.
Overall safety program will be improved.
Operations will achieve greater compliance with government regulations.
Detailed records will be available to support litigation.
Incident losses will be better controlled.

Source: American Institute of Chemical Engineers.

TABLE 14-7
FUNCTIONAL BACKGROUND FOR AN AUDIT TEAM

Auditing techniques
Applicable regulations
Materials and waste management
Operations and process engineering
Management information systems
Organizational and group behavior control
Objectivity

---

by management. Individual team members should be able to function well as a group and remain objective in dealing with the audit. One of the main responsibilities of the team leader is to control the team's activities while still being responsive to the needs of management.

The actual composition of the team will depend upon the nature, type, and size of the audit. The leader should be a person who is independent of the system being audited. One person should be responsible for the collection and processing of data. A modest audit could be adequately performed by a single individual. Additional personnel could provide depth and completeness to the audit (see Table 14-8). The ideal audit team should be composed of outside personnel who can maintain an objective, arms-length relationship with management. Internal audits should use employees from other facilities or operations whenever practical. Team members should be selected well in advance of the audit to ensure their availability and allow the leader to plan how to use them.

The team leader should contact the facility manager at least a month in advance of the audit to confirm the date and begin the preaudit planning process. Key facil-

TABLE 14-8
BACKGROUND OF PERSONNEL NEEDED FOR THE AUDIT

Operations
Maintenance
Process engineering and control
Safety, health, and environmental
Chemistry
Auditing experience
Technology consultant
Historical and natural resources

---

ity personnel should be available and normal site operations should be occurring during the week of the audit. The team leader should confirm with the facility manager in writing the time, date, place, and availability of personnel and records for the audit.

Information about the facility should be gathered and studied by the team before the facility visit. This should include the plant layout, unit processes, materials used, products and wastes generated, and organizational structure. Laws, regulations, permits, policies, guidelines, and procedure applicable to the facility should be reviewed. The records from prior audits should be reviewed and the status of past problems should be checked.

This preaudit information should be reviewed so that auditors can ask intelligent questions at the facility. The audit team needs this information in advance to develop a logical audit plan. This plan should outline each step of the audit, how it will be accomplished, and who will perform it. Priority items, like past audit findings or permit violations, should be particularly reviewed. Each team member should be assigned responsibilities that match the specific audit tasks. Checklists are helpful and encouraged in both planning and carrying out the audit. [10]

Proper documentation should be in place to ensure that the audit is being performed in accordance with previously agreed-to management guidelines and established program requirements. Once the on-site audit has been completed, the findings should be recorded and reported to management. The report should provide information about the facility to management, initiate corrective action, and provide written documentation of the audit and findings. Management expects the audit to be an accurate portrayal of its operation. After the deficiencies in the report have been corrected, management expects to be in compliance.

The audit should note both positive and negative aspects of the facility. Deficiencies, violations, permit status, good management practices, and innovations should be included. Action plans should be used by the facility to correct the negative aspects of the audit. This depends on management commitment and allocation of necessary resources to make corrections. Each deficiency requires an action plan that states what corrective action is to be taken, within what time frame, and by whom. Management should provide a written explanation for any deficiency that will not be corrected.

The audit program manager and the facility manager should work closely to implement the action plans. Ideally, the audit manager will oversee this process. If the facility manager is to be responsible for all action plans, the audit manager should check the status at the facility monthly or quarterly.

Depending upon the results of the last audit, there should be a periodic audit of all of the facilities in the organization, usually every 2 to 5 years. Ideally, one or more people from the previous audit will be on the new audit team. This is particularly necessary if numerous deficiencies were found in the past. All records should be retained in order to meet regulatory requirements. Longer records retention is usually desirable. Some organizations have adopted a policy of retaining records for the life of the facility.

THE INCIDENT INVESTIGATION TEAM

The incident investigation team must identify the root cause or causes, make recommendations to reduce or eliminate recurrence, and implement the recommendations. This team does not assign blame or carry out disciplinary action; that is the responsibility of management. This team should be independent and objective. It should not cease its efforts until the true root cause has been identified. Otherwise, similar incidents may happen. [3]

The investigation team leader should be a good manager with a high degree of technical expertise. The leader must remain objective, particularly in dealing with highly sensitive information with serious implications. An investigation team leader must be able to deal more openly with the public, news media, and employees than the leader of an audit team. Incident investigations are much more visible because something serious has occurred. Audits are more for internal use, to remedy hazards before serious harm results. They are less subject to outside scrutiny. This difference should be reflected in the selection of the team leader.

An investigation team requires the background and expertise in Table 14-9. The actual team composition may vary with the nature, type, and size of the investigation. The investigation scope, which will be affected by management, employees, government, and the public, will also help to determine team composition. [9]

The knowledge base of the ideal team is shown in Table 14-10. Some of these team members may be outside investigators, who are less likely to be biased in fact finding and can add credibility to the investigation. Such outsiders may include representatives from OSHA, EPA, local emergency planning commissions, fire departments, insurance companies, and unions. Personnel with preconceived ideas or emotional involvement in the incident should not be team members.

INCIDENT INVESTIGATION PLAN

Immediately after an incident, steps must be taken to protect victims and the environment and preserve evidence for the actual investigation (see Table 14-11). The AIChE has developed an investigation form to address all of the issues in an incident, including those listed in Table 14-12 on page 452. [4]

The investigation of an incident should begin as soon as possible—within 48 h.

TABLE 14-9
BACKGROUND CHARACTERISTICS FOR AN INCIDENT INVESTIGATION TEAM

Investigation techniques

Applicable regulations

Operations and process engineering

Organizational and group behavior control

Objectivity and unbiased perspective

TABLE 14-10
KNOWLEDGE BASE OF PERSONNEL INVOLVED IN INCIDENT INVESTIGATION

Operations
Maintenance
Process engineering
Safety, health, and environmental
Experience with the actual incident
Chemistry
Materials and metallurgy
Instrumentation and computer software
Community
Government

The evidence is most available immediately after the incident. Witnesses should be interviewed, photographs and videos taken, physical evidence identified and secured, records, log books, and instrument charts collected, and materials, streams, and residues sampled and analyzed. Since all of these items could become significant evidence, their integrity and validity need to be protected. [32]

CONDUCTING THE INCIDENT INVESTIGATION

The evidence gathered during the initial stage of the investigation could be substantial, but only a portion of it may be relevant to the incident. The data should be arranged chronologically in order to establish a time line of events before, during, and after the incident. Evidence that does not appear to fit should be regarded as suspect.

After the evidence has been analyzed and the probable root causes determined, recommendations for corrective actions should be made. There should be at least one recommendation for corrective action for each causal finding. All of the recommendations should be prioritized based upon the following criteria:

TABLE 14-11
TASKS IMMEDIATELY FOLLOWING AN INCIDENT

Rescue victims and provide medical treatment.
Stabilize the area to minimize further consequences.
Address environmental protection concerns.
Preserve the evidence at the scene.
Select team members.
Select team leader.

TABLE 14-12
ACTIVITIES OF THE INCIDENT INVESTIGATION

Gathering evidence

Analyzing evidence

Determining the root causes

Recommendations for corrective action

- Safety of the operation under investigation
- Long-term operating objectives
- Cost and benefit analysis
- Ease of implementation
- Ability to provide timely feedback
- Legal requirements
- Development of more practical alternatives

IMPLEMENTATION OF THE RECOMMENDATIONS

The most important aspect of the incident investigation is implementation of the recommendations. Without adequate follow-up the incident could be repeated. Recommendations should be clearly stated in the investigation report. The person responsible for implementing the recommendation and the completion date should be identified. The final processing and storage of all evidence and records should be well defined. Often this responsibility is vested with the safety function of the organization. A chain-of-custody record is usually a normal management control for these important documents, which should be retained for the life of the facility.

Many incidents are caused by the failure of protective coatings used to protect against steel corrosion. Protective coatings have a predictable service life under normal circumstances. When a coating fails prematurely or unexpectedly, the reasons need to be investigated. Typically, the coating has been corroded or contaminated, the viscosity or solvent evaporation rate is too high, the coating was improperly applied or exposed to ultraviolet light and excessive erosion. Field testing and laboratory studies can aid in the final conclusions and recommendations. [8]

Fire incidents are fairly common in both the industrial and public sectors. The main objective of the investigation is to determine why the fire occurred and whether the affected equipment may be returned to service. Data collection during and after a fire should focus on the fuel and ignition source, heating and cooling rates, and the temperature and time that equipment was exposed to the fire. A videotape of a fire in progress can be extremely useful in assessing fire damage. Photographs also provide excellent documentation. [21]

Incidents often occur at older chemical plants. Many old plants survive because of periodic modernization plans and corrective action before problems become hazardous. Management has to be committed to making necessary capital expenditures; upgrades and maintenance programs can greatly offset the deterioration process of aging plants, and thus reduce and prevent incidents. [18]

Audits and investigations should be integral parts of safety, health, and environmental protection. The more complete, detailed, and factual the information, the more likely the management commitment to corrective action. Audits and investigations can successfully eliminate workplace hazards, improve operating methods, educate and train employees, and prevent property damage, injuries, and death in the workplace and in surrounding communities. [1]

EMERGENCY PLANNING AND RESPONSE

Emergency planning can minimize losses. Top management is responsible for emergency planning, which is usually done under the guidance of the safety, health, and environmental protection sector of the organization. Planning should focus on protecting the health and safety of employees and the public, as well as property and the environment, and on restoring normal operations after an accident. [1,13,16]

A comprehensive emergency plan should cover fires, natural disasters, and other such incidents. Developing a good plan requires knowledge of the resources, operations, and capabilities of the facility and the surrounding area. [7,17] Response teams should be readily accessible to the disaster. Buildings and operations that are potential hazards should be designed so that they can be segregated from other less hazardous operations or offices. Damaged utilities, buildings, process equipment, and storage should be salvaged, overhauled, and made operational as soon as possible. Environmental contamination should be contained and mitigated, in compliance with regulations. The plan should be applicable to all anticipated emergencies and tailored to the needs of the organization.

Specialized groups should be developed and trained to respond to emergency situations such as fires, chemical spills, and medical emergencies. These teams may include local services like the fire department, police, and hospitals. [7,9] An internal and external alarm system, communication systems, shutdown and evacuation procedures, emergency power systems, and medical treatment procedures should all be present in a well-conceived emergency plan.

The complexity and refinements in the plan depend on the size of the organization, the type of facilities and operations, and the location of the operations in relation to the employees and the public. Large organizations with multiple facilities will require more people than a small company with a single operation. [27]

EMERGENCY HAZARDS

If specific hazards are targeted, they can be better managed. The particular character of a hazard can be identified based on its location, the processes involved, and work practices. [23] The planning for the identified emergency response should then be based upon the potential disaster and its threats to people, property, and the environment. [1,33]

Fire hazards are the most common emergency in the workplace because many common flammable and combustible materials and wastes are normally used in the work environment. [12] Usually, fires begin as small blazes that quickly get out of control. Prompt action by a trained person can bring most fires under control while they are still small. If a fire grows, more extensive fire-fighting forces will be needed. Emergency planning should consider smoke and toxic gases that result from the burning materials, as well as water runoff that may become contaminated, resulting in a disposal problem.

Flood hazards are present whenever an operation or facility is located in a floodplain. Protection should be provided by dikes. Recommendations should be sought from the National Weather Service, the Geological Service, and the Army Corps of Engineers. If flood conditions are anticipated, several precautions should be considered.

- Emergency power supply
- Water pump availability
- Electrical hazard protection
- Equipment, vessels, and chemical storage protection
- Potable water supply availability

Tornado and hurricane hazards can have catastrophic effects on facilities. These hazards may occur without much advance warning and at any time of day or night. Emergency provisions include warning and moving personnel to a safe location. Hazardous power and gas lines and debris must also be managed during these situations.

In many areas of the world earthquakes may strike without warning. Earthquakes may cause structural failures to buildings, bridges, and roads, as well as damage to operations and related infrastructure. In such an emergency the gas supply should be shut down and water alternatives for fire fighting and drinking should be available.

Riots, other types of civil strife, and sabotage can create extreme emergency situations that should be dealt with through government police departments and other such agencies.

Work-related accidents may occur, causing emergencies in the workplace and surrounding community. Complex chemical processes can create hazardous chemical gas and liquid releases. Such incidents are subject to federal regulations: the SARA Emergency Planning and Community Right-to-Know Act (40 CFR 355 and 370); Hazard Communication (29 CFR 1910.1200); and Hazardous Waste Operations and Emergency Response (29 CFR 1910.120). [2,19,29,31]

Emergency plant shutdowns should follow a well-conceived checklist that personnel are trained to follow. Deviations from a standard shutdown procedure may result in fires and explosions.

Plants located in areas subject to extremes of heat and cold must have special protection. For example, an operation in North Dakota requires much more protection from freezing and wind chill than one located in the Gulf Coast.

EMERGENCY ACTION PLAN

After potential emergencies have been identified, a written action plan should be developed with facility personnel and outside agencies. This plan should contain the items listed in Table 14-13. Other items may be required on a case-by-case basis.

The chain of command for the emergency plan should be as short as possible and all personnel should be able to respond decisively under conditions of high stress. The overall director should be familiar with the facility, its operation and key employees, and the local community.

The director is responsible for training personnel in emergency control and response. There should be periodic disaster drills to keep people in a state of preparedness.

The communications and command center should have maps, alarms, telephones, two-way radios, public address systems, and emergency power and lighting. The written shutdown and start-up procedure should be readily accessible to all personnel at the center.

Fire and security alarm systems should have signals that are easily recognizable to all employees. They should also interact with the local fire and police departments, whenever appropriate. Management is also responsible for providing a secure workplace and surrounding area. This requires a good relationship with local police, a well-trained security force, and communication and training of employees.

Specific people should be designated to assemble and lead employees to safety during emergencies, directing them away from hazards and preventing worker panic. Emergency transportation should be provided to move employees, provide medical services to victims, and transport emergency crews and supplies. [15,28]

First aid is usually required during emergency situations. If medical personnel and facilities are not available on-site, they should be near the facility. [6]

Industry and government emergency agencies should be contacted as soon as

TABLE 14-13
ITEMS FOR CONSIDERATION IN AN EMERGENCY ACTION PLAN

Emergency organization chart with phone numbers and addresses

Risk assessment of expected emergencies

Map of the facility layout

Identification of location of key emergency equipment, supplies, shelters, assembly areas, evacuation routes, and communications and command center

List of outside agencies with phone numbers and addresses

Identification and location of alarm systems

Shut-down and start-up procedures

Control of visitors and news media

possible. In some cases, regulations may require that government agencies be contacted.

BIBLIOGRAPHY

1 *Accident Prevention Manual for Business and Industry, Administration and Programs,* National Safety Council, Itasca, Ill., 1992.
2 Altvater, T. S., "Material Safety Data Sheets," *Professional Safety,* October 1990.
3 *Guidelines for Investigating Chemical Process Incidents,* AIChE, New York, 1992.
4 *Guidelines for Process Safety Documentation,* AIChE, New York, 1995.
5 *Guidelines for Process Safety Fundamentals in General Plant Operations,* AIChE, New York, 1995.
6 Anton, T. J., *Occupational Safety and Health Management,* 2d ed. McGraw-Hill, New York, 1989.
7 Baldwin, D. A., *Safety and Environmental Training,* Van Nostrand Reinhold, New York, 1992.
8 Burgess, R. A., and M. M. Morrison, "Uncover Mistakes via Coating Failure Analysis," *Chemical Engineering Progress,* September 1995.
9 Cheesman, J., "Transferring Accident Investigation Training to the Workplace," *Professional Safety,* October 1994.
10 Cheremisinoff, P. N., and N. P. Cheremisinoff, *Professional Environmental Auditors Guidebook,* Noyes Publications, Park Ridge, N.J., 1993.
11 Crow, P., "U.S. Industry, Government Efforts Seek to Improve Pipeline Safety," *Oil & Gas Journal,* April 24, 1995.
12 Crowl, D. A., and J. F. Louvar, *Chemical Process Safety: Fundamentals with Applications,* Prentice-Hall, Englewood Cliffs, N.J., 1990.
13 Curry, F. H., "Prevent Accidents before They Happen," *Chemical Engineering,* June 1995.
14 Gowen, L. D., "Causal Analysis a Means for Improving Safety," *Professional Safety,* November 1995.
15 Hammer, W., *Occupational Safety Management and Engineering,* 4th ed., Prentice-Hall, Englewood Cliffs, N.J., 1989.
16 Kavianian, H. R., and C. A. Wentz, *Occupational and Environmental Safety Engineering and Management,* Van Nostrand, New York, 1990.
17 Kim, I., et al., "Risk and the CPI," *Chemical Engineering,* February 1995.
18 Kirschner, E. M., "Chemical Industry Modernizes Aging Plants for Safety and Efficiency," *Chemical and Engineering News,* July 10, 1995.
19 Krieger, G. R., *Accident Prevention Manual for Business and Industry: Environmental Management,* National Safety Council, Itasca, Ill., 1995.
20 *A Guideline for Pipeline Safety,* Marathon Pipe Line Company, Findlay, Ohio, 1995.
21 McIntyre, D. R., *Guidelines for Assessing Fire and Explosion Damage to Chemical Plant Equipment,* Materials Technology Institute, St. Louis, Mo., 1990.
22 Morgan, D. J., and G. H. Swett, "Simplify Remedial Decisions by Using Risk-Based Management," *Hydrocarbon Processing,* August 1994.
23 Petreycik, R. M., "EU Cracks Down on Chemical Labeling," *Safety and Health,* March 1996.
24 Philley, J., "Investigate Incidents with MRC," *Hydrocarbon Processing,* September 1992.

25 *Helpful Information for People Who Live or Work near Our Pipelines,* Phillips Pipe Line Company, Bartlesville, Okla., 1995.

26 Reynolds, J. T., "Risk-Based Inspection Improves Safety of Pressure Equipment," *Oil & Gas Journal,* January 16, 1995.

27 Ritzel, D. O., and R. G. Allen, "Validity of the Basic Principle of Safety Management or Loss Control," *Professional Safety,* February 1996.

28 29 CFR 1910.35, OSHA, Washington, D.C., 1994.

29 29 CFR 1910.120, OSHA, Washington, D.C., 1994.

30 U.S. Department of Labor, *Accident Investigation, Safety Manual No. 10,* Washington, D.C., 1987.

31 Wentz, C. A., *Hazardous Waste Management,* 2d ed., McGraw-Hill, New York, 1995.

32 Woodward, J. L., "Structuring Accident Investigations," *Chemical Engineering,* June 1995.

33 Wortham, S., "Make Safety Audits Work for You," *Safety and Health,* March 1996.

PROBLEMS

1 What are the potential benefits of a safety, health, and environmental protection audit?

2 Why should audits be planned in advance?

3 Discuss a three-staged audit process.

4 What are the most common characteristics of successful audits?

5 What are typical safety and environmental audit protocols?

6 What is a near-miss?

7 What are the components of an MIS to prevent recurrence of an incident?

8 What is the root cause of an incident?

9 Why is the determination of root causes usually a politically sensitive issue?

10 What are the potential benefits of incident investigation?

11 Why is management commitment important to both incident investigation and audits?

12 Discuss the functional background that should be considered for an audit team.

13 What personnel knowledge should be available for safety, health, and environmental audits?

14 Why should audit team members be selected well in advance of the audit?

15 Why should prior audit findings be reviewed before an audit is performed?

16 After the audit has been completed and the audit report issued, why should there be follow-up with management?

17 What are the objectives of the investigation team?

18 Why should the investigation team leader possess both business and technical expertise?

19 What team background characteristics should be included in an incident investigation team?

20 What kinds of knowledge does personnel need for an incident investigation team?

21 What should be done as soon as is practical after the incident and prior to the investigation?

22 What are the main activities of an incident investigation?

23 Why is it important to understand the timing of the sequence of events before, during, and after an incident?

24 Why should there be at least one recommendation for corrective action for each causal finding for an incident?

25 How would you assess and prioritize the present environmental releases at a large oil refinery?

26 An epoxy protective coating failed in service. It is believed that poor resin-hardener mixing and high humidity may have caused the failure. How would you verify this cause?

27 If you were working near a natural gas pipeline that begins to leak, what would you do?

28 Six factors contributed to the 1984 tragedy at Bhopal, India. Discuss the contribution of each factor.

    a Water contamination of the methylisocyanate storage tank

    b Failure to maintain the design intentions of the emergency flare and scrubber system

    c Failure to execute established emergency response procedures and open the bypass valve to dump the storage tank into an adjacent standby tank

    d Failure to install an isolation blind prior to maintenance work, allowing a flow path for the water contamination

    e Reduction in shift operators from 12 per shift to 4 per shift

    f Reduction in shift supervisory staff from 3 per shift to 1 per shift

29 The *Challenger* space shuttle disaster resulted from a combination of management system failures and physical system deficiencies. The O-ring seals were known for their substandard performance prior to the 1986 fatal launch. In order to remedy this potential problem, it was decided to reassemble the O-ring system after each mission and to pressure-test the O-ring system. The test pressure was increased from 100 to 200 psig because of concern about O-ring failure. It was later discovered that this quality assurance pressure increase actually increased the failure rate of the backup sealing putty. Discuss the QA and performance trade-offs for the O-ring system.

30 The average oil refinery or chemical plant operating unit is generally designed to last 15 years without a major refurbishment. How would you go about verifying the remaining life of the unit during its expected life?

31 A major fire in a hydrocarbon processing unit killed four people and caused extensive property damage. An overfilled tank spilled pyrophoric liquid, which ignited. What are the possible root causes of this disaster?

32 The fire department used 100,000 gal of water to fight a moderate-size industrial fire. A storage tank with 0.2 percent solution of chromium in water ruptured during the fire, spilling 700 gal, which became mixed with the fire water. All of the contaminated fire water went into the local sewer system. What was the chromium concentration of the water that went into the sewer? Is this an environmental hazard based upon a regulatory standard of 5 ppm chromium maximum?

INDEX